Health and Inequality

Health and Inequality

Geographical Perspectives

SARAH CURTIS

First published 2004
Reprinted 2006, 2007

SAGE Publications Ltd
1 Oliver's Yard
55 City Road
London EC1Y 1SP

SAGE Publications Inc.
2455 Teller Road
Thousand Oaks, California 91320

SAGE Publications India Pvt Ltd
B 1/11 Mohan Cooperative Industrial Area
Mathura Road, New Delhi 110 044
India

SAGE Publications Asia-Pacific Pte Ltd
33 Pekin Street #02-01
Far East Square
Singapore 048763

British Library Cataloguing in Publication data

A catalogue record for this book is available from the British Library

ISBN 978 0 7619 6822 1
ISBN 978 0 7619 6823 8 (pbk)

Library of Congress Control Number 2003109263

Typeset by Photoprint, Torquay, Devon
Printed and bound in Great Britain by Athenaeum Press Ltd., Gateshead, Tyne & Wear.

For Brian

Contents

List of figures

List of boxes

Preface

This book has been written for readers with a range of different interests. Students undertaking school or university courses in geography will probably come to this book with some knowledge of the geographical ideas it contains, but may be less familiar with topics related to health. For them, I hope this book will show how geographical ideas can be applied to the important and fascinating subject of health and health inequality. Other readers, studying public health and health policy, for example, may be familiar with the health themes addressed here but will be able to find out from this book why geography is important for health and how a geographical perspective can help to elucidate issues of health inequality. The basic question addressed in this book is about why *places* are important for health. The discussion is designed to show how we can view places as made up of different 'dimensions' of the physical and social landscape that are important for health and health variation. The examples considered illustrate and explain how a wide range of theories and methods used in health geography complement each other and bring particular strengths to the field.

The book is mainly written with readers in mind who are following undergraduate and postgraduate programmes at university. However, some of the topics are likely to be useful for school students, especially as the geography of health is now appearing in school curricula. For students who are at a more introductory level, or teachers planning introductory courses, it may be helpful to look at the sections provided at the end of each chapter, which suggests further reading, identify the fundamental learning objectives addressed by each chapter and give some examples of key questions. Some of the boxes through the text can be used to help focus attention on specific examples. Students may be less concerned with the more detailed references to research publications given throughout the text. The references are included to help the more advanced reader who may wish to cover some topics in more depth. The book is intended to be relevant for readers from various different countries, and the discussion is international in scope. Although it is predominantly focused on cities in the developed world, comparisions have been made with some material relating to rural areas and to developing country settings.

The structure of the book is based on a theoretical framework for research in health geography. The first section explains five different types of 'landscape', relating to biological, physical and social environmental factors which are important for health. These chapters review the results of geographical research on the significance for health and health inequality of these landscape perspectives. The emphasis on space and place, and on geographical differentiation, demonstrates the contribution from health geography. The discussion also shows how health geography interfaces with other disciplines in medicine, natural sciences and social sciences, and argues that an interdisciplinary perspective is essential for useful research on public health.

One could imagine places as made up of all these different types of landscape, overlaying each other and interacting to produce health inequality. The second section of the book therefore focuses on specific aspects of health and disease (mental health and illness, and tuberculosis) and also on health policy. These aspects can be understood geographically by taking a perspective which combines the five types of landscape. This has particular relevance for policy and action to address urban health inequalities within particular local areas, as well as at the wider national and global scales. The book illustrates how geographical perspectives contribute to the design and evaluation of interventions to address urban health inequalities.

Acknowledgements

I would like to thank all my colleagues, friends and family who have given me advice and support as I prepared this book. Brian Blundell and Diana Curtis have, as ever, given endless help and encouragement and cheerfully put up with all the anti-social behaviour that goes with book writing. Professor Ian Rees Jones, St George's Medical School, London, is an inspirational collaborator and helped to shape my early thinking about the framework of this book. Several people have generously given time to review and comment on earlier drafts, and I have tried to reflect their good advice in the final version. For this, I am especially grateful to: Dr Michael Almog, New York University; Dr Priscilla Cunnan, Imperial College, London; Professor Antony Gatrell, University of Lancaster; Professor Wil Gesler, University of North Carolina; Professor Victor Rodwin, New York University; Dr Mary Shaw, University of Bristol; Dr Glenn Smith, Royal Free Hospital Medical School; Professor Susan Smith, University of Edinburgh; and François Tonnellier, CREDES, Paris. Robert Rojek and all the staff at Sage Publications have been both patient and supportive throughout the production of this book. I would also like to acknowledge all my colleagues at Queen Mary, University of London, who are a wonderful group of people to work with and who create the best kind of healthy academic environment.

1 Introducing Geographical Perspectives on Health and Inequality

In societies around the world there is debate about the inequalities in health and health care that we observe among populations and among places. Research seeks to improve understanding of the *causes* of health variation. Such knowledge should provide evidence about how best to influence the causes of health inequality and produce health gain for human populations.

These questions are preoccupying geographers as well as researchers in other disciplines, and this book aims to show how geographical perspectives on health help us to understand health variation. The book aims to review and develop the theoretical and empirical basis for geographical analysis of health inequality. Questions to be addressed include:

- Why is it useful to understand health inequality in terms of characteristics of *places* as well as *people*?
- Why is population health in some parts of cities relatively poor?
- What are likely to be effective strategies for reducing health disadvantage for these populations?

A number of theoretical strands need to brought together to inform our understanding of the health of disadvantaged populations. Research in geography and related fields has begun to provide more integrated theoretical frameworks, as well as a large body of empirical evidence.

The discussion in the following chapters is also intended to illustrate and explain how different methodologies in the geography of health and health care can be applied in research on urban health inequalities. These methodologies cover a wide range of quantitative and qualitative approaches. These approaches are often pursued rather separately by different groups of researchers, but the discussion here aims to bring out the complementarity between them, as well as the particular strengths of each.

The discussion will be international in scope, using examples from different parts of high and low income countries. However, given the concentration on urban health, much of the discussion will focus on cities in high income countries which have the greatest degree of urban development, but which also show a good deal of intra-urban variation.

This chapter reviews some concepts of health and health care that are important for the discussion here. It also introduces some important aspects of health variation from a geographical perspective. The last part of this chapter introduces a general conceptual framework for discussion of the geography of health, which involves five types of landscapes that are important for health inequality. This sets the scene for a more detailed discussion in Section I of this book, where Chapters 2 to 6 aim to illustrate the theoretical and methodological strategies that health geographers have used to investigate these different types of landscape. In Section II, Chapters 7 and 8 illustrate how these different conceptual landscapes together produce whole systems of geographical variation in particular types of health problem, such as mental illness and risks of infectious disease (specifically tuberculosis). Chapter 9 discusses the relevance of these geographical perspectives for policy.

VARYING INDIVIDUAL AND SOCIAL CONSTRUCTIONS OF HEALTH AND ILLNESS

The concept of 'health' is open to differing interpretations. The biomedical perspective on health focuses on presence or absence of diagnosed diseases, but broader definitions include the idea of health as 'a state of complete physical, mental, and social well-being and not merely the absence of disease or infirmity' (WHO, 1946: 100).

Health can also be viewed as a socially constructed phenomenon, having different meanings for different people. Our understanding of health relates strongly to our individually and collectively constructed ideas of identity and the nature and significance of the body. Many authors have discussed results from studies in fields such as anthropology, sociology and social geography which have demonstrated that individuals vary in their perception of what counts as healthy or unhealthy and in their definitions of illness (e.g. Cornwell, 1984; Fitzpatrick et al., 1984; Helman, 1984; Donovan, 1986a; Calnan, 1987; Eyles and Donovan, 1990; Miles, 1991; Stainton-Rogers, 1991; Scambler, 1997; Butler and Parr, 1999). People use different conceptual frameworks to understand health and often individuals work with more than one of these frameworks, producing complex and variable frames of reference. Frameworks for understanding and explaining health include:

- the idea of health as *balance*, or illness as 'imbalance';
- the notion of the body as a *machine'* and of illness as 'malfunction' of the machine;
- the idea of *locus of control* (the perception of the degree of control the individual has over his or her own health);
- health or illness seen as the outcome of *fate* or *divine will*;
- ideas about health providing *freedom* or 'release' to do as one pleases, or as functional ability to carry out key roles, such as work as an employee or a homemaker;
- the concept of health as *resilience* against threats of infection or hazards;
- ideas about *access* to the means for good health, such as health care and a reasonable standard of living.

These 'health beliefs' are held by different individuals, in varying degrees. However, there is a tendency for some elements of health perception to be shared and reinforced collectively among people in the same society or social or ethnic group. This gives rise to what may be seen as culturally specific aspects of health belief, which may be typical of particular social and geographical settings.

VARYING ACCOUNTS OF WHAT DETERMINES HEALTH AND ILLNESS

Perception of the determinants of health (factors which produce health and illness) is also socially constructed and variable, depending on the temporal, social or geographical context. This has resulted in varying views on how to promote health and treat illness.

The *social model* of health sees the health of an individual person as the outcome of a range of *socio-economic* and *political* determinants, as well as medical care (e.g. Whitehead, 1995). Commentators such as Dubos (1960) and McKeown (1979) have made the case that medicine has a relatively small impact as a determinant of health of populations. They argue that aspects of living conditions, associated with varying levels of social and economic development, are more significant in determining whether or not people become ill. This perspective places strong emphasis on an understanding of the social and physical environmental factors influencing health. The geography of health shows how these vary from place to place, often in association with levels of socio-economic development, and this helps to explain population health differences between different parts of the world.

Other authors have argued that *health care* should be considered a crucial determinant of health in the population, and that differences in access to effective health care can have major impacts on health (Schofield and Reher, 1991; Bunker, 1995). According to this view, variations in the consumption of health care, in all its forms, contribute to differences in health between populations and individuals. Geographical variation in the provision and use of health care is therefore an important focus for the geography of health. These different perspectives are reflected in a variety of different medicines. The relationship of biomedicine to 'complementary' or 'alternative' therapies is increasingly being recognized as a matter for debate by health care professionals (Saks, 1992; Sharma, 1992; British Medical Association, 1993; Cant, 1999). Lay views, held by people in wider society, on how best to care for health also need to be understood as socially and individually constructed, and subject to variation and to debate.

Curtis and Taket (1996) argued that 'health care' extends beyond professional medical services to include the range of formal and informal services and activities which are intended to maintain good health and treat illness. These include action by lay people to care for their own health and provide informal care for others in their family, their social circle or their wider community. Thus, for example, Milligan (2001) has examined the role of the voluntary sector in geographies of care. Health care is delivered in a variety of settings, or 'spaces of care', ranging from institutions such as hospitals to community facilities or people's homes and public places. These have been the subject of geographical studies which can help in understanding the processes involved in health care and can contribute to strategies to make health care more effective.

The discussion in this book also shows that there are many *non-medical services* and resources which are not primarily intended to affect health but which do so indirectly, because they influence the determinants of health. These include, for example, services providing education, housing, access to employment and to social and leisure opportunities, sources of consumer goods, social support mechanisms operating in families and communities, as well as measures to control the physical quality of the environment.

ECOLOGICAL VERSUS ATOMISTIC VIEWS OF HEALTH

A large part of the body of knowledge concerning health variation comes from studies of health in geographically defined populations, focusing on aggregated groups of people. However, geography also helps us to understand the health of individual people, for example, by exploring the interaction between people and places.

There is ongoing debate over the value of *ecological* perspectives (using aggregated data for populations and their environment) as opposed to *atomistic* analysis (focusing on information at the level of individuals). Critics of the ecological approach argue that it may result in the ecological fallacy (wrongly attributing the average characteristics of the aggregated population to individuals within the population). Associations observed at the ecological (population group) level between health status and other factors may not mean that these variables are necessarily associated at the level of the individuals within the group. On the other hand, a perspective that concentrates only on associations observed at the individual level may also have limitations. A purely atomistic approach may not correctly identify processes operating at the collective level of whole communities or over whole areas (Schwartz, 1994). A geographical perspective argues for a consideration of how individuals interact with their wider environment. It seeks to develop theories to explain this interaction, which are the subject of this book. It often involves analyses that combine ecological data for areas with atomistic data for individuals.

SPACES OF RISK FOR HEALTH AND GEOGRAPHICAL PERSPECTIVES ON THE STRATEGY OF PREVENTIVE MEDICINE

Beck (1992) suggested that in modern societies inequality is conceptualized in new ways. Socio-economic variation has often been construed in terms of class divisions in the distribution of wealth. Beck argued that the idea of the 'class society' is now reinterpreted in terms of the 'risk society', in which hazards are unevenly distributed between social groups. This is associated with growing public concern about issues such as the impact of environmental pollution and degradation, global environmental change, and threats to security presented by war and violence.

Media coverage of accidental chemical and nuclear emissions, floods and droughts, international conflicts and terrorist actions reflects these concerns.

Many of the hazards of concern in the risk society are hazards to human health. However, we cannot usually be *certain* that a person exposed to a particular hazard will necessarily suffer damage to their health. We need instead to understand that someone exposed to a hazard will have a different *probability* (or risk) of health damage than a person who is not exposed. It can be difficult to assess these risks. Furthermore, understanding of risk is socially constructed (Löfstedt and Frewer, 1998). Shrader-Frechette (1990) has argued that assessments of risk made by scientific experts are not value free and that interpretation of risk in social policy needs to be sensitive to lay understandings of risk. Lay people, according to Shrader-Frechette, consider risk not only in terms of the probability of fatality, or simple economic criteria, but also in terms of how well the risk is understood and whether exposure to the risk is unequal for different populations.

Box 1.1, for example, illustrates statistical probabilities of death, for the average individual in a western society, due to different risk factors. Different people do not always rate health risks in a way that is straightforwardly related to the probability of health effects expressed in these terms. For example, some people who are concerned about the health risks posed by nuclear installations may smoke cigarettes. In some ways this might seem irrational: the statistical likelihood of dying in a given year due to cigarette smoking is much greater than the risk of health damage due to a nuclear accident (see Box 1.1). On the other hand, concern about the statistically smaller risk posed by nuclear installations might stem from the idea that the unlikely event of a nuclear accident could have extensive impacts, which would be difficult for society as a whole to manage.

In geographical terms we can think about *spaces of risk*: different combinations of health determinants coming together in different ways, resulting in variation in the risks to health for population in different places. Part of the role of health geography is to describe and explain these spaces of risk. Smallman-Raynor and Cliff (2001) have discussed what they call *epidemiological spaces*, which correspond to the geographically varying risk of specific diseases. Their example focuses on the use of time-space mapping of the geographical diffusion of contagious disease considered in more detail in Chapter 6. Spaces of risk for health may also be defined in terms of factors which are important for non-infectious diseases, for injury, or for good health and wellbeing. An understanding of spaces of risk requires consideration of a range of

Box 1.1 Statistical and perceived risks to health

Different groups of people may rate risks differently. Slovic (1987) compared the rankings given to 30 different activities and technologies by different categories of respondents. The results in Figure 1.1 illustrate the different average ranks recorded for a group of college students and a group of experts, for selected risks.

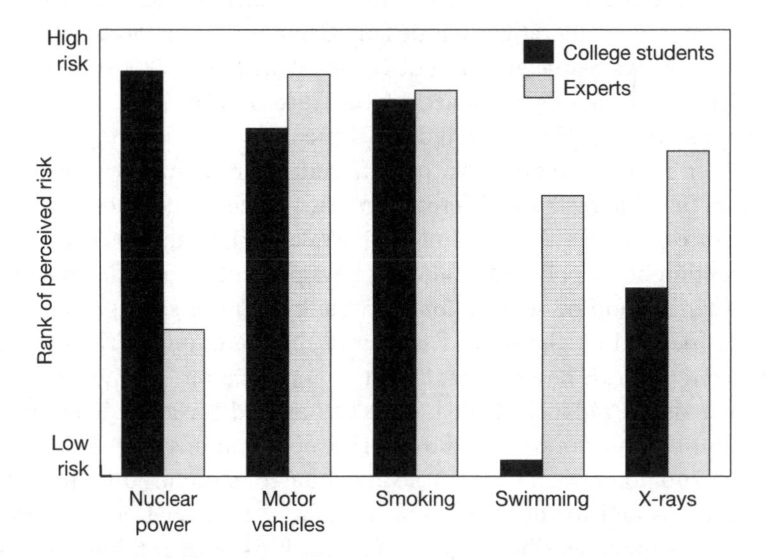

FIGURE 1.1 Perceived risk of activities and technologies assessed by college students and experts (after Slovic, 1987)

For some types of risk the two groups made very different average assessments. For example, experts ranked nuclear power installations as a less important risk than did college students. Students ranked smoking as a serious risk, but less so than nuclear power. (However, as shown below, in information published by the British Medical Association, the statistical probability of dying due to smoking is much greater than the risk from nuclear power plants.) Experts considered swimming and X-ray exposure as higher ranking risks than did college students.

Rankings by experts correlated with statistical evidence of mortality risk, but lay respondents like the college students tended to give greater weight to factors such as potential for major catastrophe, or risk to future generations.

The British Medical Association (1990: 28) reported the following risk of an individual dying in any one year due to specific causes:

smoking 10 cigarettes a day	1 in 200
accident on the road	1 in 8000
accident at home	1 in 25,000
accident at work	1 in 43,000
hit by lightning	1 in 10,000,000
release of radiation from a nuclear power station	1 in 10,000,000

physical, social, economic and political processes operating in different places.

As discussed above, risk is variably interpreted and socially constructed, and this is also true for perceived spaces of risk. Thus individuals, and the social groups to which they belong, will vary in their perception of health risk in a place. For example, work by Elliot et al. (1993) and Eyles et al. (1993), on perceptions of risk associated with proximity to incinerator facilities in Canada, has demonstrated differences in views between different groups. Hinchcliffe (2001) analysed the British government's response to the risk of BSE (bovine spongiform encephalopathy). He illustrated the difficulties of translating uncertainty and controversy about risk in different places into practical policy. Competing interests of different agencies affected decisions on how to control or manage the risks of BSE. Wakefield et al. (2001) discuss the perception of air pollution, showing how perception of risk was linked to whether it could be sensed (for example by sight or smell). The response also depended on the shared values in the community. Thus the social environment can be important for the appreciation of physical environmental risk. Craddock (2000a: 164) has argued for a feminist and post-structural approach to the study of risk of conditions such as HIV/AIDS, which should be seen as 'historically situated, structured by institutions, households and nations' and also influenced by global, as well as local, political economies. She suggests that health risks are influenced by social ideologies and cultural codes and by power relations (specifically gender relations) prevailing in particular times and places, which affect the entitlements and empowerment accorded to different groups in society. This supports similar arguments put forward by other authors, reviewed in Chapter 3, concerning the importance of power relations for health risks and inequalities.

An interesting aspect of these spaces of risk is that they combine information about the *people* occupying places and about *places* themselves. This is a very different perspective to that commonly adopted by biomedicine, which is concerned more with health risks for individual people. Rose (1992) has proposed a 'strategy of preventive medicine' which focuses on populations rather than individuals. He argued that: 'A preventive strategy which focuses on high-risk individuals may offer substantial benefits for those individuals but its potential impact on the total burden of disease in the population is often disappointing' (1992: 73). This is partly because it is not only those with the greatest risk of illness who suffer health problems. Much of the total burden of disease is carried by larger groups of other people in the population, for whom the individual risk is less. Community strategies, which concentrate on reducing risks for everyone rather than just finding and treating the 'high risk' individuals, will, it was argued, have more overall impact on health.

Community strategies will often be aimed at geographical areas where the 'target communities' are living, and they need to be based on an understanding of the interactions between populations and their environment that affect health. Thus an understanding of the geography of health in terms of 'spaces of risk' for health becomes important to the successful implementation of Rose's strategy of prevention.

VARIATIONS IN HEALTH AT DIFFERENT GEOGRAPHICAL SCALES

The geography of health is centrally concerned with the question of health variation in populations in different parts of the world and how we may understand these. Figure 1.2 shows life expectancy data for selected countries in 1999. The worst life expectancies in countries shown here are among the lowest in the world, averaging less than 40 years (e.g. Sierra Leone, Niger and Malawi). These countries have average life expectancies which are only half of those in countries such as Sweden or Japan, which are among the best in the world. Generally life expectancy is worse for men than for women, but the sex differences also vary between countries. These figures are indicative of very significant inequalities in health and death rates at the international scale.

Within countries there are also strong inequalities in life expectancy. Figure 1.3 shows male and female life expectancies reported by Raleigh and Kiri (1997) for 20 of the 105 English health authority areas in 1992–4. The health authorities illustrated are those which are ranked in the top 10 or the lowest 10 areas nationally on male life expectancy. These are compared with the average life expectancy for England as a whole. There was a gap of more than 6.7 years in average life expectancy for men between the best and the worst ranked health authorities. Female life expectancies are higher than for men, and areas rank differently on male and female life expectancy. A gap of 4.7 years separates the top and bottom ranked areas for female life expectancy. Cambridge had the best male life expectancy (76.6 years) and was ranked third best for female life expectancy (81.1 years). Manchester was ranked worst in England for male and female life expectancy at 69.9 and 76.7 years respectively).

There are also big differences in life expectancy and mortality within cities. For example, Figure 1.4 shows information on life expectancy for London boroughs in 1997–9, from a report by Fitzpatrick and Jacobson (2001). Life expectancies for men show the strongest differences and

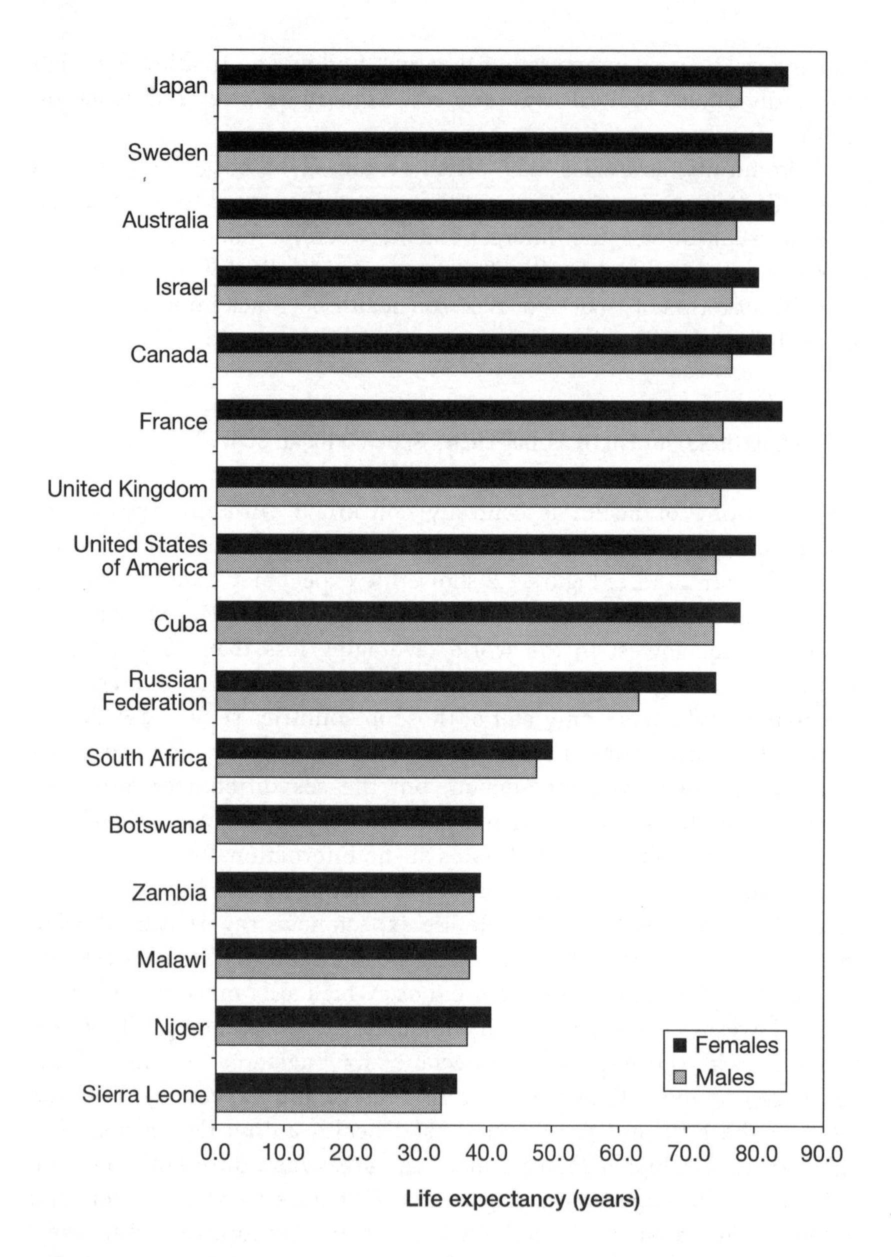

FIGURE 1.2 International variations in life expectancy, 1999 (WHO, 2000a, Table 2)

there is more than 5 years difference in the average life expectancy for boroughs with the worst male life expectancy compared with those with the best.

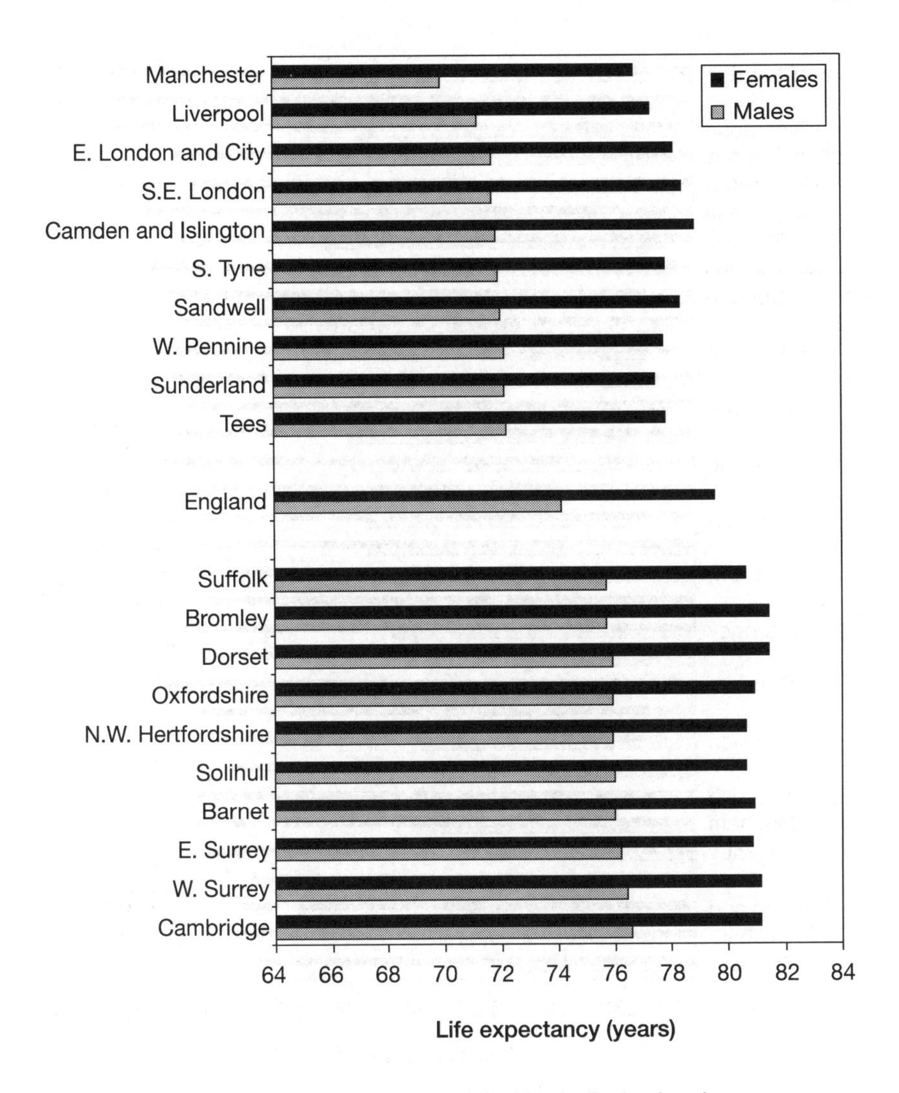

FIGURE 1.3 Life expectancy for area health authorities in England and Wales, 1992–4 (Raleigh and Kiri, 1997)

Thus we see variations in health between geographical areas at international, national and regional scales. These inequalities are evident even in countries like England, where national indicators of health are relatively good. The degree of inequality between areas is variable for different groups in the population (for example we see here that the situation differs between men and women). This book considers how health geography can help us to understand these variations and inform strategies to tackle the health inequality which lies behind the figures shown here.

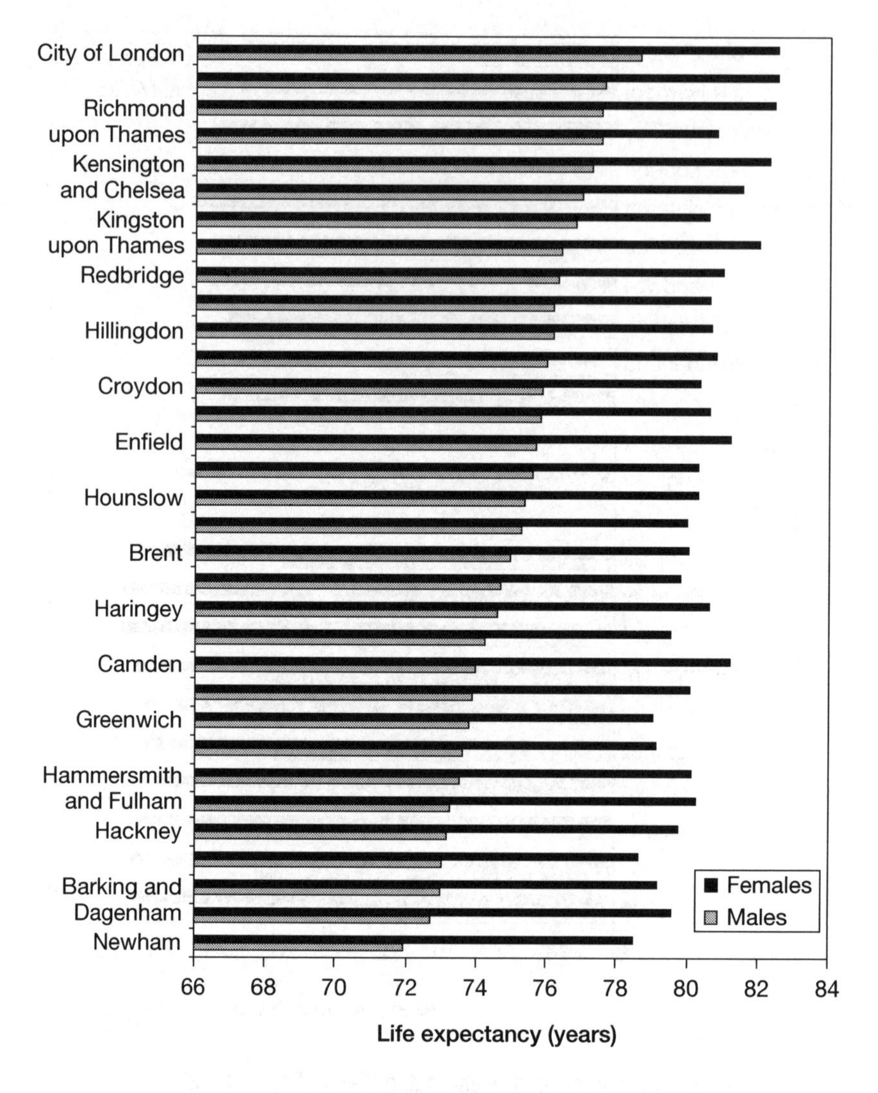

FIGURE 1.4 Life expectancy in London boroughs, 1997–9 (Fitzpatrick and Jacobson, 2001)

HEALTH AND HUMAN DEVELOPMENT: THE EPIDEMIOLOGICAL TRANSITION

An important concept in health geography is the link between health and human development. Authors such as Omran (1971) have proposed the epidemiological transition model, which describes the relationship between human development and the changing importance of different causes of illness and death in the population. The model has provided a

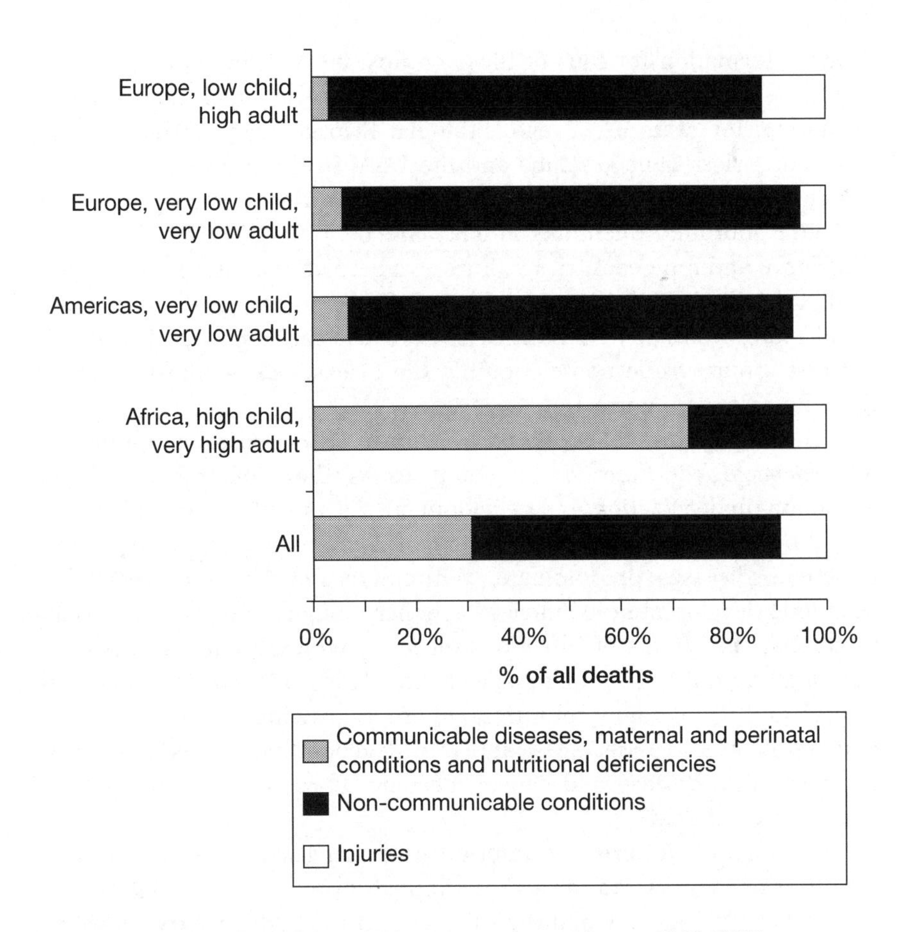

FIGURE 1.5 Deaths by type of cause for groups of countries, *c.* 1999 (WHO, 2000a)

useful framework for considering global geographies of illness and mortality. The significance of different types of disease for population health is variable in space and time and this is associated with overall levels of illness and death in the population. Low income countries, with low levels of socio-economic development, have higher levels of morbidity and mortality due to infectious diseases, while high income countries have a larger proportion of deaths and illness arising from non-infectious diseases such as cancer or cardiovascular disease.

These differences are evident today for groups of countries analysed by the World Health Organization (WHO). For example, the *World Health Report* presents information on mortality and disability, around 1999, in the population of groups of member states, classified by region and by relative levels of child and adult mortality (WHO, 2000a). Figure 1.5

shows information for four of these groups: 20 African countries where mortality is relatively high among children and very high among adults (including, for example, Congo, Ethiopia, Eritrea, Kenya, South Africa); three countries (Canada, Cuba and the USA) in the Americas with very low child and adult mortality; 26 European countries with very low infant and adult mortality (including France, Greece, Ireland, Israel, the United Kingdom); and nine countries in Europe where child mortality is low, but adult mortality is high (mainly in Eastern Europe, and including Hungary, Kazakhstan, Russia). Infectious diseases cause the majority of deaths in the first group, while non-communicable diseases cause most deaths in the other groups (Figure 1.5).

The original model proposed by Omran (1971) postulated a process of *epidemiological transition* in countries as they undergo social and economic modernization and development. As a result, countries make a transition from the situation typical of low income societies, where infectious diseases predominate. With increasing levels of social and economic development, countries have increasing resources and capacity to combat the effects of infectious diseases, which often cause particularly high mortality in younger age groups. This leads to a situation more typical of high income countries, where a growing proportion of the population live to older ages, and the 'degenerative diseases', such as cardiovascular disease and cancer, become more common and are the main causes of death.

Omran's model broadly describes the changes seen in high income countries as they passed through the industrial revolution during the last part of the 19th century and the early part of the 20th century. However, this model of epidemiological change does not appear to apply so straightforwardly to lower income countries undergoing social and economic transitions today. Many of these are seeing increases in the burden of morbidity and mortality due to non-infectious diseases, while they are still unable to control infectious diseases (e.g. see discussion in Curtis and Taket, 1996; Picheral, 1989, 1998; Vaguet, 2000). Furthermore, even in countries where infectious diseases have been well controlled in the past, we have seen, in recent decades, a re-emergence of these diseases, and also the emergence of new infectious diseases (including illness associated with infection with HIV (human immunodeficiency virus)). Thus the epidemiological transition appears to be a continuing, rather than a completed, process in developed countries, and it may be possible for it to 'reverse'. Smallman-Raynor and Phillips (1999) consider trends in mortality in 20th century Europe 1901–75. They note a convergence between areas in the comparative prevalence of degenerative diseases and growing spatial variability in the prevalence of classical infectious diseases, which may reflect a fifth phase of epidemiological transition. The resurgence of infectious diseases is diverting health care resources that could have been used to tackle degenerative diseases. Other research has suggested that trends in mortality due to non-communicable diseases in late 20th century Europe were not convergent for all countries. Warnes (1999) commented on international variability and trends in mortality in old age, showing, for example, that from 1960 to 1990 mortality rates due to cancer, cardiovascular disease and stroke in the UK became progressively worse than those in Switzerland. While in general the social and economic development of societies is associated with overall improvement in the health of their population, it is debatable whether societies with the greatest levels of average development and wealth always enjoy the best levels of health and wellbeing for all their people. Chapter 4 considers the association between human health and the distribution of wealth, associated with the geographically variable impacts of economic growth and circuits of capital.

THE IMPORTANCE FOR HEALTH OF URBANIZATION AND GLOBALIZATION

This book focuses especially on health variation in urban areas, although some consideration is also given to conditions in rural areas. Reasons for this concentration on health inequalities in cities include the increasingly urban profile of the world's population, and the growing concentration of health disadvantage in parts of major cities. The trends we observe in cities are arguably most indicative of new and future trends in the factors affecting public health, raising major challenges for societies over how to make human development sustainable in public health terms.

Human populations are becoming increasingly urbanized in all parts of the world. At present about half the world's population (3 billion people) live in cities. The recent United Nations (2001) report on the state of the world's cities considered rates of urbanization in groups of countries with varying levels of social and economic development measured using the UN human development index (HDI). This showed that in more privileged (high HDI) countries more than three-quarters of the population will be living in urban areas by the year 2015. However, the projected rates of increase in urbanization are greatest in poorer countries, with medium or low HDI (see Figure 1.6). These countries, which have hitherto been predominantly rural, are expected to have 40–50% of their population living in urban areas by 2015. After 2015, rates of urbanization are expected to level off, but by that time the majority of the world's population will be urban dwellers.

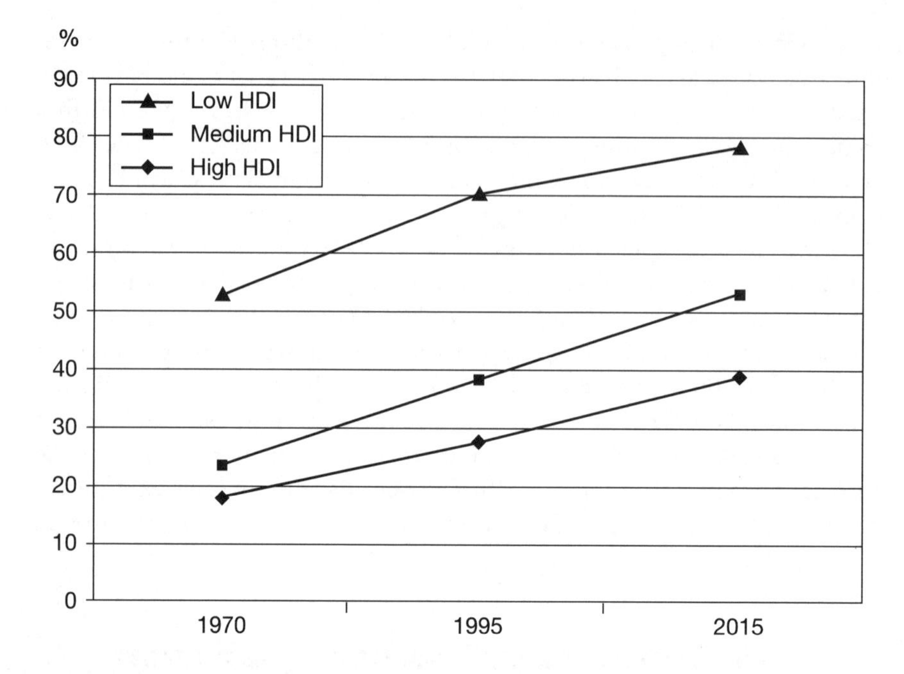

FIGURE 1.6 Urban population as a percentage of the total population in countries grouped by the UN human development index (United Nations, 2001: 10)

Another feature of current trends in urbanization is the emergence of 'mega-cities', with very large populations (over 10 million). Figure 1.7 shows recent and predicted trends in population size for some of the world's largest cities. Comparisons can be difficult because of the problems of standardizing the criteria used to define urban areas, but the diagram suggests that the projected rate of growth is much greater for some cities than others. Cities such as Tokyo, Mexico City and the New York agglomeration, which were the world's largest cities in 1990, will grow rather slowly. Although they will retain their high global ranking position in terms of city size, they will be joined by the year 2010 by cities such as Dhaka and Lagos, which will have grown much more rapidly. These cities will each incorporate more than 18 million people, and will therefore become much larger than some other major cities of the world. In the 1970s, places like London and Paris were among the world's 20 largest cities, but by 2010 their population size will not exceed 10 million and they will therefore rank lower.

Another important feature of city growth is the change in the characteristics of cities as a result of processes of globalization (reviewed,

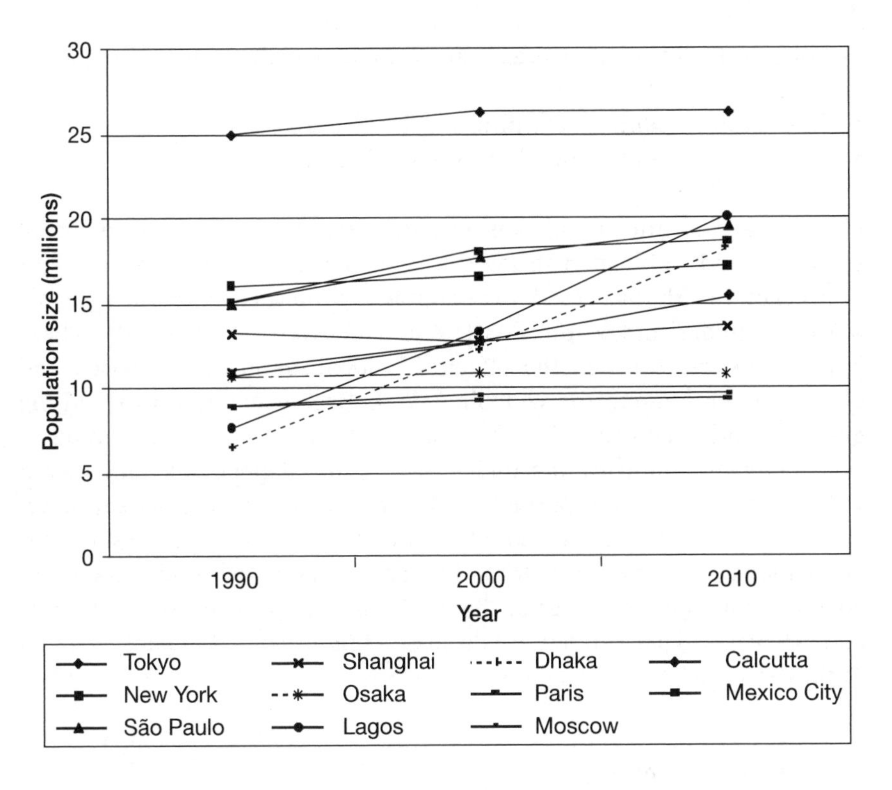

FIGURE 1.7 Trends in population size for selected large cities (United Nations, 2001: 11)

for example, in: Thrift and Amin, 1994; Massey and Jess, 1995; Clark, 1996; Waters, 1995; Keil, 1998). These include:

- compression of the time taken to cross the globe;
- increased global communication;
- global circuits of capital;
- increased impact of cultural influences from around the world.

In global cities, like London, Paris, New York or Tokyo, these trends are particularly apparent. These are large in population size, but not necessarily among the largest. More important for their global city status are features including the following:

- centres of financial/commercial control of international business and the capitalist economy;
- international transport and communications;
- significant numbers in the population who are cosmopolitan, internationally mobile and ethnically diverse;

- comparatively high average incomes supporting many people in wealthy lifestyles;
- international centres for culture and the arts and sciences;
- international centres of education and learning.

Modernization, industrialization, urbanization and the associated social, economic and political restructuring which these produce have major impacts on health, discussed, for example, by McMichael and Beaglehole (2000), Bettcher and Wipfli (2001) and Woodward et al. (2001). These impacts operate through the effects of globalization on the social and physical environment, the widening gap between rich and poor and the global spread of consumerism. The ease and speed of global communications between countries and their growing interdependence and shared vulnerability are also important for health and wellbeing. Factors influencing health in a distant country now often have repercussions for populations closer to home. In particular, it is now well understood that conditions in poor countries of the world have impacts on rich countries and vice versa. These effects are discussed in more detail in later chapters in this book.

SUSTAINABLE CITIES?

There is growing concern about how to make globalizing economies and urbanization sustainable, and debates over the impact of urban living on population health and welfare are often central to this discussion. Selman discusses the nature of sustainable development and cites the definition offered by the Brundtland Report (World Commission on Environment and Development, 1987): 'development that meets the needs of the present without compromising the ability of future generations to meet their own needs' (1996: 10–12). The United Nations Conference on Environment and Development (1992) has set out a number of key areas to be addressed, summarized as *Agenda 21*. These include action with the aim of providing or improving: adequate shelter for all; human settlement management and land-use planning; integrated environmental infrastructure; sustainable energy and transport systems; sustainable construction (see also Selman, 1996: 109–12).

The full range of issues involved in sustainable urban development is therefore very broad (for example, see the collection of readings edited by Satterthwaite, 1999a). *Agenda 21* incorporates both a global perspective and a local agenda. Sustainability is therefore seen to depend partly on local context and on the actions of a range of stakeholders at the local level, and regulation and coordination of these is often viewed as a special responsibility of local governments. Five categories of environmental action are important in cities and clearly indicate the centrality of environmental health objectives, according to Satterthwaite (1999b). These include: what he calls 'the *sanitary agenda*', concerning control of infectious and parasitic diseases; reduction of chemical and physical hazards; achievement of a high quality urban environment, for example, in terms of green space and cultural heritage; minimization of transfer of environmental costs to other areas; ensuring sustainable consumption that meets the needs of city dwellers without undermining environmental capital. Chapter 6 considers geographical perspectives on some of the physical environmental factors which are significant for environmental health. Urban public health depends on sustainable strategies for the social as well as the physical environment, so the socio-economic and political processes discussed in other chapters of this book are relevant to sustainability. Selman (1996) points out that it is debatable whether sustainable development, as defined by the Brundtland Report, is compatible with the idea of continued economic growth at the rates experienced globally in recent decades. Discussion in Chapter 4 returns to the issue of whether global economic growth is beneficial for health and equity in health.

THE SIGNIFICANCE OF THE 'URBAN PENALTY'

The urbanization trends shown above mean that the health of urban populations will become an increasingly important issue for all parts of the world over the next 10 to 20 years. This is all the more true because the health status of rapidly urbanizing populations is often relatively poor (Harpham, 1994a, 1994b). In large urban areas of poor countries, infant mortality has declined more slowly than in smaller towns and villages and in some of these countries urban infant mortality rates have increased (United Nations, 2001) to high levels: for example, urban infant mortality rates in African countries are in excess of 12 per 1000 for girls and 15 per 1000 for boys. These countries are therefore facing the problem of the 'urban penalty'. This damaging impact on health of rapid urbanization and industrialisation was also observed during the 19th century industrial revolution in countries like Britain, where infant mortality rates were very high until urban public health measures were introduced to ameliorate the poor living conditions associated with rapid urban industrial growth (e.g. Williams and Mooney, 1994; Congdon et al., 2001). Even today, in some parts of cities in more developed countries, health is still relatively poor (as shown above for London). There is a growing body of policy and theory concerning action to address health inequalities, which focuses especially on the health disadvantage found in urban areas. This is discussed for example in Chapter 9, which considers WHO Healthy Cities initiatives and national urban regeneration programmes.

The scale of contemporary urban development, in terms of population size, density and geographical extent, encourages a focus on the significance of urban as opposed to rural health disadvantage. In urban areas, there tend to be geographical concentrations of deprived populations that are large and dense enough to have a marked effect on geographical health variation in a country as a whole. Those arguing for a focus on deprived urban areas maintain that these are the places where the burden of ill health is greatest.

HEALTH INEQUALITIES IN RURAL AREAS

Although this book is especially concerned with health in cities, many impacts of social exclusion and poverty on health, discussed in Chapters 3 and 4, are likely to be relevant in rural as well as urban areas. There are some aspects of deprivation which may be more specific to rural areas and are also important for health. For example, Gesler and Ricketts (1992), Verheij (1996) and Higgs (1999) have compiled overviews of factors which influence health outcomes in rural areas of high income countries. The health disadvantage of rural populations in low income countries often presents even greater challenges (e.g. Phillips and Verhasselt, 1994). Authors such as Haynes and Gale (1999) have shown that the sparsely distributed populations of rural areas in Britain result in rather heterogeneous populations being aggregated in small area statistics. Thus small, localized pockets of severe rural deprivation tend to be lost in the average for the wider area. Those arguing for greater resources for rural areas suggest that regional resource allocation formulae, based on area averages, often underestimate the significance of these deprivation 'hot spots' in the countryside. The pattern of rural health and its trends over time may also be significantly affected by the impact of migration flows on the composition of the population.

The reviews cited in the previous paragraph refer to evidence that, while semi-rural areas may show health advantages compared with inner cities, inhabitants of the most remote rural areas have relatively poor health. This links to the proposition that distinguishing characteristics of rural deprivation relate to isolation and lack of access to services and facilities, which may have important effects for health as well as other life chances. There are special features of rural areas, such as the lack of opportunity to benefit from economies of scale in health care provision and the costs of sparsity and isolation, which make rural health care provision relatively expensive (Watt and Sheldon, 1993). Issues of access to health care in rural areas are given some attention in Chapter 5. Chapter 6 also considers natural environmental health hazards that are important in rural as well as urban areas.

UNDERSTANDING HEALTH INEQUALITY: THE WHOLE SYSTEM PERSPECTIVE

Pratt et al. (1999) argued that to address health inequalities and achieve health improvement in urban populations it is necessary to adopt a 'whole system' approach. According to this perspective, biological metaphors, rather than mechanical metaphors, are more appropriate conceptual frameworks for understanding reality (Battram, 1998). Similarly, Capra (1997) develops ideas put forward by authors such as Maturana and Varela (1980) concerning 'living systems' which are dynamic. Maturana and Varela described the process of 'autopoeisis', by which the components of a living system act together to make and remake the whole system. Capra (1997: 156) identifies three dimensions of 'living systems': the *pattern* of relationships which makes up the system; the *structure* which comprises the physical form of the system; and the life *process* by which the system actively creates itself. A key aspect of the 'living system' is that these different dimensions cannot be understood in isolation, but need to be understood in terms of their relationship to each other and to their context. Capra argues that:

> The great shock of twentieth century science has been that systems cannot be understood by analysis. The properties of the parts are not intrinsic properties, but can be understood only within the context of the larger whole . . . Systems thinking is 'contextual' which is the opposite of analytical thinking. Analysis means taking something apart in order to understand it; systems thinking means putting it into the context of the large whole. (1997: 29)

Pratt et al. (1999) use the whole system approach to inform strategies for organizational change and participation in the public health arena because it emphasizes the need to view the factors influencing population health as complex systems. This focuses attention on the connections between the different parts of the system as well as on individual component parts:

> Complex issues such as urban regeneration, homelessness, underachievement in school or long term unemployment are influenced by the actions of many individuals, groups and organizations . . . In trying to tackle them the tendency is to break them into actionable parts. Yet . . . many of these problems refuse to go away . . . it might be more fruitful to think of them as issues for an interconnected system to tackle together. We chose to shift our attention from parts onto the whole and thus to the connections between parts – how things fit together. (1999: 3)

Taken to its logical limits, whole system theory may not be very applicable in practice because it is difficult to imagine or model systems that are very complex and extensive. This book does not try to take this perspective to such an extreme. However, as an approach to population health it is useful here because it examines the relationships which connect a variety of factors relevant to health, it focuses attention on processes which are dynamic, and it accepts that these factors and processes may be sensitive to context. This book considers some of the connections between public health and dimensions of urban systems such as: the economy; employment and working conditions; housing and the domestic environment; patterns of consumption and access to goods and services; the 'pollution syndrome'; social and political structures; and wider participation in decision making by different social groups.

These aspects have not always conventionally been considered as part of the 'health system' and there has historically been a separation between research, policy and practice relating to health and medicine on the one hand, and social and economic policy on the other. However, there is a growing realization that policies and actions relating to these various dimensions of urban life are relevant to public health. Actions and interventions outside the conventional 'health sector' should be subject to health impact assessment, and planned in ways which are most likely to reduce damaging effects on public health and promote health gain. The growth of interest in such approaches to health impact assessment is discussed in Chapter 9.

LANDSCAPES OF HEALTH INEQUALITY

Geography as a discipline is particularly well suited to the 'whole system' way of thinking, since geographers are concerned to examine the interaction between population and environment, and how this varies across space and in different types of place. The geography of health is focused on the ways that the health of populations is differentiated between places and the range of factors that explain these differences. The concept of a *landscape* has often been used to convey the idea of a system of factors and processes that interact in particular settings to produce geographical variation. The elements included in a conceptual landscape depend on knowledge of the key relationships and causal pathways producing geographical variation. It is possible to imagine different conceptual landscapes of health geography, associated with different theoretical perspectives on health variation, overlaying each other in the same place.

Ideas about space and place have been developed as central themes in geography. (However, some geographers share with other social theorists a concern that space and place should not be given disproportionate importance and should be considered in relation to other important aspects of social organization and processes.) For example, this book explores ideas of geometric and social space used in geography and considers notions of place in terms of *location* in physical space and *position* relative to other places. A place can also be considered as a *locale*, with idiosyncratic attributes making it distinctive, and it may be interpreted in terms of *senses of place,* which result from the ways that individuals and communities imbue certain geographical settings with social significance and values. Some of these aspects of space and place are objectively measurable and quantifiable while others are socially constructed, relative and more appropriately explored and described using qualitative methods.

This book examines the articulation between theory and empirical methods for understanding health variation in urban areas. It draws partly on previous discussion by authors, who argue for a renewed emphasis on the importance of *place* for health inequalities (for example, Jones and Moon (1993), Kearns and Joseph (1993) and Macintyre et al. (1993). Curtis and Taket (1996) and Picheral (2001) have also described changing perspectives in the geography of health and a re-emergence of place as a key theme in this field, viewed more recently using concepts from humanistic and cultural geography. Curtis and Jones (1998) recently argued that a perspective on the importance of place in the geography of health is informed by elements of a number of theoretical frameworks. They suggested that these questions could be expressed in terms of conceptual landscapes. Five types of landscape focus are considered in this book, each of which is linked to a particular theoretical perspective as follows:

Theoretical framework	*Landscape focus*
theories of sense of place/identity	therapeutic landscapes
theories of social and political control	landscapes of power and resistance
theories of production/structuration	landscapes of poverty and wealth
theories of consumption and lifestyle	landscapes of consumption

theories of ecological/epidemiological processes	ecological landscapes

Each of these is given detailed attention in the following chapters. The perspectives considered here are certainly not unique to geography, and to some extent this book is concerned with the ways that health geography interfaces with other disciplines in medicine, natural sciences and social sciences. Indeed, one of the important messages of the text is that an interdisciplinary perspective is essential for useful research on public health. However, an emphasis on space and place emerges particularly strongly in health geography. This book presents evidence concerning relationships influencing public health that are dependent upon the setting, defined by place, time and social context. These may have particular implications for strategy and action to tackle urban health inequalities.

FURTHER READING

For useful reviews of the scope and content of health geography, students might like to look at other texts as well as this one. The following are recent publications which I would recommend. I have referred to them at several points in the course of this book.

Gatrell (2002) provides a very good introduction to the geography of health. It is written in a way which makes it very useful and accessible for students and it is international in scope. It provides especially good coverage of themes relating to physical health and the physical environment and so it is a good complement to this book, which is more especially focused on the social environment, and considers mental as well as physical health.

Shaw, Dorling and Mitchell (2002) provide another accessible introduction to the literature on health inequalities from a social and geographical perspective.

Meade and Earickson (2000) is another very authoritative account of this field, with a particular emphasis on more 'classical perspectives which emphasize the links between geography and medical ideas about disease, especially studies from the United States.

Howe (1997) is a new edition of a text that provides a classic perspective on medical geography, with some very interesting British examples to consider.

Texts that are older, but still relevant today for the ideas they include about how to approach geography of health, are: Jones and Moon (1987) and Curtis and Taket (1996).

Students may also like to consider how ideas about health geography relate to perspectives from other disciplines. This chapter has referred, for example, to authors writing about public health epidemiology, such as Rose, G. (1992) *The Strategy of*

Preventive Medicine (Oxford, Oxford University Press), and from a sociological perspective, such as Beck, U. (1992) *Risk Society: Towards a New Modernity* (London, Sage). A good collection of writing on health inequality from a range of social science disciplines is provided by Bartley, M., Blane, D. and Davey-Smith, G. (eds) (1998) *The Sociology of Health Inequalities* (Oxford: Blackwell). For example, Chapters 1, 3 and 6 by Wilkinson and colleagues, Popay and colleagues and Shaw and Colleagues provide other perspectives on how social environment influences health.

OBJECTIVES AND QUESTIONS

Learning objectives for students reading this chapter will probably include the following:

- to understand that ideas of 'health', 'illness' and risk to health are socially constructed and therefore variable between societies and social groups;
- to appreciate problems that may arise from the ecological and the atomistic fallacies;
- to have an appreciation of the extent to which indicators of health vary at different spatial scales;
- to understand the significance of the 'urban penalty' for health in a rapidly urbanizing world;
- to appreciate the scope of a geographical perspective on health.

After reading this chapter you may want to review how far these learning objectives have been achieved by attempting to answer the following questions.

1. 'It is impossible to construct a single definition of health.' Explain the extent to which you agree with this assertion.
2. Why is it important for geographers to consider health inequalities at different geographical scales?
3. What do you understand by the ecological and the atomistic fallacies? Why are they both important for health geography?
4. 'Geography of health is an ideal perspective from which to consider the whole system of factors influencing health.' Critically evaluate this assertion.

SECTION I

Geographical landscapes of health

2 Therapeutic Landscapes: wellness, healing places and complementary therapies

---

As we saw in Chapter 1, ideas of health and illness are constructed in varying ways. This chapter considers in more detail the contrasts between biomedical perspectives on disease and alternative views of health and illness. The chapter starts with a discussion of lay understandings of the body and its capacity for socially meaningful roles and activities. It focuses especially on interpretations of health that emphasize wellbeing or a varying capacity for roles and activities which are important for participation in society. Therapies based on complementary and alternative medicines, other than biomedicine, have characteristics that make them especially sympathetic to this view of health. They tend to treat health and illness holistically, in the context of people's life experience and their social and physical environment. Some aspects of these 'alternatives' to biomedicine are therefore considered in this chapter.

Also of particular interest here are the ways that health is associated with senses of place. The relationships between the body and the socially defined spaces it occupies are considered. Geographers have explored how the experience of places can affect health and wellbeing, either in ways that are 'therapeutic' and supportive or, conversely, in ways which are detrimental to good health. Experiences in social spaces that are not medical are important for health, as well as those within medical spaces such as hospitals or clinics. This chapter considers 'therapeutic landscapes' in the home, natural spaces and 'healing places'. It also discusses the relevance of therapeutic landscapes for settings for medical treatment.

Throughout this chapter, we also see that the perception and experience of therapeutic landscapes are variable for different social and professional groups, and between societies and countries. Discourses about health and therapeutic landscapes often reflect social relations and they contribute to differentiation and inequality among social groups. The views of the most powerful groups and nations tend to dominate in debates about what affects health and illness and what are suitable settings for treatment, but they are often questioned and resisted by the alternative perspectives of less powerful groups.

THE BODY: BIOLOGICAL ENTITY OR SOCIAL REPRESENTATION?

The research reviewed in this chapter raises questions about how we understand the human body. Hall (2000) argued that interpretations of the body have often been dominated by a biological perspective which treated the physical body as distinct from the mind, and led to an emphasis in biomedicine on the physical and material state of the body, rather than on the way that this state is interpreted and represented in a social sense. Biological differences between bodies (including the presence or absence of disease or impairment) were the focus for definitions of health and illness. The social model of disease and disability, on the other hand, interprets the significance of differences between bodies as something that is socially created and represented. These perspectives emphasize that health and disability are relative, socially constructed phenomena. Socially determined factors influence the experience of wellbeing, illness and disability associated with different physical states.

However, the individual experience of the biological state of the body is also important. There are some real biological and physiological differences between, for example, men and women, or people with differing genetic makeup. People do experience variation in their physical capacity for different activities, or in the presence or absence of pain or other symptoms associated with illness, even though the impact of these is constrained and constituted by social and spatial factors (as argued by Kearns, 1993; 1994). Hall (2000) therefore suggests that a perspective that acknowledges both the biological and the socio-spatial aspects of bodily experience, and the interaction between them, can provide a more complete understanding.

HEALTH, CAPACITY AND GEOGRAPHICAL PERSPECTIVES ON DISABILITY

The interpretation of the human body as the outcome of social construction and representation puts in question ideas about what comprises a 'healthy body'. As noted in Chapter 1, health is often defined in terms of capacity to perform roles and activities. Especially important for perceived health is capacity for *salient* roles, which are seen by individuals and by their social group as important for satisfying and fulfilling participation in society (Thoits, 1991; Curtis and Lawson, 2000). These salient roles will vary among individuals and may, for example, be work related (a person's capacity for paid employment and the type of job they can do), associated with caring for a family and maintaining the home, or linked to social, sporting or creative activities. Illness or disability is particularly likely to appear significant to the person affected, and to cause distress or mental illness, if it limits capacity for salient roles.

Current perspectives in health geography have therefore begun to recognize (perhaps rather belatedly) that there is an important link to be made between, on the one hand, more conventional health geography concerned with medically defined disease, and, on the other hand, research relating to disability and impairment. Differences in physical capacity of the body may result from disease. However, variations in the physical and mental state of the body also can arise congenitally or due to external agents. Human bodies are all to some extent differently physically formed and are biologically 'programmed' to function differently because of genetic variability. Sometimes differences in physical capacity also arise due to the effects of injury or other external causes, which are not related to disease or genetic predisposition.

Differences in bodily capacity can also be seen as determined, at least in part, by the social construction of a healthy body and of the types of roles and activities which societies and cultures expect of their individual members. Oliver (1990) considered various social theories that explain the social construction of 'disability' in terms of: social progress and enlightenment; capitalist modes of production; administrative categorization; cultural and professional ideologies such as sexism, racism and medicalization; the development of political movements.

In geography, the emphasis has also been on ways that individual bodily capacity interacts with characteristics of space in the construction of perceptions of 'difference'. Golledge (1993) considered particularly the

Box 2.1 Dimensions of disability and social acceptance

The following list is adapted from a discussion by Dear et al. (1997: 471–73), who defined eight social *dimensions of acceptance*. These influence the way that biological variations in capacity are viewed socially, and the extent to which they are seen to be 'different' from socially determined 'norms'. The more 'different' a person's condition or bodily state on these criteria, the less 'acceptable' they are likely to be to others in society.

Functionality: capacity to perform certain activities and jobs.

Aesthetics: visibility of difference and whether it conflicts with accepted standards of appearance.

Established conventions: whether bodily difference interferes with established, predictable rules for social interaction.

Individual culpability: whether difference in capacity results from failure to comply with social norms.

Dangerousness: whether behaviour causes discomfort because it is seen as a risk to others or themselves or is unpredictable.

Treatability: whether a person has a condition which can be cured or reversed by relatively simple medical intervention or whether their condition is incurable or can only be treated through special interventions.

Empathy: whether others are willing or able to be sympathetic or empathetic to a disability.

Contagion: whether a person's condition is seen as a threat to others because it could be communicable.

way that the spatial and physical environment affects the disabling effects of visual impairment. Imrie (1996) considered the processes of social differentiation by which groups establish their identity and distinguish themselves from 'other' groups. He suggested (Imrie, 1996: Figure 2.2) that this has resulted, for example, in the construction of two 'categories' of people: those who are 'able bodied' (and stereotypically viewed as 'normal, good, clean, fit, able, independent') and those who are 'disabled' (stigmatized as 'abnormal, bad, unclean, unfit, unable, dependent'). Dear et al. (1997) identified eight dimensions by which people with disabilities may be socially defined as 'different' from other members of society (see Box 2.1). The extent of difference from socially defined 'norms' determines the social acceptability of biological differences, as discussed in Chapter 3.

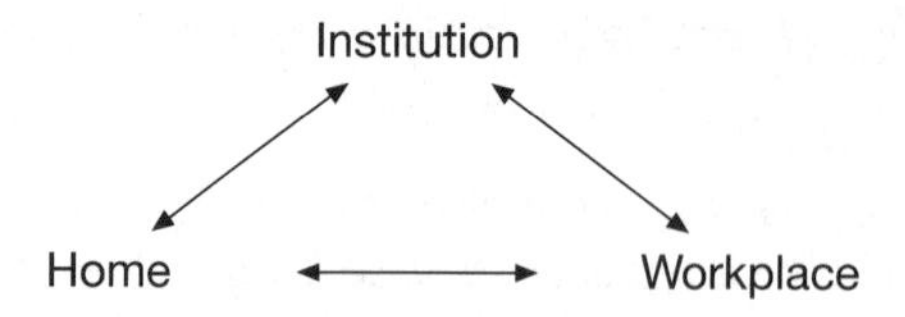

FIGURE 2.1 Typology for a 'social space of impairment' (proposed by Gleeson, 1991)

Gleeson (1999) also considered bodily capacities as culturally and historically conditioned by the social and physical environment. He employs conceptual frameworks which include ideas from Lefebvre (1991, translated from the French original of 1974). The conception of the body was seen as dependent on time and space. Gleeson (1999: Figure 4.3) interpreted Lefebvre's ideas in terms of a 'social space of impairment', shown in Figure 2.1, comprising the 'key nodes' of home, institution and workplace. The social construction of 'disability' will depend on the social relations which exist in these types of spaces.

Several of these authors (Imrie, 1996; Dear et al., 1997; Gleeson, 1999) consider the processes of oppression and exclusion which operate to disadvantage those whose health or bodily capacity differentiates them from the socially accepted 'norms'. These effects are discussed in Chapter 3.

HEALTH AND WELLBEING

This chapter aims to show how geography can move beyond ideas of health in terms of longevity or absence of disease and disability, towards more positive concepts of health and wellbeing. Positive health may include ideas of fitness and realization of the individual's full capacity for participation in society. There is debate over the relationship between these positive aspects of health and concepts of wellbeing. Wellbeing relates to aspects of mental state as much as physical condition. It is a complex construct concerning optimal experience and functioning (Ryan and Deci, 2001). It is often interpreted in terms of notions of happiness, life satisfaction and quality of life. Good health may be associated with, or contribute to, wellbeing (Fernadez-Ballesteros et al., 2001). However, other factors also seem to be important for wellbeing. For example, religious involvement, spirituality and personal meaning have been found to influence wellbeing, at least in older populations (Fry, 2000). Senses of achievement and freedom in life and work may be important for happiness in younger populations (Furnham et al., 2001). Ryan and Deci (2001) suggested that there are two broad perspectives on wellbeing which provide the theoretical basis for research. They defined the *hedonic* approach, focusing on happiness and defining wellbeing in terms of attainment of pleasure and avoidance of pain. The *eudaimonic* approach emphasizes meaning and self-realization and the extent to which a person is fully functioning. These different notions of wellbeing are not universally supported in the literature. For example, Haybron (2001) argues against the hedonic theory, since happiness does not necessarily derive from pleasure. Also Haybron argues that happiness does not depend on experience at particular points in time.

Wellbeing, or quality of life, is not an identical construct to wealth or a high standard of living. Wealth and wellbeing probably relate in rather complex ways (Sirgy, 1998; Easterlin, 2001). For example, Easterlin (2001) considered survey evidence on reported happiness and satisfaction with material standard of living (perhaps a limited aspect of overall life satisfaction). Easterlin postulated that at any point in the life cycle, an individual's life satisfaction will depend on their own experience and the expectations and norms of the social group with which they interact. For those who enjoyed rising incomes through life, happiness did not seem to increase because their expectations rose in line with changes in their lifestyle and that of their peers. This argument is interesting, from a geographical point of view, because Easterlin suggested that sense of happiness depends not only on the expectations within one's immediate family, but also on the perceptions of a wider social group with which one has contact in day-to-day settings, such as the place of work or residential neighbourhood. This suggests that social context is important for life satisfaction, as well as individual characteristics. Easterlin's work also highlights variability among social groups in life satisfaction and the processes influencing satisfaction, which may be important for our understanding of inequalities in health and wellbeing. The relevance of place for the interaction between wealth, poverty and health inequality is further discussed in Chapter 4.

PERSPECTIVES ON WELLBEING FROM DIFFERENT THERAPEUTIC STANDPOINTS

The biomedical perspective is very pervasive in studies of health, at least in modern western society, and it represents a particular view on health and illness. Biomedicine focuses on medically diagnosed diseases (recognized in terms of bodily signs and symptoms) and tends to define 'wellness' in terms of absence of, or low risk of, disease. The traditional emphasis in biomedical treatment has been on curing illness although it also has a growing role in illness prevention and health promotion. Philo (2000) has drawn attention to the apparently geographical language in Foucault's (1993: 4–15) interpretation of the development of modern medicine. Foucault discussed three types of 'spatialization' progressively imposed by biomedicine during its development. These might be interpreted in terms of spatialization of thought (or imagination), bodily spatialization and institutional spatialization. Foucault argued that medicine first imposed a particular way of thinking about illness, as an array of bodily signs and symptoms which could be used to recognize a specific disease in any person. This produced a narrow view of the individual patient in terms of a limited set of specific attributes. The second form of spatialization involved the detailed localization of disease involving a view of the human body as a rather impersonal 'space' in which disease is observed and treated. The third type of spatialization involved the concentration of medical activity and treatment in a specialized type of institution – the hospital.

Complementary medicine seems to have a stronger association with promoting wellness and wellbeing than biomedicine. Curtis and Taket (1996) and Wiles and Rosenberg (2001) reviewed studies which emphasize the role of complementary medicines in building resistance to illness and protecting and promoting good health, as well as curing illness. Coward (1993) pointed out that one reason for the recent growth of interest in alternative medicines is that they reflect the great importance which people in western societies now place on health and on the body. She also argued that there is a strong perceived connection between alternative medicine and the links made in lay culture between nature, natural lifestyles and wellbeing.

Complementary and alternative therapies tend to interpret health and wellbeing as attributes of the whole person. As noted above, the holistic view contrasts with the medical perspective on health, explored by Foucault, which tends to view the individual body as a 'case' of a disease identifiable from a specific set of symptoms which are always similar for the same disease. Coward (1993: 95) argues that, ' "Holism" is the great strength of the alternative health movement', calling for a view of health as the 'wellbeing of the whole person . . . not just a fit body, but a well mind or spirit'. Aakaster (1986) has identified several differences between biomedicine and other forms of medicine (outlined in Figure 2.2). The holistic view of disease espoused by alternative therapies emphasizes these distinguishing features.

Perspective on:	'Conventional' (biomedicine)	'Alternative' (complementary)
Health	Absence of disease	Balance of opposing forces
Disease	Specific; located in particular organs	Bodily indications of disruptive forces
Diagnosis	Morphological	Functional
Therapy	Combating destructive forces	Strengthening constructive forces
Patient	Passive recipient	Active participant

FIGURE 2.2 Attributes of 'conventional' and 'alternative' medicines (after Aakaster, 1986)

Aakaster also draws attention to the common idea in alternative medicines that health is about maintaining a good balance of the various elements in the body. In these forms of medicine, the strategy for maintaining good health or treating illness often involves restoring a healthy balance and building up the capacity of the body to resist disease. This strategy also stresses the power that a person has to exercise control over their own health. In biomedicine, the patient has conventionally been treated as relatively passive and subordinated to the expertise of the doctor. It can be argued (see Chapters 3 and 5) that disadvantaged populations are most likely to find themselves disempowered in biomedical settings, and that this may contribute to health inequalities. In contrast, in complementary therapies, the relationship between the doctor and the patient is one that empowers the patient to play an active and central role in their therapy. This could be particularly beneficial for socially disadvantaged populations. However, in many western societies, complementary therapy is relatively costly to the patient, because it is not subsidized by the state in the same way as biomedicine. Chapter 5 discusses the commodification of complementary therapies and their use as consumer products. Compared with more affluent social groups, poorer populations may not therefore have the same opportunities to use these alternatives to state funded biomedical care.

Curtis and Taket (1996), Gatrell (2002: 7–8) and Wiles and Rosenberg (2001) have presented geographical perspectives on these holistic, alternative views of health and the body. Wiles and Rosenberg (2001:220–1) suggest that complementary therapies support a contextualized (and therefore fundamentally geographical) view of the experience of health and illness. Pawluch et al. (2000) also emphasize the importance of social and ethnic influences for individual variations in perception of alternative therapies. Part of the attraction of complementary therapy in western countries is its avoidance of the reductionist perspective of biomedicine, which conventionally considers the human body out of its social and natural environmental context. The value placed on the individual identity of the person (seen to be lacking in biomedicine) is also important for many users of complementary therapy (Luff and Thomas, 2000).

Biomedicine is strongly associated with the paradigms of western scientific thought and associated with the way of life in economically developed 'western' cultures. Minority ethnic populations in western countries may view traditional medicines as part of their cultural heritage. For these groups, using traditional medicines may be a way of affirming an ethnic identity which is distinct from the majority culture (e.g. Donovan, 1986a; Lawson, 2000). Use of certain types of alternative medicines also has a strong link to spirituality (e.g. Astin, 1998; Engebretson, 1999; Lawson, 2000) and may be valued because these alternative therapies allow people to express the links between physical and spiritual wellbeing which are important to them.

Variation between countries in access to biomedicine are closely associated with the international inequalities in health and human development which were outlined in Chapter 1 (this is discussed in more detail in Chapter 5). In low income countries, which find it difficult to support the costs of biomedicine, traditional medicines may offer health care that is sympathetic to local cultures and is less costly than biomedicine. However, health inequalities may result if people are too poor to have any access to biomedical care and are only able to use alternative medicines. The dominance of biomedicine in debates about health also reflects the global political and economic hegemony of high income 'western' countries. Some non-western countries have promoted alternative medicines as a way of expressing a degree of national independence from western influence. Curtis and Taket (1996) draw attention to the symbolic significance of complementary medicines as expressions of other national and ethnic identities, especially for cultures with long established traditional medicines, such as those in China and India. For example, Chenhuei Chi, discussing the development of traditional medicine in Taiwan, commented that 'When both traditional and modern Western medical resources are incorporated into a national health care system, they also foster self-reliance and self-sufficiency' (1994: 308). The powerful influence of national governments over the development of traditional, alternative medicines is also illustrated in Chenhuei Chi's (1994) account, which contrasts the recent development of Chinese traditional medicine in China, where it has been strongly promoted, with the very different pattern of development in Taiwan.

THERAPEUTIC LANDSCAPES: THE SOCIO-CULTURAL LINKS BETWEEN WELLBEING, HEALTH AND PLACE

As explained in Chapter 1, geographical perspectives on health and wellbeing may be expressed in terms of conceptual 'landscapes'. The concept of *therapeutic landscapes* has been used to present geographical ideas about factors important for wellbeing. The theoretical underpinning for therapeutic landscapes comes particularly from cultural and humanistic geographies. The idea of a link between place and psychological wellbeing has a long pedigree in geography, connected, for example, to theories of sense of place and their relationship to an individual's cultural identity.

Yi-Fu Tuan (1974), writing from a humanistic geographical perspective, proposed the term *topophilia* to express the affective bonds which individuals feel towards certain places. He discussed the idea of 'spaces of care', with which people feel emotional ties. Health geographers working on these questions (see, for example, Gesler, 1993; Kearns, 1994) also frequently invoke work in geography by Eyles (1985), Jackson (1989), Harvey (1989a), Cosgrove and Jackson (1987) and Massey (1991) who extended the application of social and cultural theory in human geography. These approaches emphasized the role of place in the construction of individual senses of identity and the ways that this is influenced by the wider society to which the individual belongs.

Yi-Fu Tuan's (1974) work was especially concerned with how place relates to perceptions of *shared* cultural identity of whole societies and communities and the historical continuity of these cultural traits. More recent geographical work focuses also on how identity depends on senses of *difference,* involving distinctions between oneself and others in society. This perspective contributes to the notion of societies as complex and pluralistic. There is also more emphasis on changing and fluid senses of identity, depending on the social and spatial context.

Therapeutic landscapes consider the role of place in social relations. They concern the ways that wider social structures interact with individual human agency and perception. It is possible to see places as having a passive role as 'containers', where social processes are played out in particular settings. However, geographers also view places as actively contributing to these social processes. Places may *constrain or limit action* because they separate social groups spatially (e.g. through residential segregation), or because certain types of behaviour and interaction are imposed on people by their presence in certain types of place (for example, through expectations about what is proper behaviour in a public place). This is discussed in more detail in the following chapter concerning processes of social control. Places in the built environment may also *produce or constitute social relations* by representing and reinforcing certain aspects of social structures. Authors such as Gregory (1994: 94) draw attention, for example, to the importance of 'symbolic' space which legitimates and communicates the ideological structure of a society. Places which have culturally specific functions, such as centres of the arts, religion or government, are, of course, created by society. Once created, however, they actively contribute to the maintenance of the ideological character of a society. Green spaces and wildernesses also have a symbolic role for societies, representing socially meaningful relationships between human beings and the natural world.

Kearns (1991) and Kearns and Joseph (1993) have drawn attention to the importance for health geography of these symbolic meanings attached to places by social groups and by individuals. Kearns and Gesler (1998: 7–8) discuss various aspects of this 'cultural turn' in the geography of health. Cultural factors affect the way that human beings interpret and create physical dimensions of landscape which are important for health (e.g. the quality of the air, water or ground, or the character of buildings). These authors also invoke the idea of the individual as 'implicated in a cultural landscape and an active agent in inventing places'. These landscapes have particular meaning and symbolism for people, associated with their experience of health and wellbeing. Furthermore they demonstrate the role of place in the social structures which frame pre-capitalist and capitalist societies and which are important for health experience.

Health geography has thus recognized that sense of place may affect therapeutic landscapes of wellbeing. The idea of therapeutic landscapes has been developed by authors such as Gesler (1992; 1993), Kearns and Gesler (1998) and Williams (1999a). Some places have attributes making them beneficial to psychological and physical health. Other places have quite the opposite characteristics and may therefore be detrimental to health. Gesler (1993) and Williams (1999) define therapeutic landscapes as encompassing physical, psychological and social environments with a reputation for treatment or healing in a physical, mental or spiritual sense (Gesler, 1993: 186). A broader perspective on therapeutic landscapes would include not only their contribution to healing of illness or distress, but also their role in promoting good health and wellbeing. For example, Williams (1999b: 1) suggests that people associate certain places with 'peace, relaxation, rejuvenation, restoration'. Fullilove (1996) suggests that

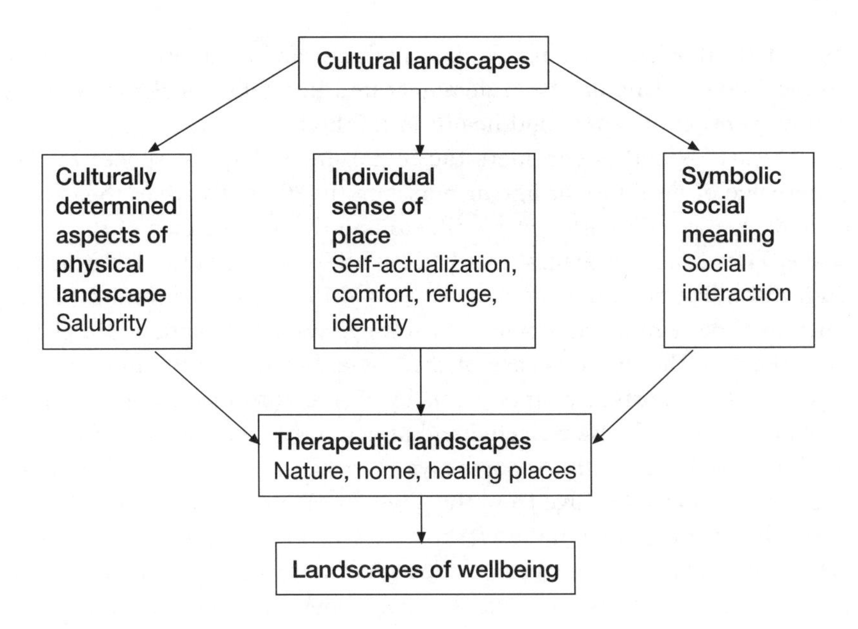

FIGURE 2.3 Therapeutic landscapes of wellbeing

place is important for senses of attachment and familiarity. This is argued to be important for mental health, as discussed here in Chapter 7.

For example, place may influence wellbeing because it is physically salubrious. Place may be linked to one's sense of individual identity, helping to realize one's personal wishes and aspirations and offering protection and refuge from external threats. Place contributes to well-being by helping to foster a sense of belonging to, and being of value to, a wider society or a particular social group. These processes are seen to operate through the symbolic role of space and also by its more material role in permitting or preventing social interaction and physical experience. These relationships are summarized in Figure 2.3.

A number of studies in geography and related disciplines have examined the therapeutic relationship between place and wellbeing. Different types of spaces can contribute to health and wellbeing, including the home space, natural spaces, places imbued with special healing properties such as spas and pilgrimage sites, and certain types of medical spaces. Examples of these are discussed below.

THERAPEUTIC ASPECTS OF HOME

The home space is important for wellbeing. It is often depicted as private, intimate, feminine and domestic (McDowell, 1983; 1993; Squires, 1994).

As such it offers special characteristics which can be important for health and wellbeing. This has been demonstrated in studies of the experience and perception of illness and health in the home.

Dyck (1998: 107) considers the importance of home spaces for the experience of disability or impairment. Smith (2000) discusses the experience of young men with chronic illness and the significance of the home space. These authors suggest that bodily differences that mark out people with disabilities or illnesses may be seen differently in different places. They may be seen as 'acceptable' in settings such as hospitals and clinics or perhaps in the private space of the home, but less so in public places. Dyck (1998) reports accounts given by disabled women, showing how experience of the home reflected and affected their experience of disability. Both Dyck and Smith report ways in which the private and intimate home space offered refuge from the challenges encountered elsewhere in public life, for example in the workplace. This idea of the home as a 'safe haven' is a common perception, not specific to people with illness or disability, and it is also a permissive and liberating place where people can express themselves more freely.

However, Dyck reported that the home space was seen by some as a place of struggle with disability, where the women she interviewed tried to retain functional independence in acts of daily living. Both Dyck and Smith discuss the problems created for people needing formal health and social care at home, when the home space is encroached upon by professional carers. The sense of privacy and intimacy at home, for the person being cared for, may seem incongruent with the home as a 'place of work' for their carer. Smith comments on the difficulties for young men with chronic illness confined to a feminized home space where they felt they could not develop masculine roles.

Mowl et al. (2000) discuss the relationship between the home space and perceptions of the ageing body. Stereotypes of the ageing body are often associated with frailty and mental and physical decline, changes in physical appearance and behaviour, and reduction in the capacity to sustain roles and activities which are important in earlier life. The older people who were interviewed by Talbot in the study reported by Mowl et al. (2000) often associated restriction to the home space with these changes. They tended to describe themselves as making efforts to avoid the stereotypical attributes of ageing. Their strategies included going outside the home, to places where they could be active, part of a social group, and where they needed to pay attention to maintaining their physical appearance, avoiding an ill-kempt image redolent of old age.

The significance of home for health and wellbeing is perhaps most starkly demonstrated by the health damaging effects experienced by people who are homeless or for whom the home does not meet their material needs. These impacts of housing and homelessness on health are considered in more detail in Chapters 3 and 4.

Thus home as a place is subject to complex interpretations, depending on the range of experiences undergone in the home and the different symbolic meanings attached to the home space. The home has both therapeutic and non-therapeutic potential and therefore may either mitigate or exacerbate health inequality for those who are relatively disadvantaged due to illness or frailty.

THERAPEUTIC FEATURES OF NATURAL SPACES

Some research on therapeutic landscapes has concentrated on the ways that natural spaces contribute to health and wellbeing. For example, Palka (1999) reports research in the Denali National Park, Alaska, which was interpreted as promoting wellbeing through access to a pristine natural environment offering relaxation, restoration and an escape from more stressful, manufactured landscapes. Visitors interviewed by Palka acknowledged a sense of the park as a healthy place which was different from other landscapes. The scale, beauty, diversity, wildness and abundance of wildlife, the remoteness and quietness of the area and its unspoilt, unpolluted natural character, were often appreciated by visitors. They reported benefits to their wellbeing through feeling more relaxed, energetic, invigorated, and able to reflect on the beauty of nature.

Bell (1999) and Bell and Evans (1997) discuss the 'redemptive' characteristics of the 'National Forest' project to create a new forest environment in a part of the east midlands in England. The National Forest would link two remnants of ancient English forest in the area, thus underlining the historical significance for British culture of trees and wooded landscape. The environmental change was expected to be associated with improvement in the physical, mental and moral status of people using it and encourage healthy and responsible citizenship. The natural aesthetic was planned to 'rehabilitate' features of the industrial and agricultural landscape in this part of England where coal mining and intensive forms of agriculture have denuded the landscape of trees. The plans for the National Forest expressed an association between rehabilitation of the natural environment, in line with principles of biodiversity and sustainability, and creation of environments which would be morally and physically therapeutic for human health (Bell and Evans, 1997: 267). It was hoped that the forest would prove economically beneficial, as well as aesthetically and morally redemptive, by encouraging new tourist activity in the area. This connection between the attractive, therapeutic character of places and their economic success is also evident with respect to other types of healing places, as discussed below.

Bell has also pointed out (e.g. Bell, 1999) that these therapeutic and 'redemptive' dimensions of landscape might depend on the values and perceptions of different groups of people using these natural spaces. If efforts to 'improve' landscapes are unsympathetic to ideas of what is therapeutic for relatively disadvantaged groups, this may contribute to health inequality. She discussed, for example, the tendency for the National Forest strategy to be dominated by the views of professional experts rather than those of local people. The advantages of 'improving' the landscape by promoting natural and sustainable spaces attractive to tourism needed to be weighed against the loss of potential benefits to welfare for local people from developing other forms of land use such as mining or landfill industries.

HEALING PLACES: SPAS AND HOLY PLACES

Gesler (1992) outlined the characteristics of places which have achieved lasting reputation for physical, mental and spiritual healing. These include:

> such natural characteristics as magnificent scenery, water and trees; human constructions such as healing temples or spa baths; contributions to sense of place such as feelings of warmth, identity, rootedness or authenticity; symbolic features such as healing myths; the incorporation of familiar, daily routines into the treatment process; sensitivity to cultural beliefs; and an atmosphere in which social distance and social inequalities are kept to a minimum. (Kearns and Gesler, 1998: 8)

Gesler (1998) considered the city of Bath in south-west England. This place has a long-established reputation as a healing place, arising from its hot springs. Its role as a healing place was most prominent at certain periods of history, especially during the Celtic, Roman and medieval periods and during the 18th century. Gesler considered factors which enhanced the reputation of Bath as a healing place, noting especially the symbolic importance of the hot mineral springs, the myths which came to be associated with the place and the fine Georgian architecture of the city. The Celts and Romans had attributed religious significance to the springs and associated them with the goddesses Sulis and Minerva who were credited with healing powers. The intense interest in classical history, mythology and architectural style in the 18th century was linked to moral symbolism concerning order, symmetry and harmony and their association with both mental and physical health and wellbeing. Not only did the hot springs support the socially and physically therapeutic role of Bath as a spa, but, Gesler suggested, the city street plan with its symmetric circles and squares invoked classical notions of perfection and proportion. The uplifting aesthetic appearance of the city created by its architectural uniformity and coherence is still part of its attraction today.

Gesler (1998) also noted, in the case of Bath, factors which have detracted from the status of the city as a healing place, including arguments often put forward by the medical establishment that the hot springs had no real healing powers, and concerns about bogus physicians exploiting the reputation of the city to practise questionable medicine. There were also interesting tensions between the success of the city as a healing place and the attributes of urban living which were seen as damaging to health. Bath grew as a city largely because of its status as a healing place. However, during its historical development, Bath, like other growing cities at the time, faced problems providing adequate sanitation, managing urban congestion and addressing unhealthy living conditions for its poorer populations. Gesler showed how, in the case of Bath, these public health challenges threatened the image of Bath as a healing place, and exacerbated health inequalities between rich and poor inhabitants.

Geores (1998) analysed the case of Hot Springs, Dakota, which showed some similarities with Gesler's description of Bath. This site was also attributed with sacred healing powers by indigenous populations (in this case the First Nation Indians who passed on myths about the curative benefits of the hot springs to white settlers). The town was developed by entrepreneurs who aimed to capitalize on the healing reputation of the springs and established other key attractions of a successful spa town such as the attractive natural landscape, comfortable hotels and leisure facilities, ready access by railway, and a concentration of medical facilities. Over time the town became known for its architectural quality and historical associations as well as its healing reputation. This example illustrated the commercialization of a healing place, as in the example of Bath cited above. There is a link here with theories of consumption and their relationship to health, discussed later in this book in Chapter 5.

Gesler's studies of Epidauros, Greece and of Lourdes, France (Gesler, 1993; 1996) illustrate similar themes and particularly emphasize the link between spiritual wellbeing and the power of healing places. Epidauros provides a historical example which Gesler studied using a variety of documentary evidence. This place was associated with the Greek god Asclepius and patients went to Epidauros seeking dream healing. Gesler suggests that many of the attributes of Epidauros as a place enhanced the sense of a sanctuary where cure or advice about treatment would be transmitted to patients in visions. Lourdes, associated with Catholic religious veneration of St Bernadette, was also a focus of pilgrimage for sick people hoping for relief of their suffering through a spiritual experience. As well as being associated with figures of religious significance, both Epidauros and Lourdes had attributes including natural surroundings or built environments that encouraged and symbolized senses of peace, redemption and religious contemplation. Both places were revered by those visiting for spiritual help and offered the expectation of cure. Both fostered some sense of fellowship and equality of access among pilgrims, and they allowed pilgrims to follow ritualistic activities, which heightened the religious experience.

THERAPEUTIC SPACES FOR TREATMENT

Geographical analysis has also focused on the ways that spaces for delivery of biomedicine can be designed according to principles of therapeutic landscapes. Gruffudd (2001) considered two modernist health centres established in London in the 1930s: the Pioneer Health Centre in Peckham and the Finsbury Health Centre. At this time, in countries like Britain, emphasis was being placed on the potential to design and manage urban environments according to modernist principles. These sought to improve wellbeing and quality of life through the application of modern knowledge and ideologies. There was emphasis on hygiene, cleanliness and salubrity as crucial aspects of modern urban design. Healthy urban design was seen to allow access to air, sunlight, parks and gardens, and social facilities. At this period, clean, white sterile spaces were being presented in a positive light by modern architects, as representative of a new order in society, inspired by modernist ideas.

Gruffudd (2001) discussed how the architectural design and organization of the health centres in London represented these ideals. The buildings were constructed of modern materials and incorporated extensive exterior glass surfaces, which could be opened up, allowing penetration of sunlight and circulation of air. The buildings were designed rationally for their various functions, with spaces for sports, social interaction and intellectual pursuits as well as medical facilities. Internal spaces were designed to be flexible and welcoming, and to encourage social contact. In Finsbury (park), entry was by easily accessible ramps, and colour schemes and reception facilities were designed to make the centre accessible and attractive for users.

Both centres were intentionally located in poor areas seen to have major public health problems, and were intended to exert an influence beyond the centre, promoting healthy living in the local community and increasing access to preventive health care. Lubetkin, the architect of the Finsbury Centre, referred to it as a 'megaphone for health', spreading its message to the surrounding area (Gruffudd, 2001). The Health Centre in Peckham developed a scheme based on a farm in Kent, where local people could be involved in producing their own healthy fresh food.

Glanville (2001) has described a trend during the latter half of the 20th century for more utilitarian hospital design. However, today there is a growing awareness of the limitations of the 'utilitarian vision' of hospital spaces and a recognition of the need to design new hospitals in ways which will help to meet a wider range of patients' needs for personal and holistic care, as described by Corner (2000):

> The early utilitarian vision has been maintained and reinforced, creating a kind of entrapment of the person who is in need of care. Hospitals represent productivity not sanctuary. The hospital bed seems to be predicated much more around power and control than the need for rest or recuperation. Progress, cleanliness and efficiency take precedence over comfort, reassurance, information or participation. Segregation, surveillance and control predominate, and individual identity is negated. These are the architectures of treatment and not of personal care.

Francis and Glanville (2001: 62–3) set out principles for hospital design to foster such 'architectures of personal care'. Their proposals paid attention to the wider urban context within which hospitals are located, the connections between hospitals and the communities they serve, and the impact that hospitals have on the communities and environments where they are located. Emphasis was placed on designing hospitals to meet targets for sustainability in use of power and water and waste disposal. They recommended that hospitals should be significant public buildings, readily identifiable, easy to access and located in central, urban sites easily reached by public transport.

Within the hospital, Francis and Glanville (2001: 62) recommended a balance between the requirement for 'sophisticated, highly engineered' spaces and the need for 'an ambience which is calming and supportive for patients and staff', and between clinical observation and privacy for the patient. The design of hospital buildings should clearly distinguish between public, social and private spaces, provide diversity, and be easy to navigate. Public spaces should seem welcoming and offer amenities such as cafés, shops, spaces for exhibitions and events, business and communication services. Their proposals appear to advocate bringing the kinds of activities typical of other public settings into hospitals, blurring the functional division between hospital spaces and other public spaces. There is an interesting connection to be considered between this proposal and examples, discussed below, of hospital spaces being commodified and packaged as 'themed' experiences similar to those created in hotels, retail centres and leisure parks. Aesthetic features of hospital buildings were also seen to be important. Artwork, interior décor, lighting and textures are seen as important in creating 'stimulating and purposeful public spaces in contrast to the private spaces for patient contact that will be therapeutic and uplifting' (2001: 62). The design of modern hospital buildings today may also be informed by research by authors such as Loppert et al. (2001) on the beneficial links between art and health.

Some recently constructed medical buildings appear to have adopted such strategies for creating therapeutic landscapes. The 'Starlight' children's wards at the Homerton Hospital in East London, for example, opened in the late 1990s, seem designed to capture a more whimsical and playful sense of place, designed to appeal to children. Plain, rectangular corridor spaces are broken up by patterns on the floor and walls representing curving, dynamic shooting stars. Bright primary colours are used and some of the observational ward spaces are designed to look like domestic living rooms, with carpeting, shelving and books, toys and television. There are intriguing parallels (including the similarity of the name) with the 'Starship' children's facility in New Zealand described by Kearns and Barnett (1999). They saw this hospital as incorporating explicit references to a science fiction television programme and building a space that functioned almost as a theme park as well as a hospital. This strategy can be seen as an extension of the use of conventional hospital spaces for creation of imaginary spaces using unconventional methods such as theatre. Van Blerkhom (1995) has described how, in New York City hospitals, performances by 'clown doctors' offered experiences for sick children which have parallels in 'alternative' practices of shaman healers. Using music, costume and sleight of hand the clowns gave expression to patients' feelings and concerns by manipulating medical symbols and violating natural and cultural norms. This can help to alleviate distress. This example is interesting because it illustrates how behaviour interacts with environment in ways which are important for health. On the one hand, a medical setting represents spatialization of medical power and encourages the biomedical approach to illness; on the other, it is possible to 'subvert' medical space to allow other perspectives to come to the fore.

A recent newspaper article (Purvis, 2001) described the new Homeopathic Hospital in Glasgow. Features of the building noted in the article include the use of natural materials; the flowing and curved lines of walls and constructional details such as door handles were seen to emphasize 'organic' forms;and wards faced outwards so that patients could enjoy views of a 'natural' garden landscape. Lighting which imitates daylight, indoor plants and wall hangings also echo natural environments. The healing potential of music and colour were considered in the design. The account describes the building as inspired by a form of 'evidence based feng shui', drawing on a review of research demonstrating links between patient wellbeing and hospital design.

These examples show how attempts to design medical spaces that will be therapeutic for users have evolved over time, reflecting changing social constructions of the sort of setting which might be therapeutic. Modernist principles in the 1930s emphasized the moral virtues of scientific, hygienic spaces, for example, while postmodern hospital spaces in the 1990s have tried to mask the clinical and scientific functions of modern medicine by incorporating references to playful or homely spaces, and by blurring distinctions between hospitals and other public spaces in the wider community. These are designed as liminal spaces, encouraging a sense of escapism from the normal constraints of everyday life into imaginary worlds. Some key attributes also appear quite persistent over time, such as the value attached to light, colour, contact with natural environments, and healthy social interactions. It is also interesting to consider who defines the characteristics of a therapeutic setting. The dominant voices are the socially powerful ones of medical personnel, architects, policy makers and academics, for example. While there is some evidence of a growing responsiveness to the less powerful participants in therapeutic landscapes, such as patients or informal carers, their position in the debate still reflects important inequalities in health and in debates about health care.

PLACE AND THE AFFIRMATION OF DIFFERENCE

As noted above, the understanding of places, as well as the perception of what is healthy and therapeutic, are open to differences of interpretation, often associated with a sense of social difference and 'otherness'. This sense of difference is linked to perception of how one's own individual and group identity is distinct from that of 'others'. Being able to affirm one's own identity may enhance one's sense of self-esteem, social value and social standing, especially where this involves expression of resistance to alternative identities imposed by powerful 'others' in wider society. This confers a sense of empowerment and may be beneficial to health for populations that are disadvantaged and at greatest risk of poor health. Some places contribute to this sense of empowerment and affirmation symbolically, by representing important cultural attributes. Cultures typically identify, for example: sacred places linked to their religious beliefs; places which are sites of historical events that have shaped the culture; and areas identified territorially as belonging to their culture or representing their 'home space' (see also Chapter 3). They can also create the setting in which social processes are played out, enhancing the sense of belonging to a group, and of difference from 'others'. Such interpretations of the therapeutic character of place also need to be understood in the light of theories of social control, discussed in Chapter 3.

Kearns (1991; 1998) provided a good example of this in his discussion of the links between place, identity and community resistance in the region of Hokianga, northern New Zealand. The population of this rural area is predominantly Maori, and Kearns (1998: 234) described the Hokianga health services as having 'grown to reflect local tradition and identity'. The area was designated a Special Medical Area because it was geographically isolated and had high levels of deprivation, making it unattractive for private medicine. Special measures to ensure adequate medical provision included salaried doctors and health care that was provided free to users. The area was served by health centres which developed an informal function as social centres for members of the Maori community to meet in ways which enhanced the sense of community. Kearns described how service reforms threatened to centralize services and introduce managed competition likely to change the health service provision in this Special Medical Area. The response was organized opposition on the part of the local community in Hokianga, culminating in the establishment of a separate Hokianga Health Enterprise Trust, with special responsibilities for health care provision in the area. The trust aimed to maintain provision that was responsive to the medical and cultural needs of the community in the area. Kearns discusses the symbolic importance of this initiative. He interprets it as an 'assertion of difference' (1998: 244) and discusses the ways that sense of place, and health care provision in that place, were associated with a distinctive cultural and community identity. When national health policy threatened to undermine this sense of difference by changing the local organizational framework for health care, a struggle developed to retain the Trust, demonstrating local resistance and affirmation of 'otherness'.

CONCLUSIONS: THERAPEUTIC LANDSCAPES AND HEALTH INEQUALITY

This review of geographical perspectives on therapeutic landscapes shows how far geography of health has moved away from a purely biomedical perspective. As a result, in countries like the UK and Canada, the relevant research groups of the Royal Geographical Society with the Institute of British Geographers, and the Association of Canadian Geographers, have renamed themselves to embrace a preoccupation with geography of *health* rather than solely with *medical* geography. Geography has demonstrated a growing interest in alternatives to conventional biomedicine, which are often more sensitive to the interaction between people and the social and physical setting in which they experience health and health care.

This broadening perspective has been accompanied by a preoccupation with a number of themes that have also been important in human geography more generally. These include theories concerning the links between place, identity and culture, and ideas about the body. They explore how socially constructed bodily experience varies between different geographical contexts. The symbolic, as well as the material, nature of space and place is fundamental to these perspectives.

The approaches discussed here are also notable for the shift in the methodological, as well as theoretical, basis for research. Qualitative, intensive studies are typical of work in this field. Gesler (1993), Baxter and Eyles (1997) and Curtis et al. (2000) have discussed the characteristics of these qualitative methods, which allow exploration of individual lives and places and the meaning of individual health related experience. Such approaches contrast with the more quantitative, extensive strategies used in some other 'branches' of health geography, such as those discussed in Chapters 4 and 6.

The research considered in this chapter has exemplified processes which contribute to the production of health inequality. Different health states and bodily capacities are valued differently by societies in general, and a person's social and economic standing is affected by their health and the way that it is perceived by others. There are strong links between perception of wellbeing, health and health care on the one hand, and individual and cultural identity and sense of social position on the other. Much of the evidence discussed here suggests that therapeutic landscapes have particular importance for psychosocial health, and in Chapter 7 we return to the question of therapeutic landscapes in relation to variation in mental health in particular. Access to social and physical landscapes that are experienced as therapeutic may mitigate the disadvantage resulting from illness. Therapeutic landscapes may also promote positive aspects of psychosocial health by boosting self-esteem and a positive sense of identity, and life satisfaction. Access to therapeutic landscapes for people whose social position or cultural affiliation causes them to feel undervalued or excluded in wider society may help to reduce health inequalities. However, the examples discussed above also show that landscapes in the home, the community and institutional settings often have potentially non-therapeutic attributes that may exacerbate health disadvantage.

The discussion above showed that complementary and alternative therapies are often viewed as contributing to wellbeing using strategies different to those employed in biomedicine. Access to, and use of, complementary therapies may contribute to health inequality in varying and complex ways, depending on the setting. The research reviewed above suggests that it is likely to be empowering and beneficial to health to have access to a range of alternative therapies as well as biomedicine. Wealthy (and generally more healthy) populations often use complementary and alternative therapies, in addition to biomedicine, to widen the repertoire of health care available to improve their wellbeing. Less advantaged groups may have access to a more limited range of therapies. These variable patterns of health care consumption and their relevance for heath inequality are discussed in more detail later in this book.

FURTHER READING

Good reviews and collections of essays on different aspects of therapeutic landscapes include: Kearns and Gesler (eds) (1998); Williams (ed.) (1999).

Readers may also want to explore more of the geographical literature concerning the body, the ways that it is interpreted socially in different spaces, and the power relationships involved. A good edited collection of essays, by authors from Europe and North America, which has taken a fresh look at these questions is: Butler and Parr (eds) (1999).

In relation to issues of architectural design of health care facilities, there are clear connections between this area of research and the work of commentators from other disciplines. A good discussion of these issues from an architectural perspective, for example is: Francis and Glanville (2001).

OBJECTIVES AND QUESTIONS

Students reading this chapter will probably have learning objectives which aim to improve their understanding of the following ideas:

- the importance of social relations, and power relations, as experienced in different spaces, for our perception of the body and its abilities;

- the perceptions of health and health care which are fundamental to biomedicine and how these contrast with perspectives from other medicines which are widely used in both high income and low income countries;
- aspects of natural and built environments which may be viewed as 'therapeutic' and the essential characteristics typical of 'healing places';
- social variability in perception of therapeutic properties of places and settings and social inequality in the experience of therapeutic landscapes.

Students might want to check how their knowledge has advanced by using the material in this chapter, the further reading indicated above, and also material on therapeutic landscapes discussed in Chapters 7 and 8, to answer the following questions:

1. What aspects of the physical, social and symbolic environment may help to create a 'therapeutic landscape'?
2. Varying strategies have been used, at different points in time, and in different societies, to create therapeutic spaces in health care facilities. Discuss what these tell us about (a) varying perception of therapeutic settings, and (b) social and power relations between different social groups involved in the design and use of these facilities.
3. To what extent is there a difference between the concepts of 'health' and of 'wellbeing'?

3 Landscapes of Power: surveillance, exclusion and social control

---

This chapter considers the significance of geographies of power for variation in health. The thought, choice and action or *agency* of individual actors is seen to interact with the power *structure* of the society in which they live, producing mental and physical health differences between more and less powerful social groups. Theories concerning landscapes of power explain the importance for these power structures of the space–time context in which people live their everyday lives. Processes involving control of resources, territoriality and surveillance all contribute to landscapes of power. The examples discussed here demonstrate how socio-geographical contexts contribute to processes that differentiate social groups in terms of their empowerment, and generate inequalities in health experience. This chapter discusses illustrations from the literature on landscapes of power in hospitals and community care settings, before moving on to consider a wider range of landscapes of difference and exclusion. The examples show how landscapes of power impact on the health of certain socially defined groups. The health related experiences of men and women, of minority ethnic groups, and of Travellers, homeless people and refugees are given special attention here.

STRUCTURATION AND SOCIAL CONTROL IN GEOGRAPHIES OF HEALTH

One conceptual interpretation of these processes is provided by structuration theory. Giddens (1984) linked ideas from geography and from other social sciences in his discussion of *the constitution of society*, in which he set out a theory of structuration. He emphasized the importance of interaction between individuals and their socio-geographical context, arguing that: 'in the social sciences, all explanations will involve at least implicit reference both to the purposive, reasoning behaviour of agents and to its intersection with constraining and enabling features of the social and material contexts of that behaviour' (1984: 179). He described how the agency of individual actors shaped, and was shaped by, social structures. These structures comprise principles of organization of society, rules and systems for allocating resources and institutional features of social systems in space and time.

Giddens was particularly interested by ideas of time–space geography proposed by authors such as Hägerstrand (1952; 1975; 1982) and elaborated by others such as Thrift (1982; 1992), Gregory (1981; 1989) and Pred (1981). For Giddens, key concepts included the 'life path' of routine, day-to-day activity in time and space, and the space–time trajectory through the life cycle. This type of time-space perspective was represented in the form of a three-dimensional diagram of individual daily time–space paths. A hypothetical example, based on Hägerstrand's space–time path diagrams, is shown in Figure 3.1. This illustrates the pattern of daily movement of three people in the same family. They each share the same space in the family home for part of the day, while at other times they are at places of work or at school in different parts of the city, or in leisure facilities where they interact with different groups of people and occupy different spaces. This type of perspective provided a contextual framework for what Giddens (1984; 3) called the *durée* – a continuous flow of conduct through the lifespan.

Hägerstrand's work was also important for Giddens because it considered the constraints which limit individual behaviour in time and space. These constraints arise partly because human movement and perception are subject to the physical limitations of the human body and the normal lifespan. People have a limited capacity to do more than one task at once, and movement in space is also movement in time. Furthermore, no two human bodies can occupy exactly the same place at the same time. Other constraints on, and opportunities for, human interaction arise from the context: for example, physical attributes and the social and economic organization of space. There are 'coupling constraints' (limited opportunities for some individuals to be together in time

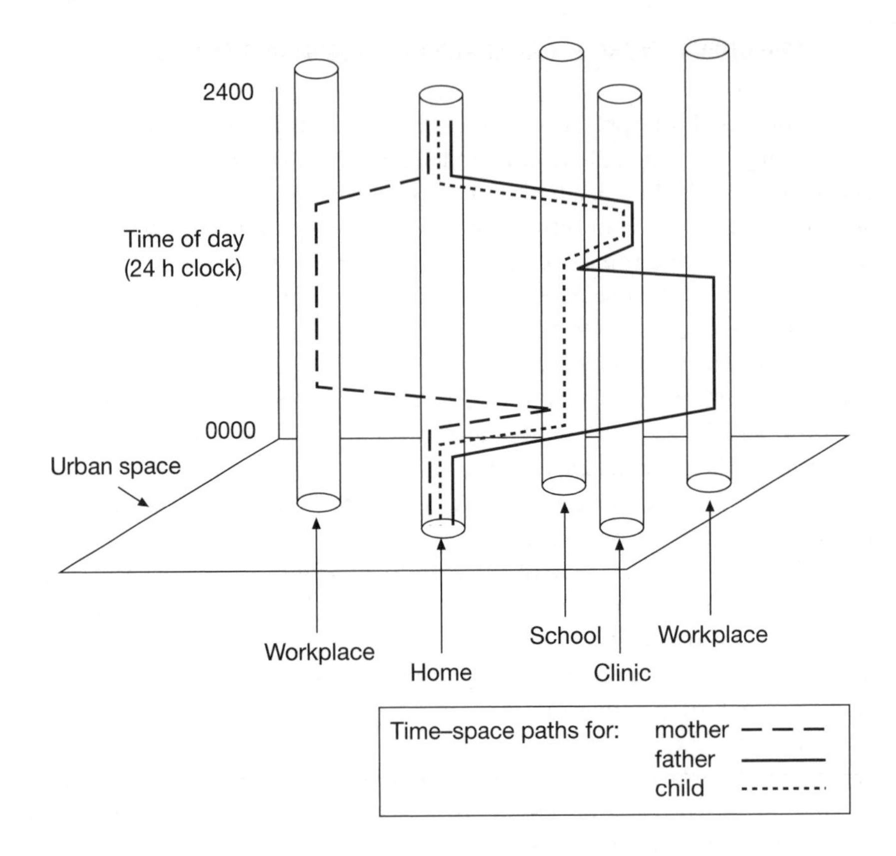

FIGURE 3.1 Hypothetical time–space path for the working day of a family of three people

and space because their space–time paths may not often meet) and, conversely, the potential for 'bundling' of time–space paths for groups of individuals whose routine does involve being in shared time–space locations (e.g. colleagues sharing a work space, family members in their household, students at the same school).

Hägerstrand's later work (1982) introduced the idea of a 'diorama' or socio-geographical contextual background which helped to form the ways that individual life paths and action were played out. Individual actors, capable of individual agency, move through daily paths in time and space which bring them into routine contact and interaction with other actors in a variety of *locales* (e.g. work, home, school). Most actors have a good understanding of the processes which maintain the social structure of the society in which they live (the conditions of reproduction of society). Everyday social practices of these actors interact with these social structures in ways which are not always obvious to the actors themselves. In doing so, they help to maintain and shape the social structures, but the social structures also frame the actions of the individuals. We therefore see a complex interaction between individual thought and action and the broader social context in which individuals live, which is strongly influenced by the geographical context of social interaction. There are links here with the ways that geographers have used the ideas proposed by Lefebvre (1991) discussed in Chapter 2.

Giddens (1984: 171, 258) discussed the idea of *time–space distanciation* through which socio-economic structures extend over time and space. Time–space distanciation operates partly through patterns of coordination of allocative and authoritative resources. Allocative resources include natural environmental resources, technologies and goods which are produced. Authoritative resources include administrative structures, human mutual associations and the organization of life chances and opportunities. The ordering of these allocative and authoritative resources in time–space puts constraints on the life paths of individual actors, which help to maintain the power relations between dominant and subordinate groups in society. Giddens explains that:

> some of the most bitter conflicts in social life are accurately seen as 'power struggles'. Such struggles can be regarded as to do with efforts to subdivide resources which yield modalities of control in social systems. By 'control' I mean the capability that some actors, groups or types of actors have of influencing the circumstances of action of others. (1984: 283)

This view presents cities, and nation states, as 'power containers' within which this ordering of allocative and authoritative resources takes place, sustaining the power relations between different social classes in society. An important feature of the concept of 'struggle', strongly emphasized in examples considered below, is that while dominant groups in society, acting within existing social structures, exercise social control over less powerful groups, nevertheless subordinated groups are also able to employ some strategies of resistance and challenge to the constraints imposed upon them.

The mechanisms by which socio-geographical spaces can operate as 'power containers' have also been explored by other authors. We can identify processes of territoriality and surveillance as key to the exercise of power. Sack (1986) discusses human territoriality – a biologically and socially rooted human trait. This takes the form of a geographical expression of social power by the use of land, the organization of human activity in space and the meanings given to places. An area becomes a territory when its boundaries are used by some individuals or groups to control access for others. This control may be established by classifying areas (e.g. designating land as private property), announcing or marking territorial zones (e.g. with signs and boundary markers) and enforcing boundaries by, for example, physical barriers or restraints, legal sanctions and physical force (e.g. exercised by military, police or security personnel).

Surveillance is a crucial mechanism for the successful wielding of power, whereby dominant social groups control information about subordinate groups and exercise control over space. Dandeker (1990) summarizes ideas from Giddens (1984) and Weber (1968) concerning the aspects of modern societies which make such surveillance more possible. These include durable state administrations which control the 'means of violence' (military and police forces); goods and services produced and distributed in a system based on large scale enterprise; modern, rational technologies such as computerized databases and telecommunications; rational, well organized bureaucratic systems. Foucault (1979) used the example of the prison as an analogy of processes operating in wider society. He used the concept of *panoptic* surveillance in the carceral setting, where discipline is exerted through supervision of the activities of prison inmates, who are unable to see their observers. Inmates therefore tend to behave as though they were under constant surveillance and discipline their own behaviour to correspond with the prison regime. Foucault argued that similar strategies for the exercise of power through observation operate in other settings. Thus the concept of the Panopticon provides an interpretation of strategies of surveillance employed in many situations. An obvious example is the widespread use of closed circuit television security systems in public spaces. This perspective represents society as a 'disciplinary cage' which is constraining but also offers safety and security.

The ideas discussed above can be used in various ways to interpret geographies of health. Selected examples are considered here to show how medical control is exercised in different time–space settings, and how dominant social groups control space in ways which influence their own health as well as that of other, subordinated and excluded, groups. We can observe these processes within medical settings and outside in the wider community.

LANDSCAPES OF POWER IN HOSPITAL CARE

Several authors have given accounts of surveillance as an important aspect of health care settings such as hospitals (Philo, 1989). Patients are removed from their normal life space in the community to be treated as inpatients in hospital. There, in Foucault's (translated) parlance, they are subject to the *medical gaze*: analytical scientific observation to identify illness and monitor the course of treatment. The process of institutional spatialization of health care in hospitals, outlined in the introduction to Chapter 2, has facilitated this controlling 'gaze' by separating illness from the 'natural' setting of the family and the community and bringing numbers of sick people together, allowing them to be viewed as cases of disease, compared, classified and treated consistently (Foucault, 1993: 109–11). The conventional hospital institution has provided facilities where patients are treated in specifically designed spaces (hospital wards) where medical personnel are able to observe them. This has developed to the maximum degree in intensive care situations, where monitoring equipment is also employed. The organization of the hospital day has been arranged largely around routines for medical observation, especially the ward rounds carried out by doctors, the most powerful medical professionals in the hospital. All patients' time–space paths, as well as those of personnel working in the hospital, are expected to fit into these routines.

The conventional regimes in hospitals have been criticized because, they caused some patients to feel their sense of autonomy was weakened or threatened in ways which they found distressing and disempowering. There are reports of patients being forced into a passive and subordinate role, or, in other cases, resisting medical authority and control over their bodies and their actions in time and space. For example, the exercise of control and surveillance in mental asylums has been the subject of a number of geographical studies, considered in Chapter 7. In a different example, Smith (2000) demonstrated this type of situation in the cases of young men with chronic physical illnesses. Smith sums up the experiences of men he interviewed as follows:

> Key aspects of the environment were the power relationships between patients and medical staff, the clash of different time–space requirements within the spatial organization of the hospital and the lack of control of both body and physical space . . . Within the hospital the men were often denied emotional expression and support. The biomedical view also meant that men's feelings about their bodies and sexualities were often ignored during treatment. (2000: 191)

Hospital care models today are less restrictive than in the past, and more attention is paid to empowering the patient in the process of care. Nevertheless, patients are still required to cooperate with particular time–space regimes in the process of medical care. The strategies of control extend beyond the space–time routines of hospital wards. The British system, for example, rations limited hospital health care resources by means of the gatekeeper role of general practitioners (see Chapter 5) and also through the operation of waiting lists, controlled by hospital administrators and consultant doctors. The patient's position in the queue depends on the medical urgency of their case, assessed by a doctor. The time they are required to wait for hospital treatment depends both on this and on the local pressure of demand. Box 3.1 illustrates the geographical variability in access to care which results from this process, and cites evidence from a National Audit Office (2001) report on inappropriate manipulation of waiting lists by hospitals. This example demonstrates a struggle between patients, local NHS administrators and central government for control over the rules and systems for allocating resources, as described by Giddens. From a geographical perspective, it is interesting that this struggle is played out differently in different areas of the country, resulting in variability from place to place in access to care.

LANDSCAPES OF POWER IN COMMUNITY CARE

In contrast with the hospital setting, which removes individuals from their normal life paths, community based (ambulatory) health services seek to integrate medical care in the locality of the normal, day-to-day life space of the individual. Nevertheless, measures to protect public health are carried out through systems of medical surveillance and control of individuals. Patients are encouraged (and in some cases required) to make use of preventive or curative treatment made available in certain places by certain medical personnel. Use of health care requires registration, confirmation of eligibility and recording of care received. In the process, a good deal of information about the individual patient is gathered and stored for use by the health service administration. Gatrell (2002: 43), discussing time–space geographical perspectives used by Schaerstrom (1996), makes the point that community based health care facilities have particular locations for service delivery, such as clinics, and are open at fixed times, forming part of the structure interacting with the actions of both patients and medical staff. Thus, for example, in order to meet government targets for immunization of infants, parents and medical staff have to agree many individual appointments for parents, who modify their usual time–space routines in order to bring their child to the immunization clinic. In each case this may require a series of negotiations between parents (over competing time space commitments at home or in the workplace) and medical staff (in relation to clinic schedules, appointment times and working hours in NHS facilities). Gatrell discusses how this process illustrates individual human action being constrained by, but also constituting, NHS structures for the delivery of care. Chapters 7 and

Box 3.1 Waiting times for hospital care in Britain: NHS 'gatekeepers' controlling flows of patients in time and space

There has been much debate in Britain recently over the geographical variability in length of waiting lists under the National Health Service (NHS). Patients with similar medical need have a variable time to wait for treatment, depending on their area of residence. This has led to descriptions of access to the NHS as a 'postcode lottery'. The hospital that patients attend is largely (though not absolutely) determined by which hospitals have contracts with the NHS to provide for patients in their area. The length of waiting times generally has also been of concern and the government has tried to bear down on waiting lists in the NHS by using waiting times as one of the performance indicators against which local health services are assessed. In 2001, health authorities were required to reduce the numbers of people on waiting lists and it was expected that no patient would have to wait longer than 18 months for a hospital appointment. In June 2001 the maximum waiting time target was further tightened to 15 months, even though some hospital trusts had not been able to meet the 18 month target for all patients. Data collected by the Department of Health (2001) from NHS hospitals were used to monitor waiting times. The data in Figure 3.2, for example, show a varying situation among health authority areas within one region in 2001. Of patients living in Cornwall or south west Devon, for example, the proportion having to wait more than 6 months for hospital care was greater than the national or regional average; in contrast, larger proportions of patients living in Wiltshire and in Dorset were treated within 6 months. This type of situation gave rise to concern about inequalities of access to care between areas. It is also an interesting example of information on time–space flows of patients being employed by central government as a method of performance management in hospitals in order to try to control the degree of variability.

The use of a limited set of performance targets to exert central control over the operation of a large organization like the NHS is often problematic, however, as various methods can be used to manipulate activity in time and space at the local level in order to subvert the intentions of central managers. Hospitals were found to be employing various strategies to improve their performance ratings. Some were 'legitimate' as far as the NHS was concerned, such as trying to make appointment systems and throughput of patients as efficient as possible. Other 'valid' approaches have involved trying to increase capacity through the use of private hospital beds for NHS patients, the acquisition of private hospitals for the NHS, or sending NHS patients to hospitals in continental European countries. These are mechanisms which involve relocating the source of care in space, and manipulating the movement of patients, in order to overcome local supply constraints on waiting time.

Other strategies to reduce waiting list statistics had perverse effects in that they were unlikely to shorten the real waiting time of individual patients. These were 'illegitimate' strategies which seemed to illustrate local resistance to central government control. For example the National Audit Office (2001) named nine NHS hospitals which had been using inappropriate methods to contain their waiting lists. These methods included: unwarranted suspension of patients from the waiting list;

continued ⇒

Box 3.1 Continued

inappropriate adjustments of waiting lists records; delays in getting patients needing treatment onto the waiting lists. One hospital was holding non-urgent patients back from the waiting list and knowingly offering patients appointments during their holidays, which they would not be able to take up. In some cases patients who had been waiting for a longer period were unjustifiably removed from the list or their records were altered.

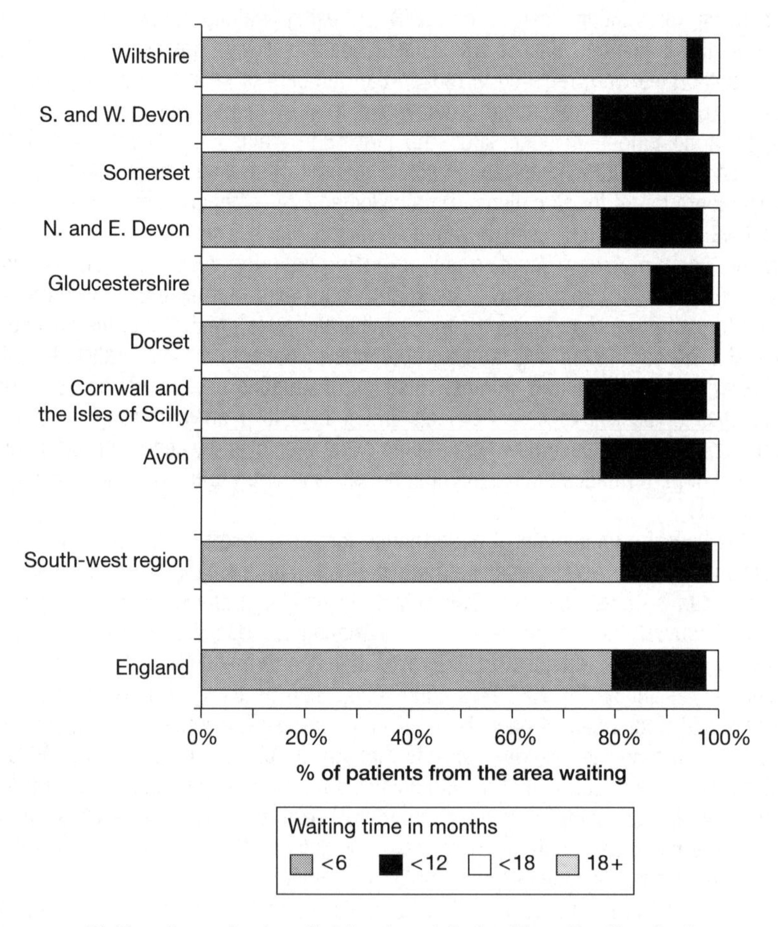

FIGURE 3.2 Waiting times for hospital treatment in health authorities in the south-west region of England, September 2001 (Department of Health, 2001)

These cases illustrated the exercise of control in ways indicative of power struggles at various levels in the NHS hierarchy. Locally, administrative gatekeepers

continued ⇒

Box 3.1 Continued

controlled patients' waiting times and access to health care in ways that produced inequity, since the management of patient flows included 'illegitimate' strategies. These examples illustrate an interesting power struggle between central government, aiming to impose a particular target, and local administrations and officials, circumventing this by varying the rules. They also demonstrate the geographically variable impact of these processes on patients, who are often in a relatively weak position to influence their waiting time for care.

8 discuss treatment of people with severe mental illness and with contagious forms of tuberculosis. These are both categories of patients over whom health services seek to exert control to ensure that they receive medically necessary treatment, even in some cases when the patient is reluctant to consent, or finds it difficult to follow the medical regime.

Medical and public health interventions may be more effective if they are sympathetic and responsive to the normal time–space paths of the patients they serve. For example, Cravey et al. (2001) use an action space perspective in their research on the geographies of beliefs about chronic illness in a rural multi-ethnic area of North Carolina, USA. They have suggested that better understanding of socio-spatial knowledge networks (SSKNs), based on the social geographies of the local populations, will help in the design of health promotion to reduce the prevalence of chronic disease. Their research in Siler City combined data from observation, interviews with key informants and interviews with samples of the population using relevant 'sites' such as community centres, churches, clinics, markets and informal meeting places. The interviews were designed to elicit information on key places in each informant's action space, and global positioning systems (GPSs) were employed to assist in the mapping of these. One objective of the research was to identify spaces where, actually or potentially, people would exchange information on diabetes. Some places could be identified as 'nodes' of an SSKN: the foci for interaction of a group of people which users viewed favourably. These were most likely to be fruitful sites for health education and promotion interventions. Other possible sites might be less effective, even though they might be providing diabetes information, because they were rated unfavourably or were inconvenient to access, so people were reluctant or unlikely to visit them. Some places would be too isolated from, or irrelevant to, residents' action spaces and so would not be useful as intervention sites. The action spaces of different social groups varied, so that the most effective intervention sites for different 'target groups' in the population might also vary, depending upon which nodes were most central and preferable in the SSKN of each group.

LANDSCAPES OF DIFFERENCE AND EXCLUSION

Geographical perspectives on power relationships and their relationship to health extend to a wide range of social groups and geographical settings beyond the confines of medical environments. Dorn and Laws (1994) and Imrie (1996) discuss the importance for the geography of health of studies which focus on bodies and the places they occupy as *sites of oppression* of people whose social position is subordinated and marginalized, or as *sites of struggle* between more and less powerful groups in society. Gleeson discusses the 'socio–spatial patterns and relations through which impairment is oppressed by dominant power relations' (1999: 54). Butler and Parr (1999) have edited a collection of essays exploring the relationships between space, health and socio-cultural, political and economic relations. These include the relations operating in the medical domain, but also those in wider society. Their book draws attention to the significance of power relations for the social construction of difference and 'deviance', resulting in the social construction of differences in physical and mental abilities in terms of 'disability'. Functional definitions of health and well-being are important to the way that health is understood, so processes of medicalization, stereotyping and social stigma contribute to geographical health inequalities and to the geography of health (1999: 14).

These accounts emphasize the reproduction of social inequality through the reinforcement of a sense of difference from 'the other', the struggles for power between these differentiated groups, and the creation of social and spatial 'distance' between them. Butler and Bowlby (1997) presented arguments from the social model of disability (considered in Chapter 2) and from feminist perspectives, suggesting that inequalities between people associated with illness or disability are based in social relations rather than in differences in biological capacity. They also discussed the social meanings of bodily images presented in public places, arguing that society generally tends to reject from public spaces those whose bodily presentation is seen to be 'deviant' or socially 'unacceptable' (1997: 419–21). Dear et al. (1997) outlined dimensions of acceptability (see Box 2.1). They argued that social rejection or exclusion of individuals on the basis of their 'disability' will depend on perceptions of these dimensions. Rejection will be most forceful when biological or psychological differences result in greater 'loss of control'. The authors seemed to imply both loss of control of bodily functions or behaviour on the part of the individual, and also loss of control of the individual by society, resulting in transgression of the expected patterns of appearance and behaviour.

Public spaces, and the time–space paths required for full participation in society, are often intolerant of differences in biological capacity, and this has the effect of drawing attention to, and excluding, people who are not able to conform to what is seen as the 'acceptable norm'. Butler and Bowlby (1997) illustrated this from research with participants with visual impairment. Dear et al. (1997: 455–7) also explained processes of *partitioning* which set up social and spatial boundaries, differentiating people who share membership of the same social group from 'other' groups. This sense of 'the same' and 'the other' is important to the ways that people construct their identities, and it also perpetuates group differences in power and access to resources.

Much of this debate concerns the health related experience of groups that are frequently subordinated in society. These include women, children, elderly people, minority ethnic groups, people with chronic illness or disabling conditions, or those who are marginalized because of their behaviour, such as illicit drug users. The health experience of all these groups is importantly influenced by their control over economic resources, since they are often disadvantaged by material poverty. The health effects of material poverty and socio-economic position are considered in greater detail in Chapter 4. The examples discussed below relate especially to the varying degree of empowerment enjoyed by different groups in different geographical settings and the extent to which individuals are able to realize their individual and group identity in ways that are important for their health. The following sections illustrate these processes as they affect groups defined by gender, by ethnicity or by their status as Travellers, homeless people or refugees.

GENDER AND HEALTH

Gender differences in health, associated with the socially constructed roles and social positions occupied by men and women, should be distinguished from sex differences between males and females due to biological differences (e.g. Arber, 1999). Gender differences reflect the varying social, economic and political power positions of men and women, and the ways in which men and women express their gendered identity. There is a large literature on differences in health between women and men (reviewed, for example, by Miles, 1991; Arber, 1999;

Annandale and Hunt, 2000). These authors discuss a 'paradoxical' situation, in that the typical lifespan of males is shorter than for females (see Chapter 1) but women have been thought to experience greater levels of morbidity and disability.

Some authors, however, have questioned how consistent this pattern really is. In fact, gender differences in health are more apparent for some types of health problem than for others. Macintyre et al. (1996: 621) concluded from careful analysis of survey data that women's greater levels of morbidity across the lifespan were only consistently found for psychosocial aspects of illness. For some physical symptoms the sex differences either were not significant or were 'reversed', being worse for men. Although it is sometimes assumed that women have a greater propensity to report illness, this is not always borne out by detailed analysis. Macintyre et al. (1999) found no evidence that men and women reported similar illnesses in different ways, or that women were more likely to report trivial illnesses. Arber and Cooper (1999) found that women in the oldest age groups are more likely to experience functional disability, but are less likely to report their health as poor.

Popay et al. (1993) have argued that where there are differences in illness reporting between men and women, this might be because they respond differently to survey questionnaires, and that surveys tap different aspects of health experience for men and women. Women may, for example, have a varying health experience due to the pressures of their complex social roles that involve balancing paid work and responsibilities in the home. Also it may be more socially acceptable for women to acknowledge psychosocial distress than it is for men. Differences in experience of illness between men and women are also argued to be due to the socially and economically subordinated position of women generally, and Arber and Cooper (1999) show that controlling for social class helps to explain some of the difference in illness experience between older men and women because women are more likely to be poor.

Furthermore, although men have a dominant role in many societies, this is not always advantageous to their health. This may explain why their life expectancy is worse than for women. Cameron (1998) and Courtenay (2000) discuss the health disadvantages associated with masculine roles and stereotypes. Cameron summarizes the *hegemonic masculine role* as follows:

> the need to be different from women (no sissy stuff), to be superior to others (the big wheel), to be independent and self-reliant (the sturdy oak) and to be more powerful than others, through violence if necessary (give 'em hell). The gender order and hegemonic version of masculinity frames relations of

> inequality, showing that men dominate both women and other groups of men. (1998: 128)

While men adopting this powerful, hegemonic gender role may experience some health advantages, especially through their greater control over society's resources, this role is also often associated with behaviours and use of spaces which carry relatively high risks for illness or injury. This may partly explain, for example, why men are more likely to suffer illness or injury due to accidents (road traffic accidents, sports injuries, effects of violence) or risky behaviours (smoking, alcohol and drug misuse). They are also less likely to admit to illness, so that they may not seek health care when appropriate. Some of these greater risks are associated with the socially constructed spaces which men and boys tend to occupy (for example, being allowed unsupervised freedom of movement outside the home when children; socializing outside the home as adults in places like bars; being engaged in physically demanding jobs in relatively risky settings such as mines or theatres of war).

Valentine (1999) described the experiences of a miner whose own embodiment of physical force and competence in physical work and sport was important to his masculine self-image. After an accident left him paralysed from the chest down, this man found it difficult to come to terms with the loss of some of these abilities. His account, presented by Valentine (1999: 174–8), explained how, through a contact made during his treatment and rehabilitation, he took up basketball, eventually playing for a national squad. The example shows how individual experience was framed by, and also constituted by, the changing spaces occupied by this man. After his injury his life path changed. The mine as a place of work was replaced in his life space by medical spaces for treatment and rehabilitation. Basketball, as a new activity, resulted from social contacts made with other people undergoing similar rehabilitation at the same health care facility, and therefore occupying the same medical spaces. Subsequently, while carrying on this new sporting activity, this man broadened the spaces he could occupy. He was able to reassert his capacity to participate in sport and use spaces for physical activity and leisure. Valentine commented on how, in this way, his sense of masculine identity and empowerment was reconstituted by challenging social discourses that present bodily impairment as weakness and incompetence.

Some studies do seem to indicate that the relatively subordinated role of many women in society has implications for their own health and that of their families. Dyck (1999) and Moss (1999), for example, discussed the workplace experiences of women with chronic illness. The experience was evidently variable, dependent on factors such as class positioning, seniority in the workforce, and the attitude of employers. However, the difficulties faced by these women reflected the constraints put on them by a 'masculine' work ethic (emphasizing capacity for full time work and avoiding expression of physical weakness). In some cases, incapacity for certain work roles due to physical illness seemed to be construed as weakness associated with gender. Chouinard (1999) also described the dominance of male power in the 'disability movement' in Canada. Disabled women's activism therefore needed to overcome the cumulative, marginalizing effects of gender discrimination, as well as discrimination in respect of differences in ability.

Another illustration of apparent associations between gender roles and health was reported by Curtis and Lawson (2000) in a study of the health experience of African Caribbean men and women in inner London. Like Macintyre et al. (1996) they questioned whether sex differences in illness are as systematic as is sometimes supposed. The pattern of illness reported in population surveys and censuses in Britain varied between men and women, and between ethnic groups. The patterns of difference seemed to depend partly on the type of questions asked about morbidity. Figure 3.3, from a national survey in Britain, shows that the

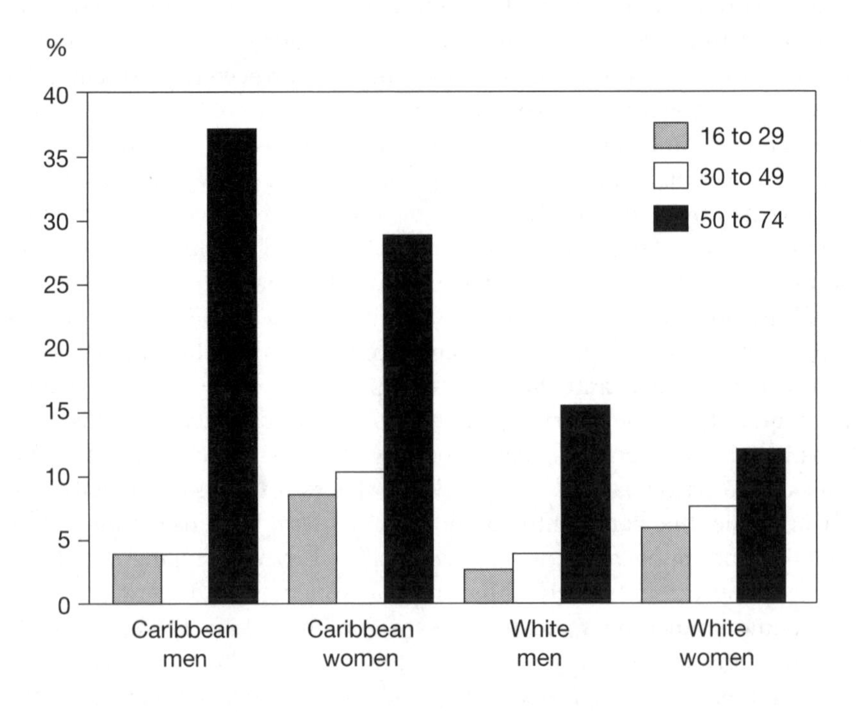

FIGURE 3.3 Proportion of 'white' and 'African Caribbean' respondents reporting health as 'poor' in a national survey in England (Rudat, 1994)

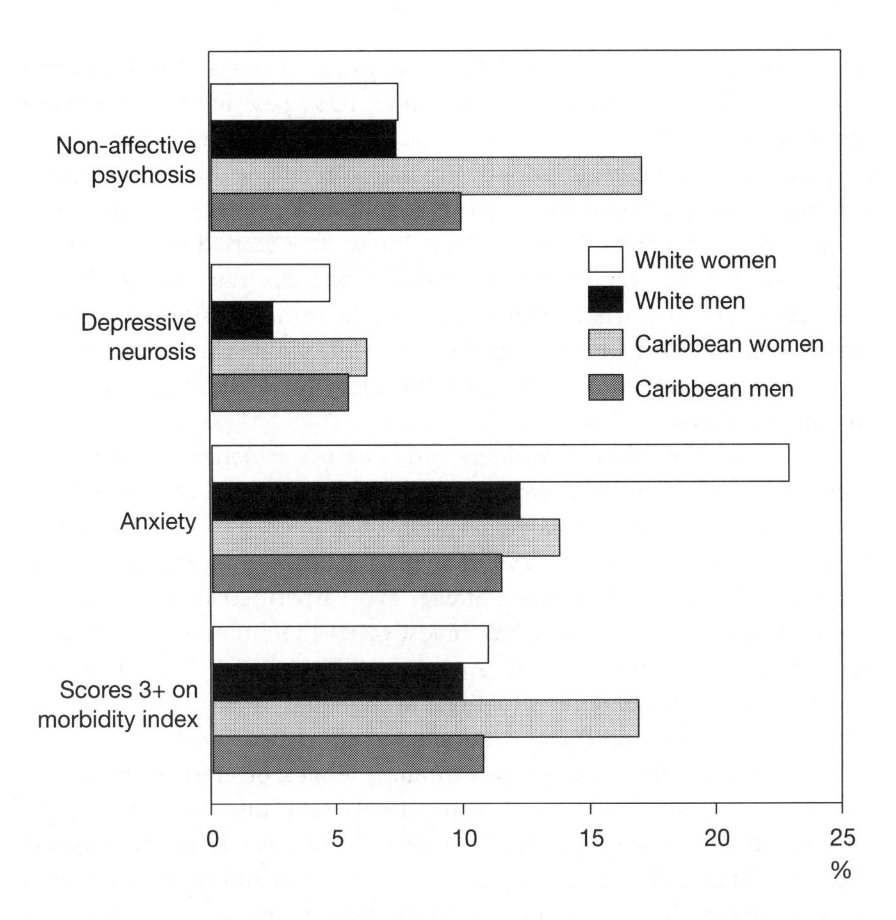

FIGURE 3.4 Prevalence of illness for men and women in two ethnic groups in the National Survey of Black and Ethnic Minorities in England (Nazroo, 1997)

sex differences in proportions reporting their health as poor were variable according to age and ethnic group. Figure 3.4, based on results from a different national survey, shows varying disparities between sex and ethnic groups, on three measures of psychosocial health and a composite indicator of reported morbidity.

Several sources, however, suggested that in systematic surveys, British Caribbean women were more likely than men to report illness. Qualitative data collected by Lawson illustrated some possible reasons for this pattern, and illustrated the impact of gendered social expectations for acceptable behaviour and the power relations of men and women (Lawson, 2000; Curtis and Lawson, 2000). The differences in socially acceptable behaviour for men and women were seen to be influential.

Men wished to assert characteristics of physical fitness and resistance to illness, associated with masculinity, and this may have made them reluctant to admit to illness or distress. Women apparently were more prepared to describe experience of illness, especially if this showed them exhibiting valued feminine attributes of stoicism and adherence to expected female gender roles. They explained how they carried on working in spite of illness. The gendered pattern of salient roles (see Chapter 2) may also have been important, since men sought to demonstrate that their health would allow them to do paid work, while women were more likely to be concerned about how their health might influence their nurturing role in the family.

The social position of women varies across societies and different socio-geographical settings. Several studies in low income countries identify the relatively subordinated position of women in society as a key factor for socio-economic development generally, and this can have consequences for health (e.g. see Sticher and Parpart, 1990; Young, 1997). For example, Cunnan (2002) has investigated the situation for African women street traders in South African cities, showing how their disadvantaged socio-economic situation, and gender constraints, combine with their working and living conditions to affect their health.

The relationship between health and the lack of empowerment for women is also illustrated by Asthana (1998), who discusses the experience of women engaged in the sex industry in Madras, India. She argued that 'individual practices together with institutional and structural forces actively shape the nature of the local commercial sex industry and patterns of HIV transmission in the city' (1998: 172). She also pointed out that the sex industry in Madras reflects the influence of patriarchal family norms (which, of course, are also evident to varying degrees in many other countries). These condition women to repress their sexuality more than men are expected to do. However, in some circumstances it is tacitly acceptable for men to seek sex with prostitutes, giving rise to the sex industry in Madras, which is particularly carefully concealed. It involves a complex and geographically dispersed group of commercial sex workers, with little control over their work, or whether or not their clients use condoms, and very poor levels of knowledge about the methods by which the HIV can be transmitted.

Asthana argues that the highly unequal power relations between men and women and the specific organization of the commercial sex industry in Madras make it especially difficult to implement community based health education strategies to try to limit the spread of HIV infection. Similar arguments about the significance of gendered power relations for health risk have been put forward by Craddock (2000a) in her discussion of the gendered risks of HIV infection in Malawi.

HEALTH OF MINORITY ETHNIC GROUPS

Relationships between health and ethnicity are also often interpreted as manifestations of variable empowerment of different groups in society. Several authors have argued that it is important to consider such explanations in order to offer a more theoretically based account of health variations in relation to ethnicity (e.g. Ahmad 1993; 1996; Bhopal, 1997; Nazroo,1998; Karlsen and Nazroo, 2000). Bhopal suggested that, without a strong theoretical basis, research on health differences between ethnic groups is 'black box epidemiology' (demonstrating associations without offering adequate explanations). As it does not offer a better understanding of *why* patterns of health differ between ethnic groups, black box epidemiology risks being racist, unethical and ineffective in providing answers to questions about how to reduce health inequalities between groups. (Similar issues are also discussed in Chapter 6: see the discussion of Figure 6.5.)

Nazroo (1998) offered a theoretical structure for interpreting the complex concept of ethnicity, comprising aspects of structural socio-economic position and of cultural identity. Karlsen and Nazroo (2000) reported analysis of data from a national survey of minority ethnic groups in Britain that was used to demonstrate how different dimensions of ethnicity relate to health. They identified five dimensions of ethnicity relating to: *nationality*; *ethnic or race* aspects (such as skin colour or descent from a particular group); *traditional* senses of belonging to a socially and culturally distinct group (manifest, for example, in dress, language, levels of intermarriage with other groups); *community participation,* reflecting how far social networks are restricted to the same ethnic group; and *racialization*, including experience of racial discrimination at work and of racial abuse. Aspects of ethnicity associated with socio-economic circumstances and racialization (which reflect the structural position of the group within the wider society) were found to be more strongly associated with health variation than dimensions of ethnicity associated with cultural identity.

Nazroo (1998) has therefore questioned the importance of 'cultural' differences for health inequalities between minority ethnic groups and the majority population and has emphasized the need to consider other elements of the structural disadvantage faced by ethnic minority groups, such as their experiences of racism or concentration in particular geographical locations. Institutionalized racism (whereby socio-economic structures systematically disadvantage minority ethnic groups) results in socio-economic disadvantage, widely experienced by minority ethnic groups in Britain, and leads to their geographical segregation into parts of the city where material living conditions are relatively poor. This influences their health experience. Nazroo (1997) showed that much of the difference in reported illness between ethnic groups in a British survey was associated with inequalities in their wealth and living conditions.

Similarly, Curtis and Lawson (2000) analysed factors associated with reported health for 'white' and 'African Caribbean' Londoners in the Sample of Anonymized Records from the British census in 1991. The socio-economic differences between individuals were strongly statistically associated with varying propensity to report illness. Figure 3.5 shows that the greatest disparities in reported illness for men and women were between different socio-economic groups. (These were classified using a composite indicator of standard of living, and distinguished in Figure 3.5 as 'high' or 'poor' living standard.) For women, and within similar socio-economic groups, there were still some differences in illness reporting, independently associated with ethnicity. This may suggest cultural variation between ethnic groups which is not explained by socio-economic disparities. However, for men there was little difference between ethnic groups within the same socio-economic category. This finding supports the view that power over material resources is a crucial factor in the health inequality between social groups. This is explored in more detail in Chapter 4.

Evidence from qualitative research by health geographers in East London has also shown that health related practices and health experience among informants from minority ethnic groups reflected their experiences of racism in British society. Illustrations are presented in Donovan's (e.g. 1986a) qualitative study of health experiences of West Indian and Asian informants and Lawson's (2000) reports from her research with African Caribbean respondents. Some of the informants in these studies attributed psychosocial distress to experience of racism. Also, Lawson's work in particular features a strong theme emerging from the comments of her informants, concerning their strategies to resist the impact of racism on their health. Resistance strategies included drawing on support from social networks, spiritual and religious faith, and finding positive ways to express their individual and ethnic identities. As noted in Chapter 2, the use of complementary and alternative medicines among ethnic minority groups may be, in part, an expression of this resistance and assertion of ethnic identity. Both Donovan (1986b) and Lawson (2000)

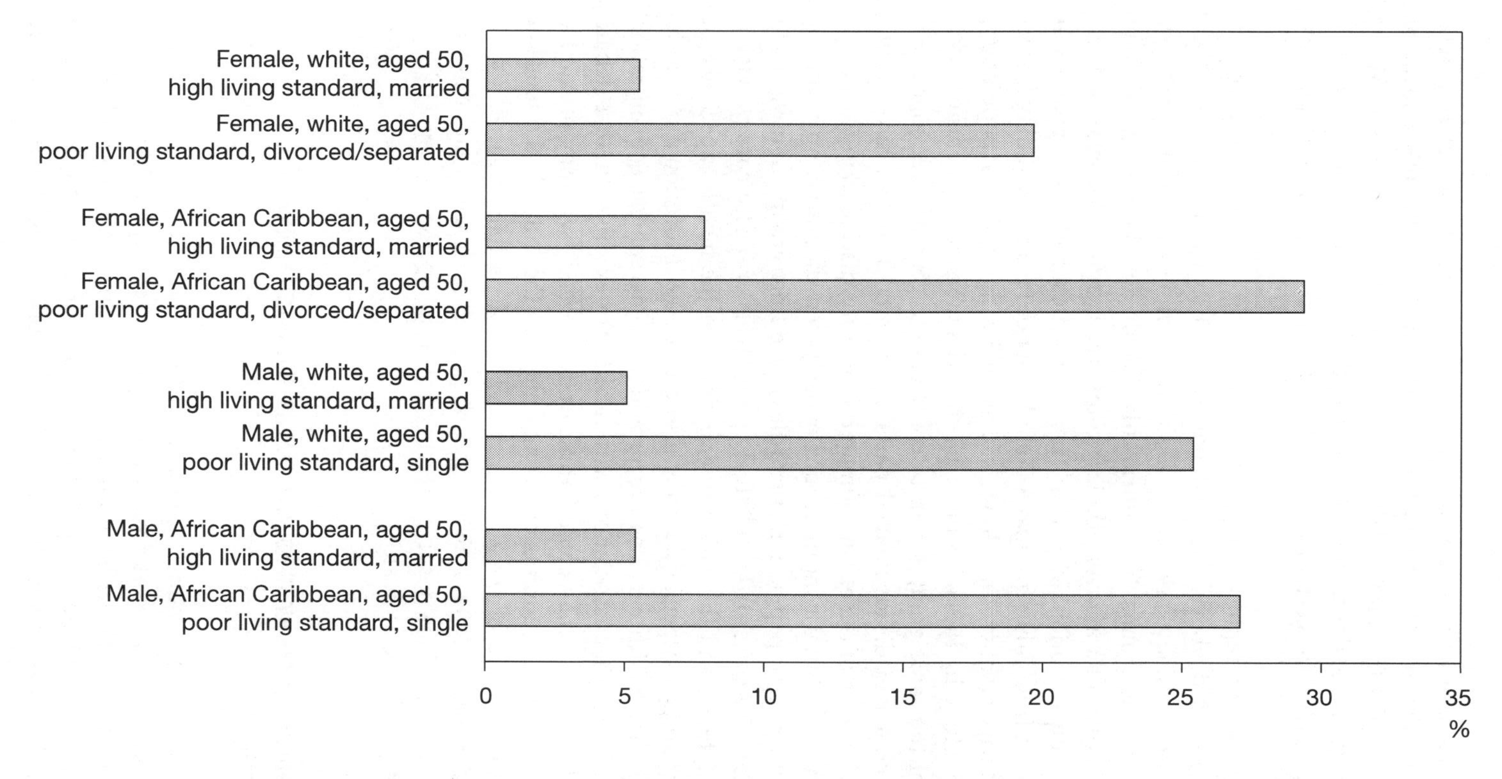

FIGURE 3.5 Proportions reporting long term illness in the 1991 census in London: comparisons by sex, ethnic group and indicator of socio-economic position (Curtis and Lawson, 2000)

showed that alternatives, seen as part of the minority ethnic culture, were sometimes preferred to 'mainstream' biomedical treatments associated with the majority culture. We also saw in Chapter 2 how the Maori community in Hokianga sought to assert their local community identity by campaigning for certain types of health service organization (Kearns, 1998).

From a geographical perspective, and using the theoretical frameworks discussed above, material poverty and experience of racialized aspects of ethnicity are partly associated with space–time paths of different ethnic groups. In countries like Britain and the USA, these may be distinctive for minority ethnic groups, for example, because of their marked concentration in certain parts of cities and their residential segregation from other groups. A complex set of factors, including historical patterns of migration flows, exclusion by the majority population, discrimination in the housing market and tendencies to cluster with others of the same group, seem to contribute to geographical concentration in poor and disadvantaged parts of the city. This can have health impacts because of the effects of material poverty, discussed in Chapter 4. When such material conditions are experienced they are important for the health of any population, regardless of ethnic group. Impoverished areas also may suffer from relatively poor levels of health care provision (see Chapter 5) and this can also contribute to the health experience of minority groups, especially if they also encounter institutional racism in the way that health services are provided. Geographical concentration may also have other associations with health. For example, there is debate over whether there may be an 'ethnic density' effect, whereby proximity to others in the same minority group enhances the individual's sense of ethnic identity and of social support and social capital resources (Neelman and Wessley, 1999; Halpern and Nazroo, 2000; Karlsen and Nazroo, 2000). This may have positive effects on health experience, especially in terms of psychosocial health.

MOVING ON THE MARGINS: TRAVELLERS, HOMELESS PEOPLE AND REFUGEES

The life spaces of minority groups may differ from the majority population in other ways, apart from the effects of geographical segregation. This section reviews the health implications of high mobility and lack of fixed place of residence for people who also suffer from social exclusion and disadvantage associated with their mobility, and focuses on the example of Travellers, people who are homeless and people who are refugees.

The health of Travellers

Itinerant populations, variously identified as Romanies, Gypsies or Travellers, present an interesting example from the perspective of geographies of social groups who are 'on the margin' of society. These groups have often been excluded by the majority in society and their 'otherness' is strongly associated with the distinctive ways in which they occupy space. Romanies are not itinerant in all countries, having been encouraged, or forced, to settle in some places (Hajioff and McKee, 2000), but in some countries they retain elements of a nomadic lifestyle.

Nomadic groups assert their distinctive ethnic identity by lifestyles that contravene and subvert the controlling practices of dominant groups in society. Their itinerant behaviour means that the normal processes used to administer and organize populations resident in fixed locations cannot apply. The standard arrangements for providing primary health care, for example, are not appropriate for these groups. Travellers are often distrusted by the majority society and represented as deviant and dangerous. Other, settled populations, who own or occupy fixed spaces, often reject Travellers through exclusionary territorial practices. Temporary stopping places become more difficult to find and are often sited in locations which are relatively risky for health (for example, derelict industrial land and spaces lacking basic amenities).

A number of studies have examined the health of Gypsies, and the findings are of interest from a geographical point of view, although this research has not, on the whole, been conducted by health geographers. Van Cleemput and Parry (2001) report on a survey of 87 Gypsies interviewed in Sheffield, UK. Their responses to a standard questionnaire on self-reported health (Euroqol) were analysed. Most (but not all) measures of health for the Gypsies were worse than those recorded for respondents from semi-skilled and unskilled manual social groups in a national survey. (For example, the Travellers were more likely to report anxiety and restriction of mobility and usual activities.) The Gypsies also reported worse health, on several indicators. However, when Gypsies were compared with an age and sex matched group of patients registered with a general practice in a highly deprived area of Sheffield, the differences were less extreme. The authors suggested that poverty and poor living conditions may account for much of the difference in health experience between Gypsies and the general population.

Similar conclusions are borne out by a number of other observers of health of Gypsies in various countries. Sutherland (1992) reported that Gypsy populations in the United States have poor health on a number of indicators, for example, high levels of diabetes, heart disease, hypertension. In Europe, Smith (1997) also commented that living standards are much lower than for the majority population, and they lack housing, adequate stopping places, education, employment and essential services. Joubert (1991) recorded that babies born to Gypsy families tend to be smaller than in other groups, due to poor living conditions and hygiene, as well as poor education of the mother. Koupilova et al. (2001) reviewed the limited evidence on health of Romany populations in the Czech and Slovak republics. Their health status was, for the most part, worse than that of the majority population, partly due to their living conditions. Radicova (2001) reported on a survey of Romany settlements in Slovakia funded by the World Bank. This documented the disadvantaged situation of Romany populations in social and material terms and also their relatively poor health. It was argued that the social exclusion and differentiation of the Romanies resulted in different life chances and risk taking. A broadly based strategy was needed to address the difficulties they faced. Particular emphasis was placed on efforts to overcome exclusion from labour markets.

Van Cleemput and Parry (2001) noted that Gypsies who travelled less frequently reported poorer health. This finding could be interpreted in several ways. It may be that those in poor health were unable to travel, or that travelling was good for the wellbeing of people in this group. The authors cited other research which showed that Gypsies associated living in a house with feeling isolated, closed in and lonely.

A wish to avoid stigmatizing Romanies may contribute to the lack of information about them, making their situation less 'visible'. However, Hajioff and McKee (2000) review evidence about the health situation of Romany peoples and comment that much of the literature focuses on control of infectious diseases in Romany populations, such as mycobacterial infection, viral hepatitis, poliomyelitis. There is much less attention given to the impact of non-communicable diseases. They conclude that much of the literature is based on concepts of contagion and is primarily concerned with protecting the health of the majority population. This observation echoes Donovan's (1984) discussion of the representation of health of ethnic minority groups in Britain at the time, which she saw as tending to emphasize their higher risk of infectious diseases and conditions seen as 'exotic', 'imported' or linked to their particular racial group (e.g. tuberculosis, malaria, sickle cell disease), rather than focusing on the risks they face for non-communicable diseases which are also common in the majority population (e.g. cardiovascular disease and diabetes).

Poor access to health care is often identified as a factor contributing to poor health for Gypsies (connecting to theories of access to health care discussed in Chapter 5). Their socially and spatially excluded status makes it more difficult to ensure access to primary health care and their experiences of discrimination discourage them from using mainstream health care. There is an absence of systematic research and of advocacy to represent this group. Koupilova et al. (2001) documented lack of communication with health workers, low uptake of preventive care and lack of advocacy for this group in the Czech and Slovak republics. Edwards and Watt (1997) reported that Travellers had high levels of unmet need for dental care but this was not due to cultural barriers to seeking care. Their lack of access was associated with their itinerant lifestyle. Those living on unauthorized transit sites have been shown to have worst access to care. Smith (1997) also explained how Romany women used mainstream health services reluctantly because of previous experience of racist exploitation and unauthorized experimentation by government run health services. Feder et al. (1989; 1993) comment on the need to improve provision of primary care to Travellers and consider the role for specialized health professionals to meet their needs.

Health beliefs and behaviours of some Romany groups may be dissonant with mainstream medical views and practices. Sutherland (1992) and Smith (1997) explain that the health related behaviour of these groups is often viewed by medical personnel as chaotic and difficult to manage. Gypsies view other ethnic groups with suspicion and may consider them a source of uncleanliness or infection. There are traditional practices and beliefs which are important for Romanies, involving ideas of good and bad fortune, purity and impurity, and inclusion and exclusion from the group. Traditional beliefs linked health, luck and travel, and included use of herbal remedies. Today there is no longer access to these remedies, or the knowledge has been lost. Thus some Romanies with poor access to medical services are also unable to rely on more traditional strategies for health.

Lehti and Mattson (2001), reporting on in-depth interviews with four Gypsy women, explored knowledge about health and attitudes to health care among Gypsies. The researchers reported that women tended to report similar types of symptoms (pain, headache, depression) and so received similar treatment, though the implication is that they were actually describing different health problems, so this might not have been medically appropriate. They tended to attend the surgery in a group of relatives or friends, which might make consultation between patient and doctor seem more complicated. Such behaviours go against the grain of medical spatialization described by Foucault, which requires cooperation of the patient in expressing details of their condition in terms that correspond to medical diagnostic criteria. The interviews showed that strict systems of social hierarchies and social rules influenced health beliefs and health behaviours of Travellers. Young women were seen to be especially vulnerable. Polaschek concludes that 'until the effects on the health care system of inequalities in power between groups in society are addressed we cannot ensure that the needs of persons from minority cultures will be met' (1998: 457).

On the other hand, Ojanlatva et al. (1997) reported how health workers sought to improve cultural competence using information from documentary sources and interviewed a Romany woman. They point out that the differences between groups like Romanies and the majority population might be exaggerated. Some minority cultures may adopt the lifestyles of the country they live in, even though their lives are not organized geographically in the same way as the majority.

Thus while population mobility in itself is not necessarily a risk for health, when it is associated with social exclusion and expulsion by more powerful groups it can be damaging. The health of other highly mobile populations lacking a fixed residential location and subject to social exclusion can also be interpreted in terms of power relationships in social space. People who are homeless or are refugees fall in this category.

Homelessness and health

Studies in high income countries have repeatedly demonstrated homelessness to be associated with particularly severe health disadvantage (e.g. Bines, 1994; Best, 1999). Researchers with a geographical perspective are among those who have examined the impact of homelessness on health in urban settings. Jones and Curtis (1997) cite evidence of relatively high levels of homelessness in London. Takahashi and Wolch (1994) report on the health situation for homeless people in Los Angeles, highlighting the problems they face in access to employment and services as well as to housing. Padgett et al. (1995) discuss the high rates of use of hospital emergency rooms by homeless people in New York City. Lack of access to alternative sources of health care contributed to this pattern. It was also associated with relatively high prevalence rates of injury and victimization (actual and feared experience of robbery, violence and assault), as much as 30 times that in the general US population. Similarly, Ensign and Gittelsohn (1998) report perceptions of health risks among young homeless people in Baltimore, which emphasized violence and victimization. Studies in New Zealand (Kearns et al., 1991) and Sydney, Australia (Dartonhill et al., 1990) have also described the marginalized position of homeless people. There are significant links between homelessness and mental illness, which are explored in Chapter 7.

Although they often constitute marginalized minorities in the urban populations of wealthy countries, the presence of homeless populations can have appreciable impacts on local indicators of population health. Brimblecombe et al. (1999) showed that some relatively affluent parts of the city of Oxford, England, had poor health indicators, which were found to be due to the location of hostels for the homeless. It is often the case, however, that wealthy communities are successful in excluding homeless people. Takahashi (1997) discusses the rejection and exclusion by the wider community experienced by homeless people, resulting from social stigma attached to homeless people and their lifestyles. Stigma is expressed by a majority view emphasizing non-productivity, dangerousness and personal culpability of excluded groups. It is suggested that *'stigma not only creates a definition of acceptable and non-acceptable individuals and groups, but also creates a powerful cognitive map of acceptable and non-acceptable places'* (1997: 904). These 'non-acceptable' places are associated with degraded urban environments and concentrations of people who are destitute and ill. Takahashi describes a web of interactions between these excluded populations and their impoverished environment. Places *'inherit the stigmatization of persons and groups'* and persons and groups can *'become stigmatized through their interactions with devalued places'* (1997: 911).

In urban areas of low income countries, extreme poverty affects larger proportions of the population. Many who have relatively fixed homes live in slum conditions. Some studies suggest that, while homeless populations do show health disadvantage, they may not be much worse off in health terms than other poor populations. For example, Gross et al. (1996) showed that among children living on the street in Jakarta, Indonesia, around half were stunted in growth, but that this was no worse than the situation for poor urban children housed in the slums of the city. Wasting (indicative of poor current diet and conditions such as diarrhoea) was not very common among these homeless children. They suggest that children living and working on the streets had a somewhat better nutritional status, and they stress the resourceful survival strategies practised by the children. Also, conditions in Jakarta may have reduced the impact of homelessness compared with other cities. Night-time temperatures were always relatively high and food was relatively inexpensive, for example. Similar findings were reported for homeless boys in Nepal by Panterbrick et al. (1996), who comment that street life may represent a relatively successful response to urban poverty for these boys. It seems likely, however, that this is only true for those who are initially relatively strong and resourceful, and that only the strong survive in these difficult conditions. These studies seem to contrast with information on the homeless in high income countries and they may illustrate how risk factors for health of excluded groups may operate differently in different types of setting. They also underline the significance for health of resistance strategies practised by groups in marginal socio-spatial positions.

Health of refugee populations

The situation for refugees has received limited attention by British health geography, although it raises some interesting issues from a geographical perspective. A review of a large body of research on health of refugees in London (Aldous et al., 1999) estimated around a quarter of a million people in the city had been through the asylum seeking process within the last 15 years. London's status as a global city, with widespread communication links to many parts of the world that are in conflict, makes it a common destination for asylum seekers. The authors noted that many of the health problems of refugees were associated with their poor living conditions and were not specific to their refugee status, except in so far as their status caused them to be at risk of material poverty. They did, however, have relatively high prevalence of certain health problems associated with the conditions under which they had left their countries of origin. These included the physical and mental after-effects of war, torture and traumatic displacement and journeying to Britain. Together with other recent migrants from poor countries, they shared relatively high risks of communicable diseases which were prevalent in their countries of origin and these risks were heightened by problems of inadequate access to health care and health surveillance services. Refugees are often in a very precarious position in London, lacking permanent homes, and are highly mobile once they arrive in the city, which makes it more difficult to coordinate services. Watters (2001) also discusses the resilience of refugees and the need to support them by building on their powers of resistance.

Geographical studies of refugee health in other countries also highlight the significance of forced displacement for health. Kalipeni and Oppong (1998) consider the refugee crisis in Africa using a political ecology approach, based on the writings of Giddens (1984), Atkinson

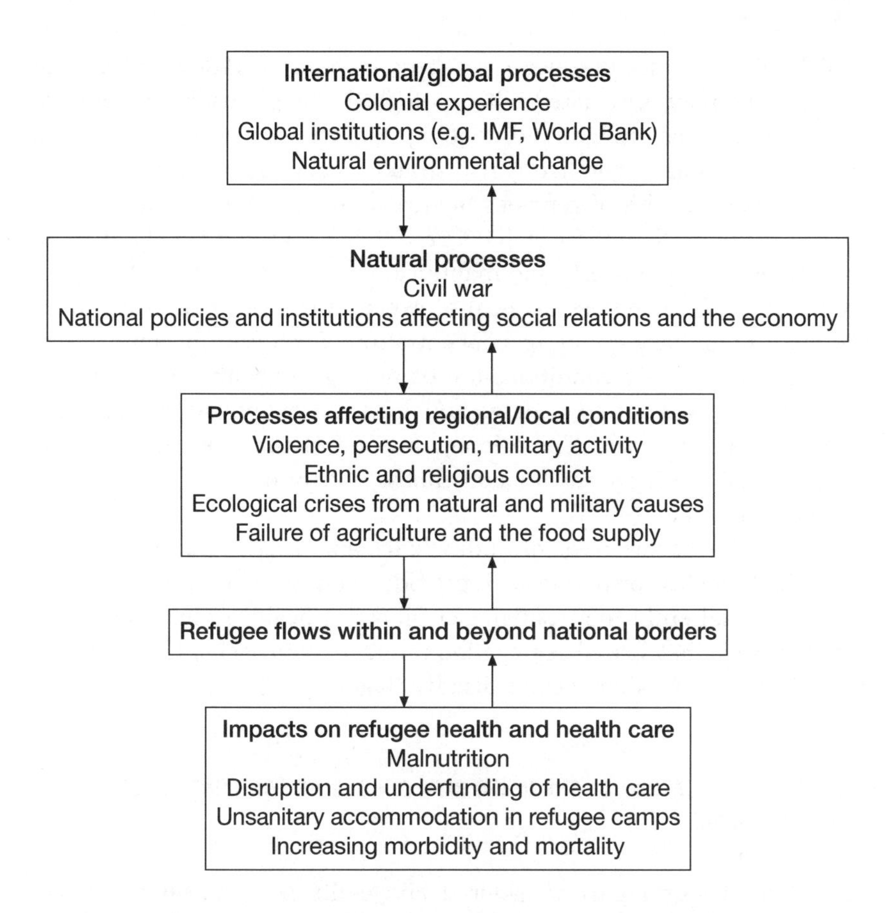

FIGURE 3.6 Political ecology perspective on the factors influencing health of refugees (after Kalipeni and Oppong, 1998; Wood, 1994)

(1991) and Wood (1994). Three-quarters of sub-Saharan African countries are affected by forced migrations of this type. While there are no precise estimates, millions of refugees are involved, and the largest numbers of women and children forced to move are estimated to be those fleeing countries such as Rwanda, Liberia, Somalia, Eritrea, Sudan and Angola during the 1990s. The political ecology approach outlined by Kalipeni and Oppong (1998: 1640) involves a number of processes operating at different geographical scales to generate refugees (Figure 3.6). International processes such as historical actions of colonial powers and present actions of global institutions like the International Monetary Fund and the World Bank contribute to the destabilization of political economies. National processes include government policies and ethnic and religious conflicts (often rooted in conditions during the colonial era). At the regional and local level, processes producing refugees include breakdown of agricultural productivity, associated with ecological change resulting from natural crises such as drought and flooding, as well as contamination due to impacts of military 'scorched earth' strategies and laying of landmines.

Effects on health of refugees in Africa operate through causal pathways including: disruption of food production and livelihoods of populations, resulting in poverty and malnutrition; disruption of health service provision and reduction of health expenditures as countries divert resources to military spending; concentration of refugees in camps where material and sanitary conditions are damaging to health. Morbidity and mortality increase due to diseases which are normally preventable, such as cholera and measles. There is aggravation of the risks of conditions such as HIV infection and mental illness, partly due to the trauma of violence and rape.

Thus the experiences of individuals and social groups who are nomadic, homeless or refugees all provide extreme examples of the links between social exclusion, spatial exclusion and health status. They highlight the ways that spatial segregation interacts with social differentiation and rejection to produce health disadvantage.

'DESERTIFICATION' OF THE INNER CITY: STRUGGLES FOR CONTROL OF CITY SPACE

Several of the examples considered above illustrate tensions between social groups over power to control and occupy space in the city. This process was identified as an important dynamic for public health in New York City by Wallace (1990) and Wallace and Wallace (1991; 1997). They described a process which they termed the *desertification* of parts of the inner city. Desertification resulted from a struggle for power over urban space and resources between communities in poor residential areas and politically powerful groups. This took place in areas which had been occupied by large concentrations of poor populations, which represented relatively stable, though disadvantaged, communities, able to exercise some influence on democratic processes in the city. Some of these communities were displaced and moved to other parts of New York City during the 1980s and 1990s.

Wallace and Wallace argue that the withdrawal of basic services, including fire fighting and policing, resulted in a steady decline of the fabric of these areas, in terms of both the social and the physical environment. Wallace and Wallace invoke a version of what Davis (1990),in his description of Los Angeles, called the 'fire zone', where parts of the city were literally razed due to uncontrolled fires destroying the buildings. These areas became available for redevelopment. Those members of the population who were able to quit the area did so, leaving a growing concentration of the most disadvantaged and excluded groups. These remaining populations, already at high risk of illness and injury, and lacking access to health services, experienced a steadily worsening situation in terms of public health and high rates of serious diseases such as HIV/AIDS related problems. Wallace and Wallace therefore described a process of *hollowing out*, or socio-spatial exclusion and marginalization, of communities whose social position was weakest, imposed by more powerful social groups. They argued that this created a situation which was dangerous for the public health of the city population as a whole, and not just for the excluded populations involved. Chapter 6 considers, for example, their analysis of the spread of HIV infection through the population of New York City and surrounding, wealthier commuting areas (Wallace and Wallace, 1993).

In the account offered by Wallace and Wallace, there were reflections of the time–space geography described above. Their perspective also encouraged a view of New York City as an interconnected, whole city system from the point of view of risks to public health. The model of desertification offered an explanation of the detrimental effects on public health of social and spatial polarization. It illustrated hegemonic power structures influencing the living spaces of disempowered populations.

CONCLUSIONS: LANDSCAPES OF POWER AND HEALTH

The examples considered in this chapter show how geographers and researchers from other disciplines have used ideas concerning the exercise of power in space, and its implications for population health. Figure 3.7 summarizes the relationships reviewed here. Power is unevenly distributed between social groups, and time–space distanciation, territoriality and surveillance reflect and reinforce these power relations. The constraints and potential of individual time–space paths are important for social interaction affecting experience of health and health care. Variable operation of rules and systems for resource allocation by those in authority produce spatially varying access to the resources that are necessary for good health. Socially constructed spaces and the actions of dominant groups interact to reinforce the power relations in society. As a result, there are health inequalities among populations separated from each other in space, time and social position. However, resistance by subordinated groups also operates through socio-spatial processes to contest, and sometimes to mitigate, these inequalities.

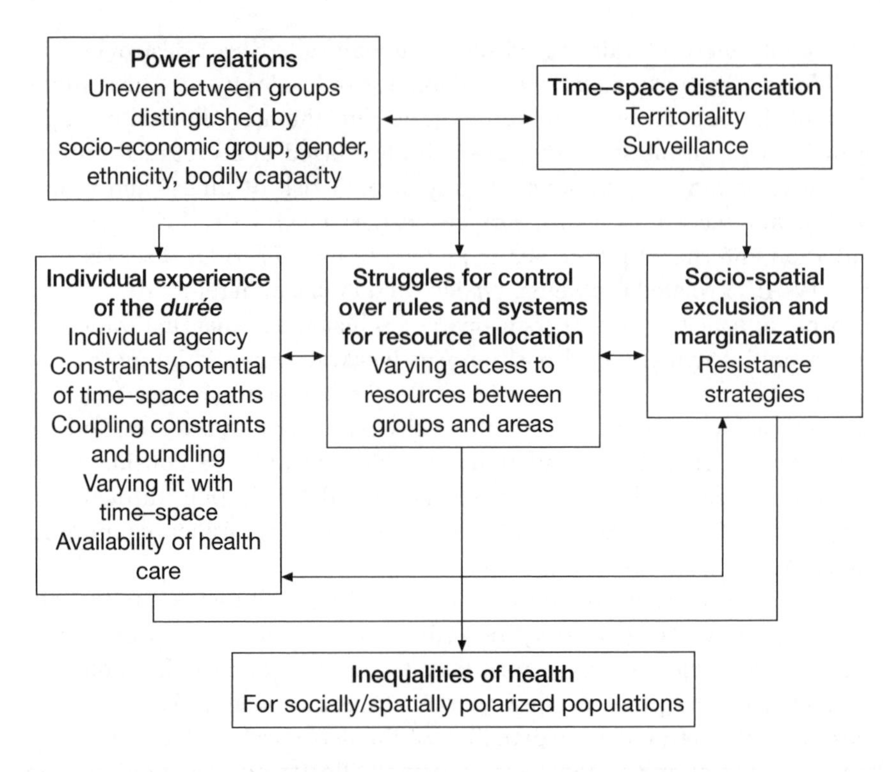

FIGURE 3.7 Key relationships in landscapes of power and health inequalities

These processes of exclusion and disempowerment, as well as the capacity for resistance, provide part of the explanation for health inequality between social groups and between populations in different places. They are also relevant for the design of health care strategies to reduce health inequalities. Spaces in which health care is delivered can create 'power containers' which exacerbate the social inequalities contributing to health difference. However, if they are located and organized in ways that break down these inequalities of power, spaces of care can offer opportunities to improve the health of marginalized populations. Action to promote wellbeing and improve health care may need to be targeted on the spaces occupied by disempowered groups and should aim to be sympathetic to the time–space paths of their daily lives.

FURTHER READING

Useful supplementary reading on the themes covered in this chapter include the following examples, which have been mentioned in the discussion here:

Sibley (1995) provides a good introduction to geographical perspectives on processes of social exclusion in a general sense (not specific to health). Also Wolch and Dear (eds) (1989) is an earlier collection of essays on similar themes, with excellent contributions by several authors, including Chris Philo. Other work by Philo provides excellent reviews and discussion of the issues raised in this chapter, for example: Philo (1997) Across the water: reviewing geographical studies of asylums and other mental health facilities. (*Health and Place*, 3, 73–89).

Butler and Parr (eds) (1999) is especially useful for the way that it crosses what has, in the past, been a divide between the work of geographers concerned with health and illness, and those concerned with 'disability'. It includes thought-provoking chapters about the experience of social exclusion and social construction of the human body.

Gleeson (1999) and Imrie (1996) also provide in-depth discussion of ideas relating to disability from a geographical perspective.

Perspectives from outside geography are also important to an understanding of the issues raised here. It is not possible to include a comprehensive list, but I would recommend, for example: Annandale and Hunt (eds) (2000) for material on gender differences; and Bartley et al. (1998) which, provides a collection of essays on health inequality with an interestingly diverse range of disciplinary perspectives.

OBJECTIVES AND QUESTIONS

I hope that students will find that this chapter contributes to their knowledge of:

- conceptual frameworks which help to explain the importance of power relations in health differences, and the importance of geographical dimensions of these power relations;
- examples of studies showing how social power relations affect the health experience of individuals and social groups who are disadvantaged by processes of social and spatial exclusion.

This knowledge should help students to answer questions like the following:

1 What do you understand by the concepts of the *durée* and *time–space distanciation*, proposed by A. Giddens (1984)? How has the work of geographers contributed to these concepts, and why are they important for the geography of health?
2 Why are 'landscapes of power' important for the health and wellbeing of patients in health care settings?
3 Discuss the significance of geographical mobility as a potential risk to health. (NB: this question could be answered in a number of ways and students might also refer to ideas from Chapter 6 to tackle it.)

4 Landscapes of Poverty and Wealth

This chapter focuses on the significance for health of geographic variations in poverty and wealth. The discussion here first considers aspects of the geography of uneven development and their relationship with health and welfare. There may be limits to the health gains associated with growth in the economic wealth of societies. This chapter reviews the debate over the importance for health of *inequality* of wealth. The discussion goes on to examine the links between poverty and health disadvantage, with a particular focus on the importance for health of conditions of material poverty in *places* and whole communities (as opposed to individual poverty). Some explanations for the associations between poverty and poor health are reviewed, and in this chapter particular attention is given to factors such as variations in social capital and the effects of material poverty associated with poor housing and poor working conditions.

GEOGRAPHIES OF UNEVEN DEVELOPMENT AND WELFARE

Many commentators see social and spatial polarization of wealth as an essential aspect of political-economic systems that dominate the world today. Within, as well as between, countries, uneven development has been widely documented and debated.

Some authors (e.g. Dollar, 2001) suggest that there is no systematic association between growth in national income and growth in inequality of income. They argue that incomes of the poorest people in developing countries tend to increase as average national income increases. However, others have observed that in many countries affected by globalization of capitalist economic systems, income inequality has become more marked. For example, Cornia (2001) reports growing income inequality between 1950 and 1990 in most countries in Latin America and the countries of the former Soviet Union, and a similar trend in Asian countries and in several of the members of the Organization for Economic Cooperation and Development. In Britain, Hills (1998) reported growing income inequality during the 1980s and 1990s, compared with the situation in 1970.

Global cities are especially influenced by these effects. While they may show particular concentration of wealth, they are also marked by striking inequalities and some authors argue that we are seeing a growing division between rich and poor urban dwellers. Sassen (1991) argued that in global cities there is a dual labour market, comprising two types of work. Core workers are essential to the functioning of multinational firms, and benefit from relatively secure and rewarding working conditions. Their situation contrasts with a 'reserve' labour force, who are engaged in work which is more marginal, insecure or poorly paid. The position of socially and economically excluded populations contrasts starkly with the situation for privileged core workers and their families (Sibley, 1995). Byrne (1997) argued that, in the past, there was a relatively 'smooth' socio-economic gradient with varying levels of poverty and wealth. However, he suggests that: 'Now things are different. There are really two sorts of lives, involving poverty and exclusion on the one hand and affluence and participation on the other. These two modes don't overlap' (1997: 34).

These social divisions are mirrored by spatial divisions. Geographical critiques of the capitalist mode of production have often focused on the view that fundamental structures in this type of political economy generate uneven development and inequalities of wealth between geographical areas (e.g. Harvey, 1982; 1989b; Smith, 1994). Socio-economic disadvantage for an individual person therefore needs to be interpreted in the light of the broader socio-economic setting within which the individual is embedded.

Research in Britain, for example, has produced evidence for increasing geographical disparities in terms of poverty and wealth. Hills (1995) cites evidence reported by Green (1994) that demonstrated persistent and growing spatial polarization of poverty in Britain during the 1980s. A measure of 'concentrated poverty' was compiled, based on ranking of small areas on indicators of unemployment, economic inactivity, living in rented housing, lack of a car and low social class. In 1991, 8.9% of small areas were classed as experiencing concentrated poverty as compared with 7.5% in 1981, suggesting that more communities were experiencing intense deprivation. The results also showed polarization in separate measures of poverty and wealth at small area level. Unemployment and inactivity levels and the proportion of adults without higher qualifications all showed increasing spatial concentration into the most disadvantaged areas between 1981 and 1991. Such uneven economic development within a country results in growing geographical concentration of populations affected by relatively low incomes, poor employment opportunities, poor working conditions, and poor housing. They may also suffer from poor access to goods and services and more exposure to hazardous environmental impacts of production (as discussed in Chapters 5 and 6). All these aspects of material poverty are damaging to health.

Uneven development results partly from the organization of production in today's global economy. The global expansion of industrial and post-industrial capitalist production has been important for the development of spatial differences in poverty and wealth at the international and the local levels. Taking an international perspective, Cornia (2001) suggests that economic globalization has been promoted by various national and international policies and actions. These include domestic deregulation of economic processes, liberalization of international trade, and privatization of key services and infrastructure that were previously organized collectively. Partly because of these changes, multinational companies have grown in importance and influence. Global flows of capital are associated with an increasing tendency for the economies of different countries around the world to be interrelated (Harvey, 1982). Within countries, international circuits of capital result in growing foreign investment, and increasing short term financial flows of money into and out of the national economy. As noted in Chapter 1, global circuits of capital are among the important drivers leading to development of global cities.

The organization of capitalist modes of production can take various forms, and Pinch (1997) considered these in terms of their relationship to the geography of welfare. Trends in the organization of production are often summarized in terms of a shift from 'Fordist' to 'post-Fordist' models. The 'Fordist' system of factory based production was exemplified by methods used by the Ford Company for car manufacture in the US in the early 20th century. It was typical of the period when extractive and manufacturing industry led economic development globally. Pinch (1997: 77) summarizes the geographical features of the classic Fordist model, which include: spatial concentration of work in large factory settings, designed for mass production; spatial division of different types of work and regional specialization; international sourcing of components; and growth of large industrial conurbations. This enabled profitable production of standardized goods at relatively low prices, affordable for the majority of people. Large industrial cities such as Manchester in Britain, and Detroit in the USA, grew rapidly as a result of these trends.

Post-Fordist models, on the other hand, describe more flexible and specialized forms of production, more typical of new technologies and service industries that have led recent economic growth. They tend to be more distributed in geographically clustered, multiple production units, with components sourced through a network of 'associated' industries located close together, often in 'new industrial spaces' in accessible but less urbanized and industrial locations. Cities like Cambridge in Britain and Silicon Valley, Santa Clara County, California in the US have seen strong development of this type.

Both Fordist and post-Fordist organizations of production are associated with the concentration of wealth and production in certain geographical areas, while other areas remain less developed. The shift of dominance from one type of model to another has resulted in 'deindustrialization', disinvestment and decline in 'old industrial' areas which were once very successful as sites for 'Fordist' style production, but are now uncompetitive internationally, and are disadvantaged compared with areas which are more attractive to the new post-Fordist, post-industrial economies.

While these processes are to some extent international, and independent of national government, regime theory (Stone, 1993; Stoker and Mossberger, 1994) postulates that the state has a role in mediating their effects. Different countries at different times operate particular regimes, or modes of regulation, in order to achieve economic stability and the reproduction of the capitalist political economy. These regimes vary to suit national conditions. They operate through the economic and social policies and regulations of national governments and other institutions. These influence the operation of markets and the relationships between various groups of capital owning classes and workers in society. This is one reason why, for example, social and economic conditions vary between countries in Western Europe, even though these countries are within the same world region. Each country operates its own mode of regulation, as well as cooperating over some aspects of social and economic policy through the European Union. At the more local level of cities, we can also envisage urban regimes: the functioning of public agencies and private interests to regulate and govern the city.

Health care can be interpreted in the light of these perspectives, as discussed, for example, by Mohan (1995) and Pinch (1997). For example, the political economy of a country, including its systems of regulation and organization, influences the availability and cost of labour to health services and the role of private sector organizations and market models in the provision of health care and pharmaceuticals. The ways that health and welfare services are produced can also be interpreted in terms of Fordist and post-Fordist models. Some features of hospital health care, for example, have strong similarities to Fordist manufacturing principles. The provision of rather standardized packages of health care, and the somewhat rigid division of labour among different types of health care worker, could also be seen as typical of a Fordist approach. The more recent growth of interest in flexible production processes has also applied to the provision of health care. Many countries have moved towards health services which are more deregulated, are organized according to managed market principles by a complex mix of public and private sector agencies, are delivered through a more varied range of facilities, and are intended to be more responsive to the demands of the consumers of care. These aspects are considered in more detail in Chapter 5.

Some accounts based on Marxist theory (critically reviewed, for example, by Mohan, 1995) also see the provision of state health care as one of the ways that societies maintain and regulate capitalist political economies in a more general sense. Collectively provided health services help to 'reproduce the labour force', ensuring 'fitness for work' of each new generation of workers. Health provision to workers and their families, funded by the state or by employers, which supports sick, disabled or frail elderly people, also helps to 'compensate' workers for their relatively weak position in the political economy and therefore maintain the political stability of the system. Collectivized health care systems, such as the British National Health Service, Western European health care based on social insurance, and the Medicare and Medicaid schemes in the USA, might all be interpreted as aiming to mitigate the impact on population health and welfare of capitalist modes of production, and to maintain capitalist societies as viable political economies.

POVERTY AND INTERNATIONAL HEALTH VARIATION

Poverty is widely accepted as an important determinant of health and is generally associated with higher levels of illness and death in the population (e.g. Marmot and Wilkinson, 1999). The processes by which poverty causes morbidity and mortality are complex and are variable in different settings. Research has demonstrated these inequalities at different geographical scales.

At the international scale, differences between countries and world regions have been presented, in a simplified form, in terms of the *cycle of poverty* and the *cycle of affluence*. These relationships are also supported by Omran's (1971) epidemiological transition model, described in Chapter 1 (see also Curtis and Taket, 1996). Jones and Moon (1987: 40) cite a conceptual framework described by Pyle (1979). Pyle contrasted the 'infectious disease model' found in poor, less developed countries with the 'chronic disease model' found in high income countries. A revised version of these models, shown in Figure 4.1, includes reference to global as well as national processes creating these contrasting conditions. Figure 4.1 also represents the interdependence of conditions in high and low income countries.

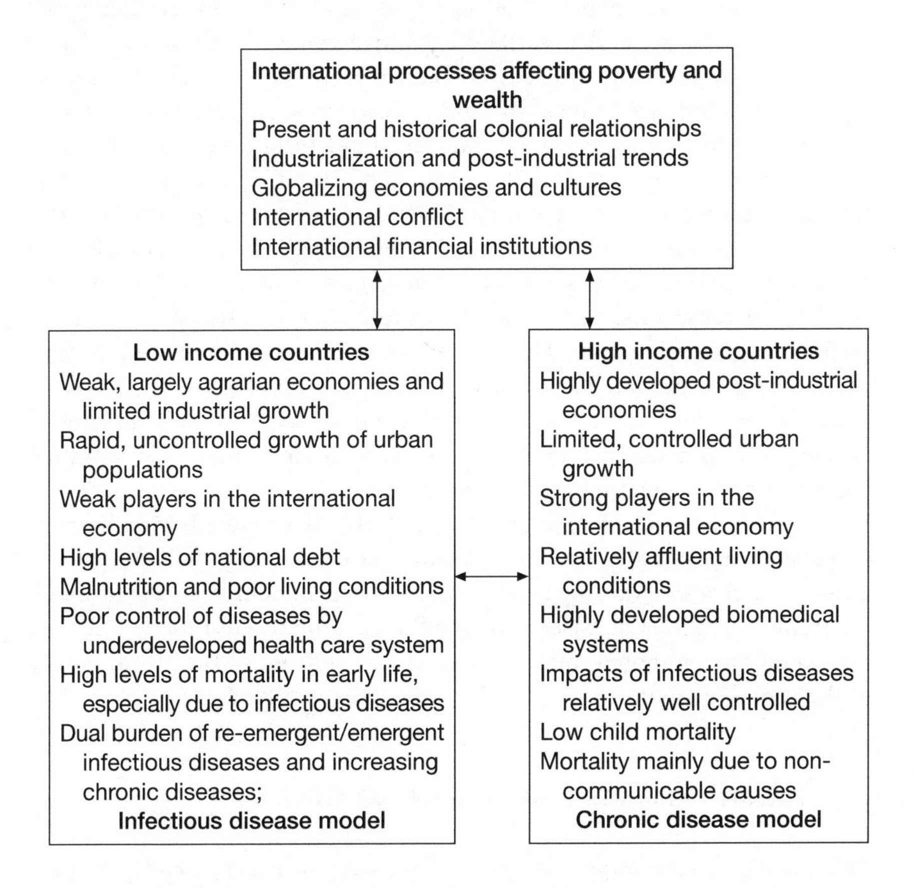

FIGURE 4.1 Models of national poverty and wealth (adapted and revised from Pyle, 1979; Jones and Moon, 1987)

The *infectious disease model* reflects the epidemiological situation found, for example, in African countries, with high levels of child and adult mortality, illustrated in Figure 1.5. These countries have conditions typical of the cycle of poverty. Low levels of economic development, poorly regulated economies and weak agrarian economies cause food shortages and malnutrition, resulting in greater susceptibility to infection and high mortality at early ages. The difficult situation for these countries is often compounded by a disadvantaged position in the global economy (undermining the operation of local markets), by high levels of national debt and by political and social instability, as presented in Kalipeni and Oppong's (1998) political ecology model discussed in Chapter 3 (Figure 3.6).

Countries in regions such as North America and Western Europe, in later stages of the epidemiological transition, and with low child and adult mortality (see Figure 1.5), are typical of those experiencing the 'cycle of affluence' and described by the *chronic disease model*. These industrial and post-industrial societies have relatively strongly developed, highly regulated economies, are often locations of powerful interests controlling global circuits of capital, and tend to be relatively stable in political terms. They have often benefited in the past from colonial relationships with poorer countries. Levels of material poverty are less extreme than in poor countries and infectious diseases are relatively well controlled. Chronic diseases are the major causes of death. The original versions of these models represented a rather conventional view of the epidemiological transition. The revised version here is intended to reflect more recent epidemiological trends such as the resurgence of infectious diseases, and the double burden faced in many low income countries of growing chronic disease mortality as well as problems in controlling emergent and re-emergent infectious diseases.

These models are important for the links they make between levels of poverty and wealth and the health status of the population at the international scale. Varying conditions in different countries, corresponding to the infectious disease model or the chronic disease model, are not completely independent, but are interrelated because of the global nature of the social, economic and political processes involved.

POVERTY AND HEALTH VARIATION WITHIN COUNTRIES

There is also a very large body of research demonstrating health inequalities associated with socio-economic position of groups *within* countries. This comes from relatively wealthy, as well as poorer, countries. For example, there is a voluminous literature on poverty and its impact on health in Britain. The evidence extends back in history for 200 years (Davey-Smith et al., 2001). In the 1980s, increasing concern about persistent health inequality in Britain led to the publication of a report summarizing the evidence, often referred to as the Black Report (named after Sir Douglas Black who chaired the inquiry). A further report was also compiled by Margaret Whitehead at the end of the 1980s. (These were subsequently published together in Townsend et al., 1988.) In the 1990s, Benzeval et al. (1995) discussed methods to tackle these inequalities, and in 1998 the Department of Health published the report of the Independent Inquiry into Inequalities in Health (1998), chaired by Sir Donald Acheson and so often referred to as the Acheson Report. The latter report was based on a series of reviews summarizing the key evidence of socio-economic health inequality (Gordon et al., 1999).

In Britain, there are few sources of data that allow researchers to relate health directly to income. Most analyses compare health for population groups defined by other indicators of socio-economic position, such as occupational social class. Class is usually categorized in groups ranging from class I (professionals who are on average more economically advantaged) to V (the most disadvantaged, unskilled workers). For example, Figure 4.2 is based on data from the Acheson Report and shows mortality among men aged 20–64 for three points in time over the last 20 years. The information for England and Wales as a whole showed a declining death rate over the period, suggesting an improving situation for population health on average. However, the data for different social groups showed that there was a gradient at each time point, such that the more advantaged social groups had lower mortality than the less advantaged groups. Furthermore, mortality for the professional groups had improved (reduced) more than for unskilled workers over the period, so that the gradient was becoming steeper with time. Similar socio-economic gradients are found for many other health indicators, such as levels of illness, general health and fitness (Benzeval et al., 1995; Department of Health, 1998).

The Black Report and subsequent reviews have discussed possible reasons for these disparities. Most commentators conclude that the differences reflect real health inequalities and are not merely statistical artifacts. The pattern results from an interaction of the *health damaging* effects of socio-economic deprivation with *'health selection'* – a set of processes by which those in poor health tend to drift down the social ladder to poorer socio-economic positions (West, 1991; see also Curtis and Taket, 1996). Some recent research in Britain suggests that the impact of deprivation on health, which is evident from birth and throughout the life

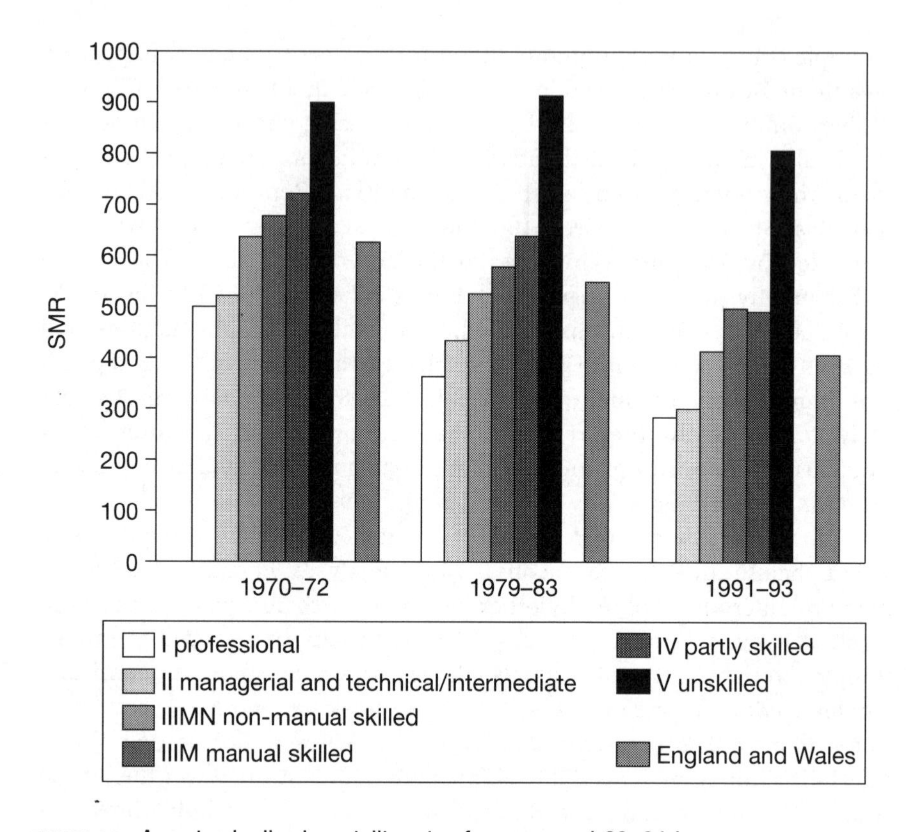

FIGURE 4.2 Age standardized mortality rates for men aged 20–64 in England and Wales, by social class differences at three time points (Department of Health, 1998: 12)

course, has a stronger effect in generating health inequality than health selection (Bartley and Plewis, 1997; Blane et al., 1999; Graham, 2000).

Similar relationships between socio-economic position and mortality exist within other high income countries. For example, Kunst et al. (1999) have examined inequalities in mortality due to all causes of death among socio-economic groups in European countries, demonstrating that manual and non-manual groups typically have different mortality experience. Figure 4.3 illustrates, for 11 countries, the differences in age standardized mortality for men aged 45–59, between those in non-manual and manual occupations (excluding the agricultural sector). Manual workers had relatively high mortality, above the middle line in the graph representing the national average. Non-manual workers had relatively low mortality, below the national average. The degree of inequality was variable between countries, being greater in countries like France and Finland

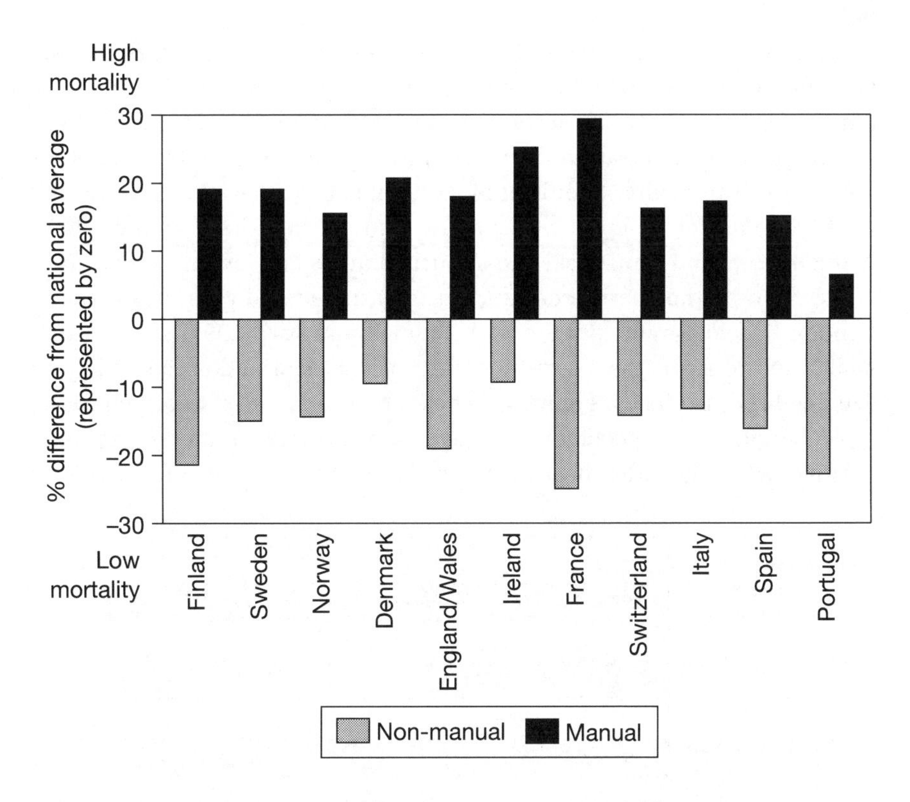

FIGURE 4.3 Age standardized mortality for men aged 45–59 in European countries (data from Kunst et al., 1999: Table 4)

than in Norway and Switzerland, for example. Some studies have used data on neighbourhood poverty to examine social inequality in mortality. For example, Bosma et al. (2001) reported data from the Netherlands showing that mortality risks were higher for people from poor neighbourhoods with high levels of unemployment. Davey-Smith et al. (1996) report a study of mortality among 300,000 white men in the US. This showed that mortality was graded in relation to the average income of the zipcode areas where the men lived and that those from lowest income areas had the highest mortality.

This health divide between rich and poor *areas* is also evident in Britain. Studies demonstrating this include those by Townsend et al. (1988) and Carstairs and Morris (1991), which showed associations between mortality and deprivation at the small area level in north-east England and Scotland respectively, with higher mortality in poorer areas. Curtis et al. (1993) and Bryce et al. (1994) showed that there was a gradient in mortality due to coronary heart disease, at the scale of small geographical units (wards) and for larger health authority areas in England in the 1980s, such that more deprived areas had higher rates of illness. Shaw et al. (1999) have also reported in Britain, in the 1990s, a strong positive association between poverty and mortality and other indicators of ill health. Shouls et al. (1996b) showed variations in mortality for typologies of local government districts in Britain in the 1990s. Affluent rural and semi-rural and suburban areas had low mortality rates, while deprived northern areas and socially mixed and polarized areas of London had high rates (Figure 4.4). Raleigh and Kiri (1997) showed that health authority areas in groups ranked by deprivation had varying average life expectancy. Figure 4.5 shows that in the least deprived areas, typical male life expectancy was over a year more than the national average, whereas in the most deprived areas, life expectancy for men was

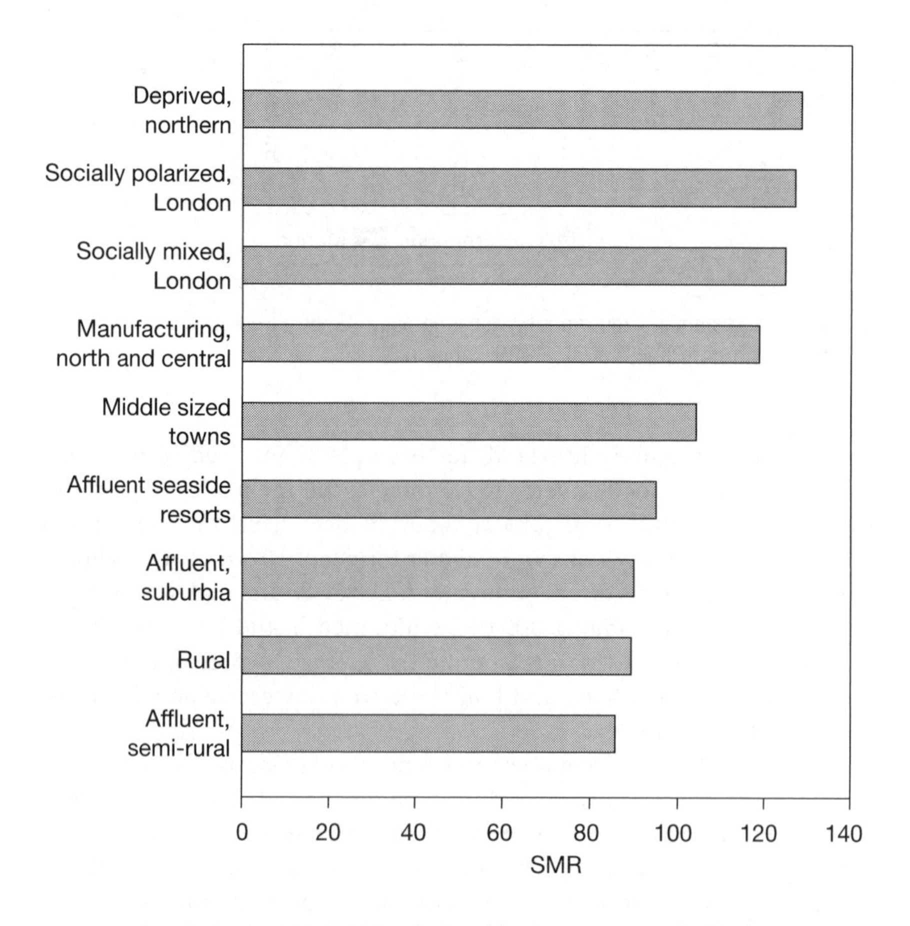

FIGURE 4.4 Local government districts clustered by type: age standardized mortality ratios for males 45–64, 1990–2 (Shouls et al., 1996b: 151)

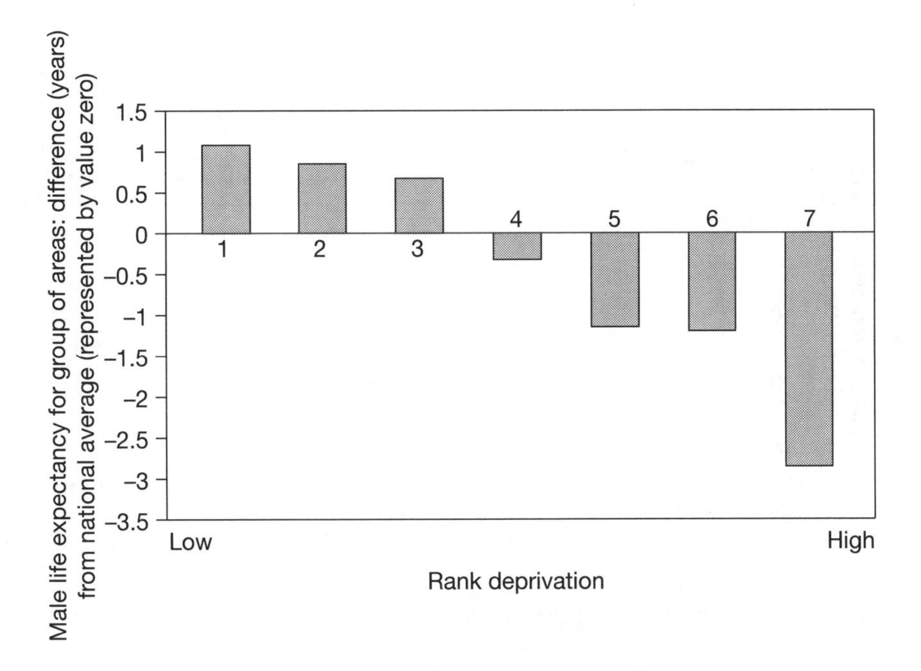

FIGURE 4.5 Male life expectancy for groups of English health authorities by rank of deprivation compared with national average (Raleigh and Kiri, 1997)

almost three years less than the national average. Congdon et al. (1997) showed that for small areas (census wards) in England in 1991, mortality and self reported long term illness were associated both with measures of deprivation (most deprived areas had poorest health) and, independently, with measures of affluence (richer areas had better health). This emphasized the idea of a *graded* relationship between deprivation and health for geographical areas, similar to that for social classes noted above.

The disparities in health between rich and poor areas of Britain have been shown to be increasing, and this finding mirrors evidence for a widening health gap between social groups, considered above. For example, Raleigh and Kiri (1997) demonstrated a widening in the differences in life expectancy. Between 1985 and 1993, the least deprived group of health authorities illustrated in Figure 4.5 showed an annual rate of increase in male life expectancy of 0.38%. For the most deprived group of areas, the annual improvement in male life expectancy was only 0.18%.

Dorling and colleagues have made several studies recording what they refer to as 'the widening gap' in geographical health inequalities

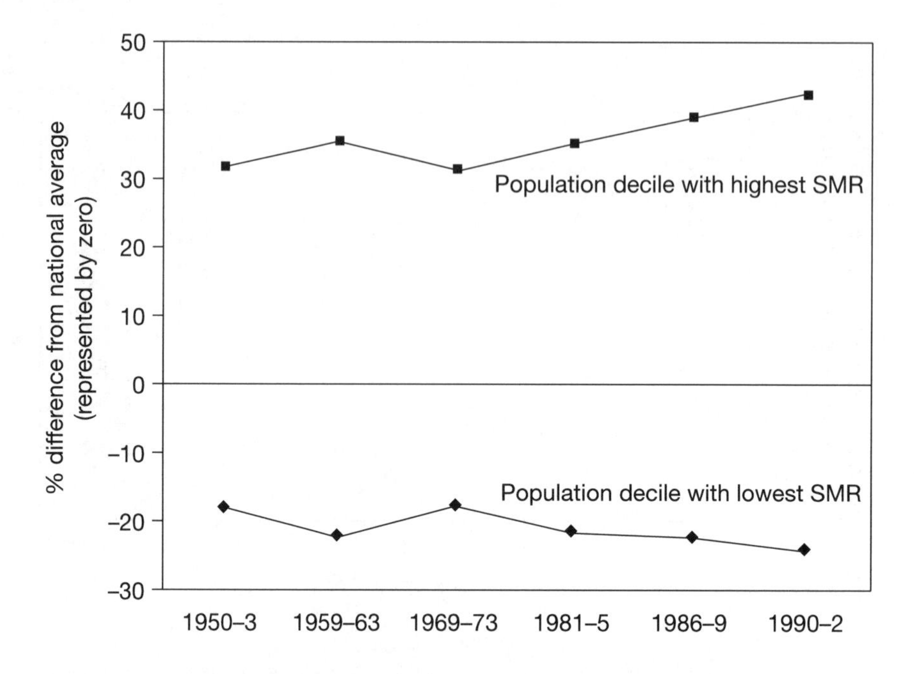

FIGURE 4.6 The widening gap between decile areas with the highest and the lowest SMRs under 65 years in Britain, 1950–92 (Shaw et al., 1998: Table 4)

between rich and poor parts of Britain (e.g. Shaw et al., 1999). One study, reported by Shaw et al. (1998), examined district mortality levels in Britain for a period between the 1950s and the 1990s. Figure 4.6 illustrates their evidence of a widening gap between areas with the highest age/sex standardized mortality and those with the lowest mortality over this period. Mortality fell in all areas throughout the period, and during the 1970s the disparity between areas of high and low mortality was fairly constant, or narrowing. However, the mortality gap widened during the 1980s and 1990s. Similar findings by Shouls et al. (1996b) also showed that the mortality gap between the different groups of districts shown in Figure 4.4 widened during the 1980s, because districts with initially high mortality showed less improvement than those with low mortality. Higgs et al. (1998) similarly reported widening disparities in mortality between more and less deprived areas of Wales over the period 1981–91.

The discussion in this chapter so far shows that comparisons of geographical areas consistently demonstrate that poor populations in economically poor areas have worse health than rich populations in economically successful areas, and this is evident in terms of both international differences between countries and local variation within countries. Evidence reviewed here from Britain illustrates these health inequalities in a particularly detailed way, and suggests that the health gap between rich and poor populations is widening. The British example shows that even in a relatively wealthy country, health inequality can be a significant factor. Growth in *inequality* may show adverse trends, even while measures of *average* health in the population as a whole are improving.

IS ECONOMIC GROWTH GOOD FOR HEALTH?

Comparisons of rich and poor countries of the world show that, in general terms, greater levels of socio-economic development are associated with better health. However, this effect is clearer for poorer countries than for wealthy nations. D. Smith has pointed out that, for a given increase of resources, the associated gains in human welfare are typically greater in poor populations than in populations that are already relatively rich (e.g. D. Smith, 1977; 1994: 139–40). This is summarized by the diagram in Figure 4.7. This relationship is significant for our interpretation and understanding of the benefits of economic growth in terms of human welfare and of principles of social justice. It has led to a debate over whether the processes of growth in the world economy that we are seeing today are universally beneficial for population health.

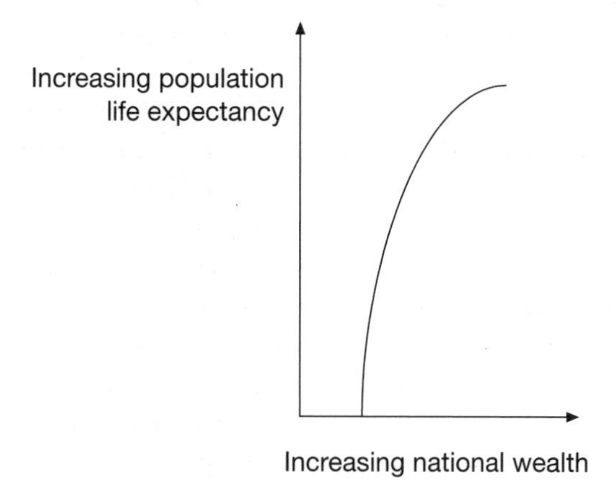

FIGURE 4.7 Typical form of the relationship between national wealth and population life expectancy (after Smith, 1994; Wilkinson, 1996)

Wilkinson (1996) also considered the curvilinear relationship which typifies the correlation between national wealth and health, shown in Figure 4.7. Analysing these relationships over time, Wilkinson took account of variations in the purchasing power of income in different countries and the fact that a given absolute income change has more impact on a low income than on a higher income. He suggested that rising national wealth *per se* does not produce health gain in wealthy countries and that in these countries a 'threshold' of income has been reached, allowing 'the attainment among the majority of the population of a minimum real material standard of living, above which further increases in personal subsistence no longer provide the key to further increases in health' (1996: 45).

There are a number of possible reasons why growth in economic wealth beyond a certain level is not invariably 'good for health'. Chapter 1 referred, for example, to the question of sustainability of development in terms of public health. If increasing economic wealth is achieved at the expense of damage to the physical or social environment, then the effects on health may be negative. Stephens et al. (2000) and Fidler (2001) have questioned, for example, whether free trade, which has been important for global economic growth, is always beneficial for health and sustainability of world resource systems. Cornia (2001), Dollar (2001) and Fidler (2001) draw attention to processes which often accompany rapid, poorly regulated economic globalization, and can be detrimental to health, such as uneven and exploitative foreign investment, unequal access to world markets, and 'undesirable' aspects of international trade such as narcotic drug trafficking. We also saw in Chapter 2 that wellbeing is not straightforwardly related to growth in individual income, though generally those on higher incomes report greater levels of satisfaction with material aspects of life.

Wilkinson (1996) suggested that one possible interpretation is that there may be changes in the nature of consumption which affect health and are associated with rising income, but which are not very well reflected in economic indicators of national wealth and price indices. Aspects of more 'affluent' consumption patterns that can damage health include over-consumption of some products (e.g. tobacco, alcohol and high fat, highly processed foods). Another example is that higher levels of car ownership may entail a sedentary lifestyle and greater exposure to risks associated with motor vehicle accidents and traffic pollution. For populations as a whole, these factors may partly offset health gains associated with greater wealth, although, in fact, in the richest countries these risks are generally worst for poorer groups in the population than for the wealthiest groups.

IS INCOME *DISTRIBUTION* THE FUNDAMENTAL CAUSE OF HEALTH INEQUALITY?

As noted above, several commentators have drawn attention to the fact that growth in average income associated with participation in economic globalization is not always associated with growth in equality of income. Increasing income inequality is argued to be detrimental to the health of the population as a whole. Wilkinson's (1996) analysis focused particular attention on this issue of the *distribution* of wealth and health *within* high income countries. Wilkinson (1996) refers to a large body of evidence of strong intra-national associations between health and wealth of different population groups (some of which is also reviewed above). He went on to argue that wealthy countries with a very uneven distribution of wealth have poorer overall health than those with more equal wealth distribution. Wilkinson suggested that aspects of society such as social cohesion and solidarity are associated with the degree of economic equality, so that social cohesion tends to be weakest where distributions of wealth are most unequal. He put forward theories concerning psychosocial pathways, which might be particularly important in causing these associations between material inequalities in society, lack of social cohesion, and poor health in the population.

Wilkinson's work has generated a good deal of debate and several other studies have been carried out in different countries to explore the question further. Judge (1995; Judge et al., 1998; 1999) questioned whether so much emphasis on one aspect of inequality (income distribution) could fully explain a phenomenon as complex as health differences. In a study which examined data on income distribution in 16 countries, Judge et al. (1998: 578) report 'only very moderate' support for the argument that income inequality is strongly associated with health status at the national scale. They comment on the difficulties of making international comparisons.

A number of studies have also examined health in relation to income distribution for geographical areas *within* countries. The evidence is variable with respect to the validity of Wilkinson's arguments. Lochner et al. (2001) reported that mortality was higher for individuals living in US states with high levels of income inequality, after controlling for individual poverty. However, Fiscella and Franks (1997) found that the association between income inequality and subsequent mortality for areas in the United States was not significant after controlling for level of income of the areas. The same authors also used survey data on health of individuals and area data on income inequality to show that individual income had the strongest effect on health (measured in terms of self-rated health, depression, morbidity assessed by a physician, and subsequent mortality). After controlling for individual income, area income inequality showed some independent association with self-rated health and depression, but not with medically assessed morbidity or mortality (Fiscella and Franks, 2000). The authors comment on various methodological constraints which might have affected these results, but their findings may also support the hypothesis that income inequality in the community is especially linked to psychosocial aspects of health.

Blakely et al. (2002) report a multi-level logistic regression analysis of self-reported health in a national survey in the USA, examined in relation to individual income and to average income and income inequality for each respondent's area of residence. Area data were considered at the level of metropolitan areas and the larger county scale. Their results suggest that self-rated general health was slightly worse in metropolitan areas with greater inequality of income, but this was partly explained by the lower *average* income in these metropolitan areas. The relationships were less clear when county level income inequality was considered. Ross et al. (2000) analysed intra-urban inequalities in income and health in United States and Canadian cities and showed that greater income inequality is associated with higher mortality levels in US cities, where inequalities of wealth are most extreme. In Canada, inequalities in wealth are less marked and there is no significant association between these inequalities and population health. This may suggest that at this finer scale of urban areas, Wilkinson's hypothesis is not universally applicable to all countries and is less pertinent to Canadian cities than to US cities.

This controversy over the importance of income inequality for illness and death highlights the need for relatively sophisticated approaches to understanding area variations in health, poverty and wealth, as discussed below.

IS POVERTY IN *PLACES* IMPORTANT FOR HEALTH?

Research has demonstrated that health inequalities are associated both with the socio-economic characteristics of *individuals*, and with levels of poverty and wealth in their residential *area*. This finding has considerable significance for health geography, since it demonstrates the relevance of *place* for health inequality.

Several empirical studies have tested whether a person's health shows statistically independent associations with the type of place and with their individual socio-economic status. (These are reviewed, for example, by Curtis and Jones, 1998.) This type of analysis makes a rather artificial distinction between geographical variation in health which is *compositional* and that which is *contextual*. Compositional variation results from the attributes of the individuals that make up the population of an area. This means that area differences in health arise because local populations vary in their composition (e.g. areas differ in the relative numbers of residents who are richer and typically healthier, or are poorer and more likely to be unwell). Contextual variation, on the other hand, arises if people with essentially similar individual characteristics show varying health in different geographical settings, in which case there is something about a place itself which is significant for health difference and which affects everyone living there, interacting with their individual attributes.

Although this distinction is made between compositional and contextual effects, they should probably be considered as interrelated, rather than completely independent. For example, the ethnic *composition* of a local population results from the ethnic identities of individual residents. However, geographical concentrations of a particular ethnic group can influence the sense of ethnic identity for a whole community. This is often manifested in aspects of the local environment such as shops, places of worship, community centres, as well as being perceived as a social attribute of a whole locality. Thus the ethnic profile of the area becomes more than the sum of the individuals, since attributes of the area contribute to the ethnic identity of the place and this can itself contribute to the individual sense of belonging to an ethnic group (see also Chapter 3). Similar arguments could be put forward concerning the relationships between the individual social class composition of a population and the sense of class consciousness and solidarity which may lead to the idea of a 'working class community' or of a 'middle class neighbourhood'.

Various statistical methods have been used to explore the relative importance of compositional and contextual effects on health difference. Some of the earlier studies used tabulation methods. For example, Fox and Goldblatt (1982) and Blaxter (1990) showed by these techniques that for people with similar individual social position (measured by social class, housing tenure or employment status), there were differences in health associated with the characteristics of their residential area (electoral ward). Mustard et al. (1999) examined survey data on reported health in Canada in relation to indicators of individual income and 'ecological' measures of income in the area of residence. They tested the hypothesis that the association with ecological measures would be more attenuated than for individual measures because individual income would have the most direct effect on individual health. However, the results showed that for most measures, health variation was as strongly associated with ecological income indicators as with individual measures.

Subsequent work has used multi-level modelling techniques which distinguish statistically between variation at different hierarchical 'levels' of aggregation. These have usually been based on data about *individual people*, combined with contextual information on the *places* where they live. The analyses generally seek to attribute variation to these different 'levels', distinguishing between health inequalities associated with individual characteristics (such as sex, age and socio-economic status) and with differences between areas of residence (often measured by socio-economic indicators for the area population). We need to bear in mind, as noted above, that it can be difficult conceptually and empirically to separate out 'individual' and 'contextual' effects. However, in terms of the statistical models used, these studies generally show that health inequality is strongly associated with individual attributes, but that these do not 'explain' all of the health variation observed. A part of health difference is also associated with area of residence, either at the scale of the immediate locality or at the scale of the wider geographical region. Typically, in areas with high levels of deprivation, the health of the population is worse than one might expect from their individual attributes alone.

Two studies of this type examined health status recorded in the Health and Lifestyle Survey (HALS), a representative sample survey of individuals in Britain. Humphreys and Carr-Hill (1991) and Duncan and Jones (1995) showed that health variation was most strongly associated with individual attributes, but there was some variation associated with area of residence. Similar conclusions have been reached in studies of self-reported long term illness, based on the Sample of Anonymized Records from the 1991 population census in Britain (Gould and Jones, 1996; Shouls et al., 1996a). Support for the existence of 'contextual effects' also comes from research on life events recorded in the Longitudinal Study, a sample drawn from the 1971 British census (Sloggett and Joshi, 1998; Joshi et al., 2000), and from the Scottish Heart Health Study (Hart et al., 1997). Haynes et al. (1997) also have shown an association between self-reported chronic illness and levels of unemployment in local labour market areas. Curtis et al. (2003) reported an analysis of a longitudinal study of a sample of the population of England and Wales. They showed that unemployment in the area of residence during childhood was associated with mortality and morbidity later in life, after controlling for individual characteristics and conditions in the more recent place of residence. This suggests that place effects may have long term health effects through the life course.

Health related behaviour also shows links to area characteristics as well as individual attributes. Duncan et al. (1996; 1999) have demonstrated, using multi-level modelling, that smoking behaviour is related to individual attributes such as sex and class, but also to the proportion of people who smoke in the local area. This suggests there may be 'local cultures of smoking'. Ross (2000) reported analysis using an ordinary least squares regression model to test whether smoking was associated with area deprivation as well as with individual income, education, race and marital status. In poor areas, men were found to smoke more.

These studies also reveal further, more complex associations between area socio-economic conditions and health. For example, Shouls et al. (1996a) showed from population census data that, although in wealthy areas people's health tended to be better generally, these areas showed a particularly strong *gradient* between rich and poor people. A similar finding was reported by Sloggett and Joshi (1998) in results from their analysis of the Longitudinal Study. In contrast, however, Stafford et al. (2001), analysing data from the Whitehall Study for a population living in London, found that gradients of ill health with poverty for individual civil servants in the study were steepest in poor areas. Graham et al. (2000) report on evidence from a Scottish study showing that although population health tended to be worse in more deprived areas, individuals' recovery from heart attack was better than in wealthier areas. Although the evidence is variable, several of these studies therefore suggest that disadvantage and the effects of illness have greatest impact on local health inequalities within communities which are generally more affluent and more healthy.

Area factors in health inequality may also operate at different geographical scales. Langford and Bentham (1996) and Congdon et al. (1997) showed, for example, that mortality and self-reported illness indicate a regional north–south divide, with worse health in the north of England and Wales, even after controlling for variability in socio-economic conditions at the small area level. Thus it is possible that broad regional factors such as climatic or other physical environmental conditions, or regional socio-economic factors, may influence health difference, as well as local differences in poverty and wealth.

These empirical studies demonstrating that the geographical context influences socio-economic health inequality are also supported by theoretical arguments. In general terms, the argument for contextual effects has been put forward in theoretical discussion by Giddens (1984) and others, considered in Chapter 3. This supports the argument that individual agency interacts with the social and economic structure of the *locale*. Research on socio-economic inequality in health has also developed more specific understandings of the causal pathways, which explain *how* and *why* poverty in an area is associated with health. These include differences between rich and poor populations in consumption of goods and services, physical environmental risks, social capital and effects of poor housing and working conditions.

Some of these pathways are discussed elsewhere in this book. For example, poor neighbourhoods typically have worse *access to goods and services* and this affects patterns of private and collective consumption, as discussed in Chapter 5. Poor areas may also have poor *physical environments* so that populations are exposed to more environmental hazards due to pollution. Geographical perspectives on these types of environmental exposures are discussed in Chapter 6.

Various authors have also examined the argument that aspects of the social environment affect the association between poverty and health. Poorly developed *social capital* in communities (often, though not invariably, associated with area poverty) may be associated with worse health experience, although the evidence is variable on this point. Social capital (the collective social resources of a community) is a complex concept, and is measured differently in different studies. For example, Veenstra (1999) examined data for individuals and found little evidence for any overall effect of social capital on health in Saskatchewan, Canada. Health was found to be more strongly related to aspects of human capital such as education and income. However, associations between social capital and health have been found in studies of geographical areas. Kawachi et al. (1997) made an analysis of survey data from 39 American states to measure levels of social capital in terms of civic engagement, norms of reciprocity and levels of trust. Each indicator of social capital showed a strong inverse correlation with age-adjusted all-cause mortality rates, suggesting a positive link between higher social capital and better health status. Siahpush and Singh (1999) reported an ecological study in Australia which measured features of social organization, such as networks, norms and trust, that facilitate coordination and cooperation for mutual benefit. Areas with higher levels of social integration were found to have lower mortality rates and higher life expectancy. Ellaway et al. (2001) reported, from a study of individuals in Glasgow, that perceptions of neighbourhood problems and neighbourhood cohesion were significantly associated with a measure of mental health (GHQ12), after controlling for socio-demographic variables. Worse mental health was associated with weaker neighbourhood cohesion, perception of neighbourhood problems, and perception of being 'worse off' than others living in the neighbourhood. Gatrell et al. (2000) have argued that deprived material and social conditions in poor areas go against residents' senses of what is 'proper' or 'socially acceptable' and that this may be damaging for psychosocial health. Graham et al. (2000) suggest that individual self-esteem may be influenced by comparisons made with others living in the same community: if these comparisons are unfavourable, this may damage self-esteem and impede recovery from illness.

The impact on health of conditions associated with aspects of *material poverty*, such as working and living conditions, is also likely to be important in understanding the ways that places affect health. A large body of evidence helps to explain the findings by Haynes et al. (1997) that conditions in local labour markets may influence health. Box 4.1 summarizes some of the evidence from research on the causal pathways linking *employment* and health. There is also a good deal of research that shows that aspects of *housing* may be important for health. Key points from this evidence are outlined in Box 4.2.

All of the explanations for health inequality reviewed here highlight the need to address health inequalities by considering not only individual poverty and wealth, but also the socio-economic structures within which people live and work. These structures cannot be very effectively influenced by individual action alone. For example, whether or not workplaces or housing conditions are conducive to good health is not just a matter for the individual to resolve, but is also a responsibility of local and national governments and other regulating bodies. Even these agencies may find it difficult to change the impact of fundamental structural processes in the political economy, which operate globally as well as locally to produce inequalities in economic conditions.

CONCLUSIONS: LANDSCAPES OF POVERTY AND HEALTH INEQUALITY

A large body of evidence has been amassed concerning links between poverty and health, including many studies with a geographical perspective. The examples used in this chapter demonstrate a range of techniques in the geography of health. Some geographical research on health variation has used statistical analysis of indicators of socio-economic conditions and population health (particularly methods such as multi-level modelling which offer scope for demonstrating contextual effects). Qualitative studies of perception also show how the material aspects of places (living and working conditions) impact on health.

This chapter has highlighted some particular aspects of geographical research on landscapes of poverty and their association with health variation. First, there is evidence for a complex association between geographies of poverty and of poor health. Population health varies geographically partly because poor and relatively unhealthy people are spatially concentrated in certain residential areas, separate from the areas occupied by affluent and healthier populations. In addition, health inequality is associated with features of places, including local landscapes of poverty and wealth. Individual poverty interacts with poverty in the wider community and, in combination, these produce effects that are usually detrimental to health. This is a finding of particular interest to health geographers because it underlines the role of places and the interaction between people and places for health difference. We need to consider 'place effects' in terms of processes which are geographically differentiated in space and time. They affect people throughout their whole life course, not just at one stage in life.

Some of the processes that explain these associations are likely to operate through the impact on health of aspects of poverty such as poor living and working conditions. These effects need to be understood not only as local problems, but as related to structural dimensions of the political economy in a country as a whole, and in the global political economy. The geographical polarizations of rich and poor and healthy and unhealthy populations, as well social group differences, are evident in some of the richest and most highly developed countries as well as in poorer countries. Furthermore, health inequalities at the fine geographical scale are clearly seen in the most advanced and economically successful urban areas within rich countries.

Wealth distribution is therefore important for health. However, it would be too simplistic to argue that uneven income distribution is the sole cause of health differences. The previous chapter pointed to other important dimensions of social inequality that also have significance for health, such as gender and ethnicity. As Poland et al. (1998) have emphasized, factors other than poverty and wealth (notably differences in access to health care) are also important for health differences, and these are the subject of the following chapter.

Box 4.1 The association between labour markets, workplace conditions and health

Labour market conditions and workplace factors can be considered as determinants of health as shown in Figure 4.8. See also, Department of Health, 1998; Bartley, 1994; Bartley et al., 1999; Karasek and Theorell, 1990; Burchell, 1994; Barnett, 1995.

Workers in depressed labour markets or in workplaces with poor conditions are likely to experience a number of 'negative' work attributes, including: low pay; lack of job security and greater risk of unemployment; low levels of job control and involvement; lack of support from colleagues; low reward to effort ratio; lack of prestige; and poor physical working conditions. Experience of unemployment and of work with negative attributes affects workers and the families and communities who depend upon their work.

Low pay produces poverty, associated with low standards of living. These include *material effects* of poor diet and poor housing and greater risk of physical effects like hazards at work (e.g. Bartley, 1994; Bartley et al., 1999).

Work with negative attributes also has *psychosocial effects* (Warr, 1987; Joelson and Wahlquist, 1987; Bolton and Oatley, 1987; Jahoda, 1992; Burchell, 1994; Barnett, 1995). Unemployment causes loss of 'traction' (the routines and habits associated with work, which help to motivate people and guide their daily activities). Job insecurity contributes to uncertainty and lack of choices and control in life, as well as disruption of life plans due to unplanned changes in employment status. Unemployment and work with negative attributes can contribute to social stigma and loss of self-esteem. Workers can also experience stress where there is an effort/reward imbalance or lack of control in work (Siegrist, 1996; Bosma et al., 1998).

The influence of work-related health determinants will vary for different people, depending on *mediating factors* such as: age; gender; ethnicity; the salience (or perceived importance) of their work for them; their personal capacity to cope with changes in their work; the amount of social support they enjoy.

The research cited also suggests that *changes in these work related conditions* can cause changes in living standards, physical hazards at work and psychosocial processes, and this results in *changes in a number of aspects of health.* Most of these studies suggest that health status deteriorates when employment status changes for the worse. Changes have been reported in *general aspects of health* (e.g. risk of death from all causes, general self-reported health) and risk of *illness and death from specific physical causes*, such as cardiovascular disease, cancers, suicide, accidents, respiratory disease (Mattiasson et al., 1990; Martikainen, 1990; Moser et al, 1987; Bosma et al, 1998). *Physiological changes* to blood cholesterol and the immune system have been detected (Mattiasson et al., 1990; Cobb, 1974). *Mental and psychosocial problems* have also been recorded, such as depression and anxiety and domestic violence (Bolton and Oatley, 1987; Hammer, 1993; Frese and Mohr, 1987; Ratner, 1983). Changes in employment status have been shown to be associated with *changes in use of health services*, sickness absence from work, pensionable disability (Vahtera et al., 2000; Iversen and Sabroe, 1988; Westin, 1990; Beale and Nethercott, 1987).

Health impacts of redundancy are most clearly demonstrated in longitudinal studies of individual workers, followed over time as they go through the redundancy

continued ⇒

Box 4.1 Continued

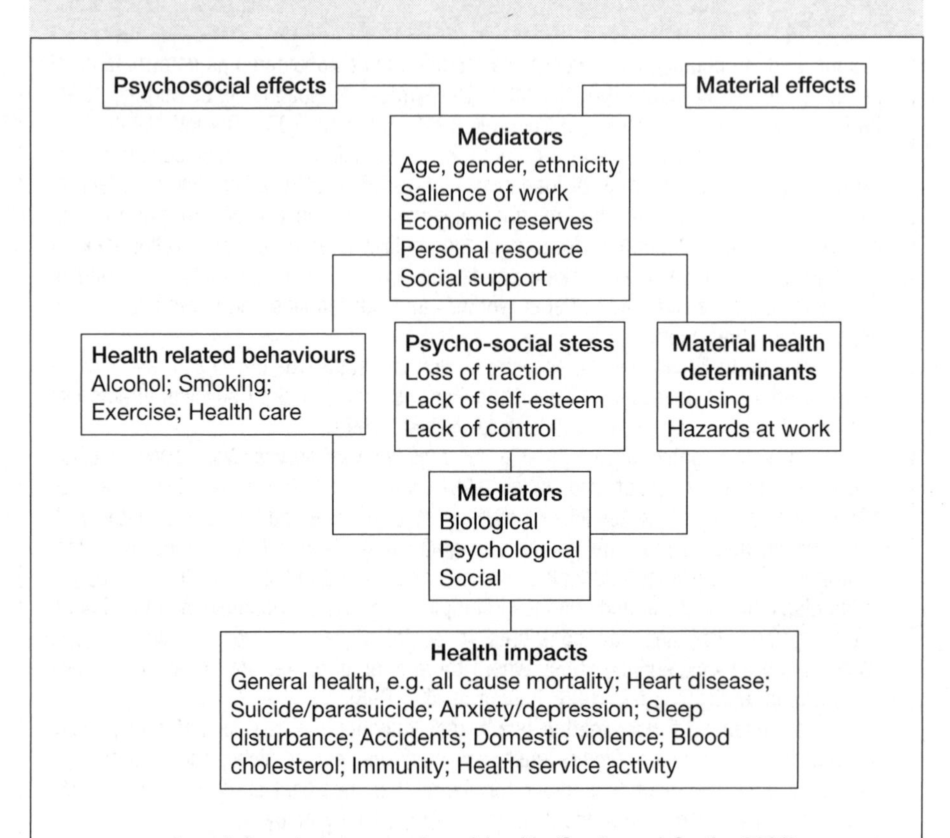

FIGURE 4.8 Associations between work and health (Coutts and Curtis, 2002)

process. These studies report changes to *physiological status*, such as blood cholesterol, that increase risk of illness such as cardiovascular disease (Cobb, 1974; Mattiasson et al., 1990). Redundant workers have increased rates of *pensionable disability* (Westin, 1990) and made increased *use of GP surgeries* (Beale and Nethercott, 1987). Redundancy is associated with *changes in rates of hospitalization* for some conditions, e.g. increased for cardiovascular disease but reduced for musculoskeletal problems, accidents (Iversen and Sabroe, 1988; Beale and Nethercott, 1987). Workers undergoing redundancy have increased risks of *stress and associated psychosocial difficulties* such as boredom, despondency, sleep disturbance or depression/anxiety (Joelson and Wahlquist, 1987; Mattiasson et al., 1990).

Many of these aspects of employment are influenced by factors that are not solely related to the individual characteristics of the workers. They are importantly affected by broad conditions in the economy, conditions in local labour markets and the characteristics of employer organizations and specific workplaces. Thus local economic conditions are significant as health determinants.

Box 4.2 Health and housing disadvantage

Several reviews have pointed to evidence of *links between poor housing and poor health* in Britain (e.g. British Medical Association, 1987; Lowry, 1991; Victor, 1997; Best, 1995; 1999; Burridge and Ormandy, 1993; Bardsley et al., 1998). Reviews have also been made in other countries (e.g. Dunn and Hayes 2000). Several different theories have been proposed to account for the *causal pathways* that may account for these associations.

There may be *physical effects* of poor housing, for example, because of damp, cold, mould spores, dust mites, household crowding. For example, several studies suggest that such conditions increase the risk of self-reported respiratory problems, although research using biometric assessment of respiratory health (e.g. using 'peak flow meters' to measure lung function) does not show the link very clearly (Hyndman, 1990; 1998).

Bad housing may also contribute to *mental illness and distress*. In Britain a large proportion of low cost housing in urban areas is in flats, often in high rise buildings. Lowry (1991: 70–1) discusses evidence that mental health of mothers and children living in such buildings may be worse, perhaps because difficulties in getting out of the home limit recreation space and produce isolation. Evidence for women in Britain suggests that high density living may be damaging to mental health (Gabe and Williams, 1993). Furthermore, Fuller et al. (1996) showed that poor mental health was associated with household crowding in Bangkok, Thailand, where there are more extreme levels of crowding than those typically found in western countries.

Dunn and Hayes (2000) have suggested an analytical model to account for the causal processes linking housing and health in a survey of two neighbourhoods in Vancouver, Canada (Figure 4.9). They tested the hypothesis that housing conditions would impact on health through their association with dwelling satisfaction and neighbourhood satisfaction. Dissatisfaction with these factors might link to health, especially psychosocial health, through processes including lack of control over key aspects of everyday life, lack of a sense of neighbourly social support, and impact on poor housing through its meaning for one's sense of identity and self-esteem. The results of their analysis generally lent support to this model, showing in particular that dissatisfaction with aspects of housing was associated with reported ill-health.

Macintyre et al. (1998) have tested the idea that links between housing tenure and health are explained by their impact on the individual's sense of self-esteem, and because housing tenure relates to income. Compared with owner occupiers, tenants were disadvantaged on several measures of health, even after controlling for income and a measure of self-esteem. This suggests that there are probably other reasons for the link between housing disadvantage and health disadvantage. These might include material effects (e.g. crowding and poor housing amenities).

The link between housing and health does not simply concern inequalities between owner occupiers and tenants, however. Smith et al. (2003) have studied the experience of home ownership for people with chronic illness. They demonstrated that such individuals are disadvantaged within the owner occupied housing sector and that this had psychological and physical impacts on their health.

continued ⇒

Box 4.2 Continued

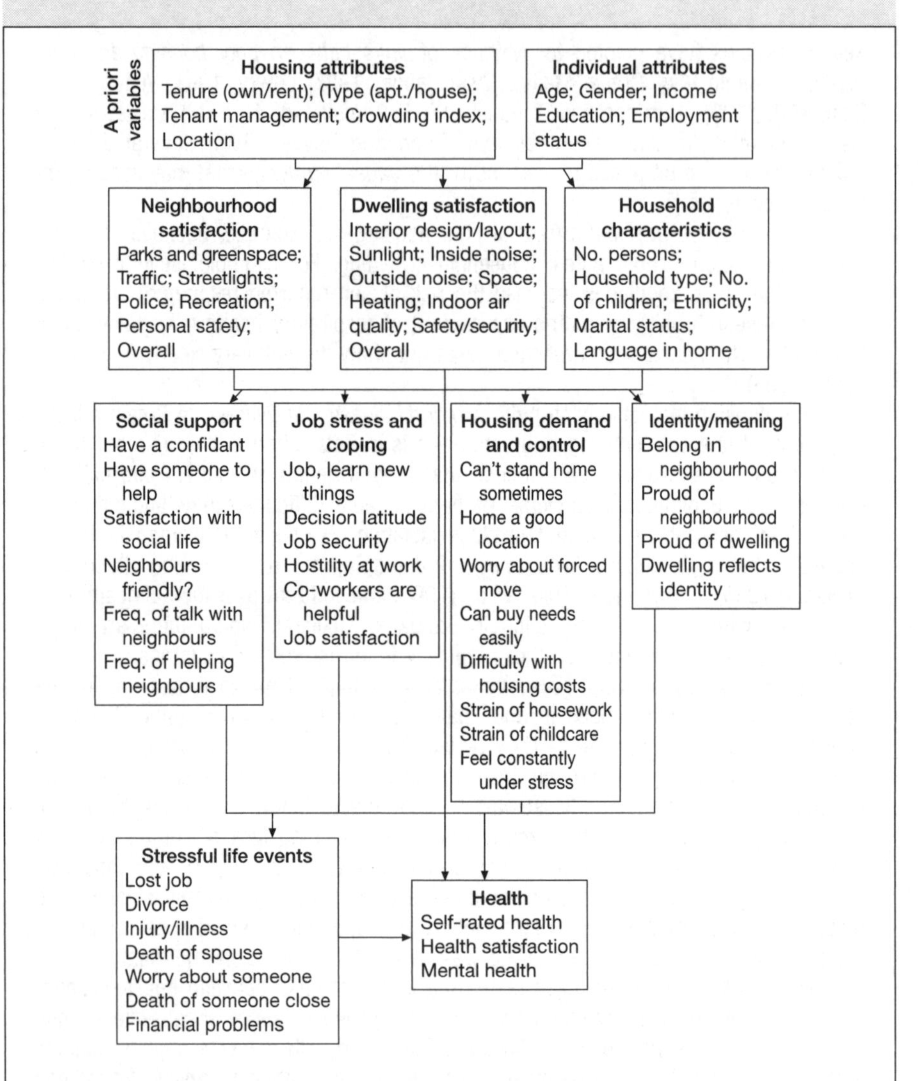

FIGURE 4.9 Conceptual framework representing relationships between housing and health (reproduced from Dunn and Haynes, 2000: Figure 1, with permission from Elsevier)

Exposure to these risks of poor housing is determined not only by *individual or household* factors, but also by wider, *contextual* variables, which vary geographically. These include conditions in local housing markets and national and local policies and action to provide social housing. In addition to the effects of *poor housing* summarized here, *homelessness* shows particularly strong relationships with poor physical and mental health disadvantage (see Chapters 3 and 7).

FURTHER READING

For readers who are new to this field of geography, a good introductory text is: Shaw, Dorling and Mitchell (2002).

The same group of authors have also produced an accessible and comprehensive discussion of trends in health inequality in Britain: Shaw, Dorling, Gordon and Davey-Smith (1999).

Many of the theoretical ideas were very well set out by Pinch (1985). A comprehensive review of concepts of social justice from a geographical perspective provided by Smith (1994).

Many of the issues summarized here in relation to globalization and health are discussed at greater length by, for example: Lee, Buse and Fustukian (eds) (2002).

Readers may also wish to refer to wider discussion of the socio-economic determinants of health, and a very useful edited collection is: Marmot and Wilkinson (1999).

In the USA, in particular, there has been a good deal of emphasis on the concepts of social capital and health that are briefly referred to here, and readers may want to refer to Putnam's seminal text on this subject: Putnam (2000).

OBJECTIVES AND QUESTIONS

Key ideas and information in this chapter, which would be useful learning objectives for students using this book as a text, include:

- the evidence for geographical inequalities in health and wealth at different spatial scales;
- the idea that poverty in places, as well as poverty of individual people, may contribute to health disadvantage and health inequality;
- the debate over the links between wealth and health and especially the controversy over Wilkinson's hypothesis that inequality of income is the crucial factor;
- arguments about how the political economy (especially the capitalist mode of production) relates to health inequality and the implications of global circuits of capital for health;
- the importance for geography of health of social determinants of health such as social capital, housing, work conditions;
- the value of methods such as multi-level modelling for teasing out the relationships between individual people, the places where they live and the health status of different populations.

This knowledge should help students to answer questions such as the following:

1 Critically review the geographical evidence, from countries such as Britain, relating to the assertion that: 'Poverty in places contributes to health disadvantage experienced by some populations.'

2 With respect to population health, should we be more concerned about wealth of nations or about inequality in wealth within nations?
3 Is global capitalism bad for health? Explain your interpretation of the debate on this question.

5 Landscapes of Consumption: care and commodification

This chapter explores the relevance of 'landscapes of consumption' for geographical health variations. It reviews examples of the literature on geographically varying patterns of use of health services and other resources important for health.

Variation in use of goods and services depends on the characteristics of individual consumers, and of their wider community. Varying provision of health related goods and services is also important. Individual inequalities of wealth help to explain differences in consumption, but other 'structural' factors are also relevant. Attention here is concentrated especially on three types of structural factors that influence landscapes of consumption in the geography of health. These are *administrative and political* structures, social structures operating in different *social milieux*, and *spatial organization of infrastructure* for delivery of goods and services.

These structures are variable in different geographical settings and they also change over time. The following discussion considers change due to three types of process: *health system reforms*, *commodification* of health and health care, and *technological development*.

In the last part of this chapter, these structures and processes are illustrated using two examples of varying landscapes of consumption. These concern use of primary medical care and access to healthy food, which are both forms of consumption that are particularly important for inequalities of health. A model of landscapes of consumption and health is proposed at the end of the chapter.

CONCEPTUAL MODELS OF HEALTH SERVICE ACCESSIBILITY AND USE

A useful starting point for consideration of varying landscapes of consumption is provided by geographical perspectives on health care use. These have been partly influenced by conceptual models of health care use proposed by authors from other disciplines, such as Aday and Andersen (1974) and Gross (1972). Curtis (1980; 1982) and Joseph and Phillips (1984) described the factors affecting geographies of health care access and use. These models combined elements of users' perceptions with information on patients' daily patterns of movement in time and space and the organization of local services.

Jones and Moon (1987) and Curtis and Taket (1996) have reviewed studies of geographical inequalities in health care and the factors affecting these, at the level of society as a whole and at the local level. Other authors have also drawn attention to the broader scale factors that affect access to and use of health care in different societies. For example, Fielder (1981) and Puentes-Markides (1992) explicitly include aspects of macro level health policy and aspects of the wider social structure and political economy in their models of health service access and use. More recently Gatrell (2002) has also considered variation in health service provision in various countries. These accounts show that a comprehensive model of health care access and use needs to incorporate a complex range of elements at different geographical scales, and that social and economic aspects of access, as well as spatial distance to be travelled to facilities, are important for heath care use.

Figure 5.1 summarizes one way of thinking about the factors influencing landscapes of health care consumption. These factors include the characteristics of *individual patients* but also 'structural' factors such as: the *social milieux,* or cultures, in which individual patients and individual health care staff are located; the *political, administrative and professional structures* influencing the funding of health care and its organization; and the *physical infrastructure* of facilities for the delivery of health care, influencing spatial accessibility. These are partly macro level structural factors, pertaining to whole societies, but they also show significant micro

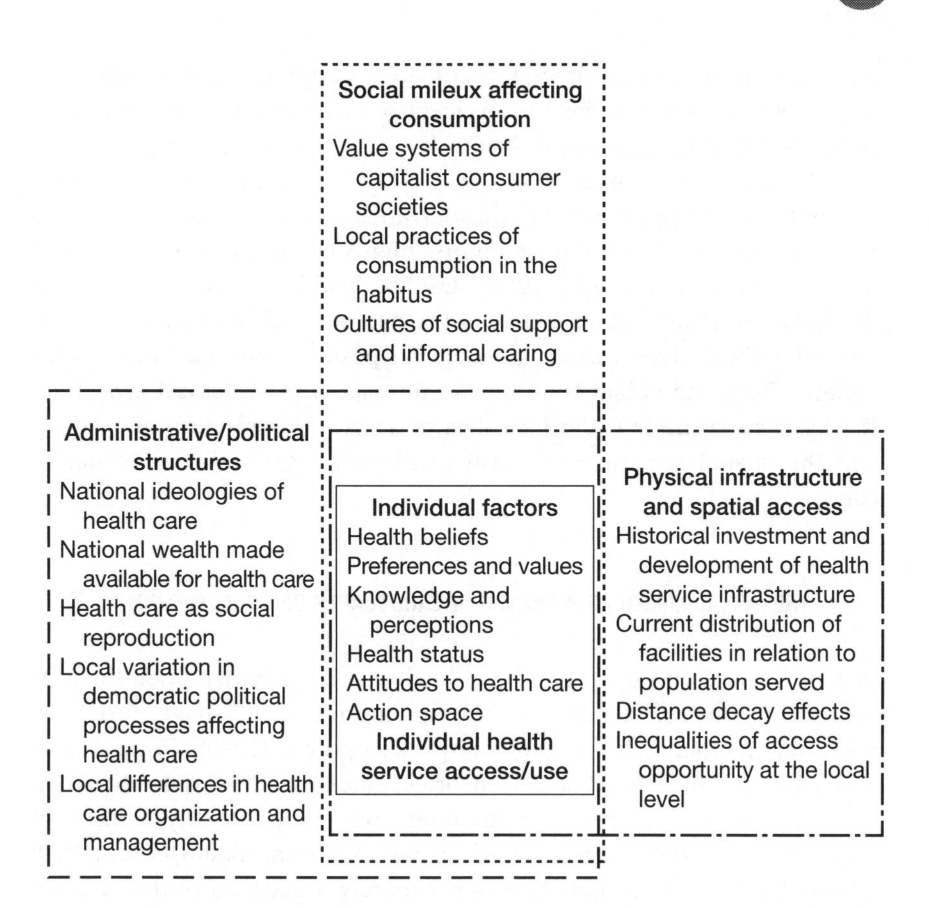

FIGURE 5.1 A schematic model of health care access and use

level, local geographical variation. Geographical differences in provision and use of health related goods and services therefore reflect differences among places, at various scales, as well as among individual people.

INDIVIDUAL FACTORS IN HEALTH RELATED CONSUMPTION

Some models proposed by authors from outside geography emphasize individual factors affecting use of health care (e.g. Rosenstock, 1966; Veeder, 1975; Mechanic, 1978). These focus attention on factors such as the individual's health status, health beliefs, perception of symptoms and attitudes to health care, as well as ability to afford the cost of health care and physical access to health care facilities. Individual aspects of social position such as age, gender, class and ethnicity are likely to be associated with these attributes, as already discussed in Chapters 3 and 4. Variations in access to, and use of, health services are often argued to contribute to socio-economic inequalities in health (e.g. Gordon et al., 1999).

These models of individual behaviour have parallels in behavioural research in geography, which consider attributes of consumers such as their purchasing power, their mobility, their knowledge and preferences (e.g. Walmsley and Lewis, 1984). Use of health related services also depends on the interaction between the patients and health care personnel, who themselves vary in their social position, attitudes and health beliefs. These individual level factors interact with structural aspects of the local social milieu, the local health service organization and policy and the spatial structures of local facilities for the delivery of health care.

THE SOCIAL MILIEU AS A FACTOR IN LANDSCAPES OF CONSUMPTION

Recent research in the geography of health care has drawn upon theories concerning the social relations of consumption which have also been taken up in other fields of geography, e.g. by Pred (1996), Crang (1996) and May (1996). These social theories emphasize the links between consumption and senses of individual and group identity, senses of difference and the experience of social exclusion. Baudrillard (1970: 103–4) argued that consumption is not merely a question of the satisfaction of individually determined needs and wants. It has to be understood in relation to a socially structured system of 'needs' which result from the system of production in capitalist societies and which also arise from the value systems of that society. Consumption is defined in terms of a system of communication – a coded system of signs which convey our sense of ourselves and of differences between social groups in society (1970: 135–6).

Theories of consumption proposed by Bourdieu (2000) have been especially influential for geographers because they offer an explanation for variability in consumption patterns between places. Bourdieu developed the idea of consumption framed by a local socio-geographical context, which he describes using ideas of the habitus and of fields. The *habitus* is a socially defined geographical space, and denotes the position of social groups in relation to scarce assets in that space. Bourdieu (2000: 101,172) also interpreted consumption practices, shaping the lifestyles of individuals within social groups, in terms of *fields*, governed by 'logic' reflecting the values (practices and tastes) of the group. Although individual consumption practices may seem to be very diverse, Bourdieu (2000: 172) argued that the logic of these fields 'harmonizes' consumption practices within the social group to which the individual belongs and differentiates these shared patterns of consumption from those of other groups. These notions of habitus and fields of logic provide a socio-geographical framework for interpretation of consumption, shown in Figure 5.2. This illustrates the interrelationships between the broader social structure, the habitus, fields of practice and taste, all producing patterns of consumption and lifestyle which reflect the individual's class membership and social position.

These perspectives on consumption are relevant to the consumption of goods and services that affect health. The pattern of health services consumption prevailing in a local area is therefore likely to depend on the social composition of the local community, the strength of social networks and social capital, and local cultures of health related behaviour and health beliefs. These factors are also important for the provision of informal health care and voluntary caring. Milligan (2000a; 2000b; 2001), for example, has reviewed the importance of the voluntary sector for care of frail elderly people and people with mental illness in Scotland. The geographical pattern of provision in this sector depends importantly on the social setting, which partly determines the availability of voluntary helpers and also the willingness of the local community to make charitable donations.

Conflicts of interest may emerge as different social groups seek to impose their preferred pattern of consumption in an area. Lifestyles of wealthy and powerful groups require consumption of certain goods and services, made available in ways which exclude or disadvantage poorer

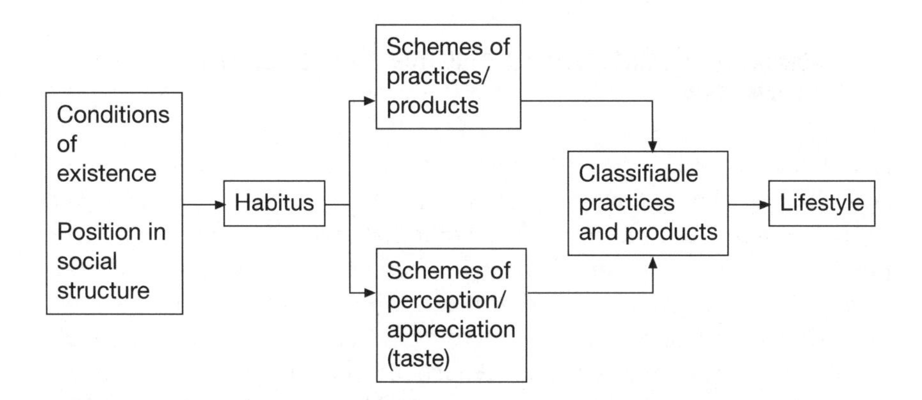

FIGURE 5.2 Structures of existence, habitus and lifestyle (after Bourdieu, 2000: 171)

and less influential members of society. Exclusive lifestyles may emphasize high cost and very secure residential areas, designer clothing, fashionable food retailers and eating places, exclusive sports and leisure facilities, or health spas or sources of alternative therapies. The exclusivity of these goods and services is maintained through their high cost, through social barriers (selective membership of clubs, for example), and through control of the siting of spaces for consuming these products and of access to these spaces. Examples include the creation of gated residential developments with private amenities, or use of security systems to control access to shopping malls.

As space in the city is taken up by facilities offering these exclusive consumption opportunities, less space is available for other patterns of consumption which are more important for less powerful groups. Thus thrift stores and charity shops may be replaced by retail outlets providing high cost goods; and poor residential areas in the centre of cities may become 'gentrified' as a result of a new enthusiasm for urban living among the elite, forcing out poor residents. Public open space, street markets and free access play spaces are replaced by office blocks with leisure space and other facilities reserved for employees only. Thus more powerful and privileged groups influence the high cost, and the constraints to access, which society places on certain styles of consumption. By doing so they limit the ability of less powerful social groups to access or benefit from these aspects of lifestyle, and this may be important for health. These processes concerning consumption need to be considered in the light of related processes of social control, explored in Chapter 3, and processes of production, uneven development and regulation, considered in Chapter 4.

ADMINISTRATIVE AND POLITICAL STRUCTURES FOR COLLECTIVE CONSUMPTION

Health care systems in different countries have often been typified according to the relative importance of the state as a provider and organizer of medical services. Usually the state plays a role in health care provision because uncontrolled free market forces will not produce a fair distribution of health services that meets the needs of the whole population. Collectivist strategies operated by governments aim to regulate health care consumption so that it responds efficiently, effectively and equitably to population needs. Jones and Moon (1987) and Curtis and Taket (1996) discuss the position of different national health systems along a continuum between collectivist and anti-collectivist models of health care. Even in countries like the USA, where anti-collectivist strategies are relatively important, public provision plays a significant part in the system. Theories of *collective consumption* help us to understand the structural processes influencing use of services such as health care.

One widely debated perspective on collective consumption was proposed by Castells (1977), who argued from a neo-Marxist perspective that the city can be seen as part of the total socio-economic system involving production, consumption and exchange. Consumption (especially collective consumption) is particularly directed at social reproduction of the political economy and of the labour force, as well as reproduction of the prevailing social ideology. This requires essential services like education and health care. Delivery of these services requires a range of physical infrastructure and institutions, such as housing, transport, schools, hospitals and clinics, sanitation facilities, recreation spaces, community centres, theatres. The free market, governing processes of private consumption, cannot be relied upon to provide access to such infrastructure for all groups in the population. The production and consumption of these services are therefore collectively organized.

As the state has to assume responsibility for these essential services, their use by consumers becomes a social and political issue. Societies normally regulate these services to make them available in proportion to need or merit, rather than allowing free market forces (of individual willingness and ability to pay) determine the pattern of use. This may be of particular benefit to poorer groups in the population that might not otherwise be able to afford services. Pinch (1985) discussed Weberian perspectives, which emphasize the role of administrative organizations in provision of welfare services and focus on actors in key managerial positions who determine the distribution of services.

When applied to health care, these perspectives suggest that state involvement in health services is a form of social reproduction of labour power, which benefits working classes but leaves the dominant capitalist classes in control. Pinch (1985) commented that, while Castells' work has been criticized, it does provide a framework which moves away from contrasting collectivist and anti-collectivist ideologies and focuses more on health care as part of the political economy. Health services also contribute to the economic system, as employers and as consumers of goods and services produced by other sectors of the economy, such as industries producing pharmaceuticals and medical supplies.

Political and administrative structures for collective consumption influence variation in health care access and use, both at the international level and locally, within countries. Wealthier countries are generally able to invest more resources in health than poor countries, and this is an important aspect of health inequality at the international scale. For example, data from WHO (2000a) in Box 5.1 shows that public expenditure on health per capita, in international dollars, is 100 times more in countries such as Switzerland, the USA and the United Kingdom than in countries such as Ethiopia and India. However, system performance can also be assessed in terms of equity of access for different population groups and responsiveness to their needs. Figure 5.3 in Box 5.1 shows that the amount of national expenditure on health services bears some relationship to other measures of service performance such as responsiveness and fairness. However, the relationship is a very imperfect one, suggesting that different countries vary in the mode of health service delivery as well as health care costs. These variations in resource use and service provision are in large part due to varying administrative and political structures influencing health policy and health service provision.

Administrative and political structures in health systems are also locally variable *within* countries. In the context of a city, Weberian theories stress the role of individual 'urban managers' who determine the direction of local policy and the deployment of resources to different areas and population groups. These managers' actions are constrained by the organizational structures in which they work. Their decisions are made partly with the aim of preserving the existing organization, rather than solely to reflect democratically expressed public views. The manipulation of waiting lists, discussed in Chapter 3 (Box 3.1), is an example of this. This perspective helps us to appreciate the significance for local health care provision of bureaucratic systems including information management, resource allocation and eligibility criteria for health care. The infrastructure is locally variable and this may affect access to health care. Penchansky and Thomas (1981) proposed the 'five As' model of health service characteristics that are significant for users of health services and may vary at the local scale: availability, accessibility, accommodation, affordability and acceptability. These are determined largely by the management and administration of care.

Varying local patterns of health service provision can be viewed as the outcome of political struggles between different classes in society and a reflection of the dominance of the capitalist socio-economic structure. Castells (1977) argued that urban social movements arise from these struggles. Pinch (1985) pointed out that alternative perspectives, from public choice theory, argue that democratic processes are the best way to resolve these struggles in a fair way. Voters are assumed to make free,

Box 5.1 International data on health service performance

The World Health Organization made a comprehensive survey of health systems in over 190 member states (WHO, 2000a). Indices were compiled to compare health service expenditure, fairness and responsiveness.

Responsiveness was evaluated using data from key informants in each country concerning seven dimensions of health care: dignity; autonomy; confidentiality (respect of persons); promptness of attention; quality of basic amenities; access to social support during care; and choice of care. Health systems were assessed in terms of variation in these aspects of responsiveness for four 'vulnerable groups': poor people, women, old people and indigenous or disadvantaged racial groups.

Fairness was assessed in terms of how far variation in estimated household expenditure on health care was proportional to differences in household incomes.

These measures are considered here in relation to indicators of annual *expenditure,* expressed as international dollars per capita population.

The measures of fairness and responsiveness tended to improve as the amount of health service expenditure increased (Figure 5.3). Thus for disadvantaged populations in poor countries, lack of health service resources may be exacerbated by poor health system performance for the most disadvantaged groups.

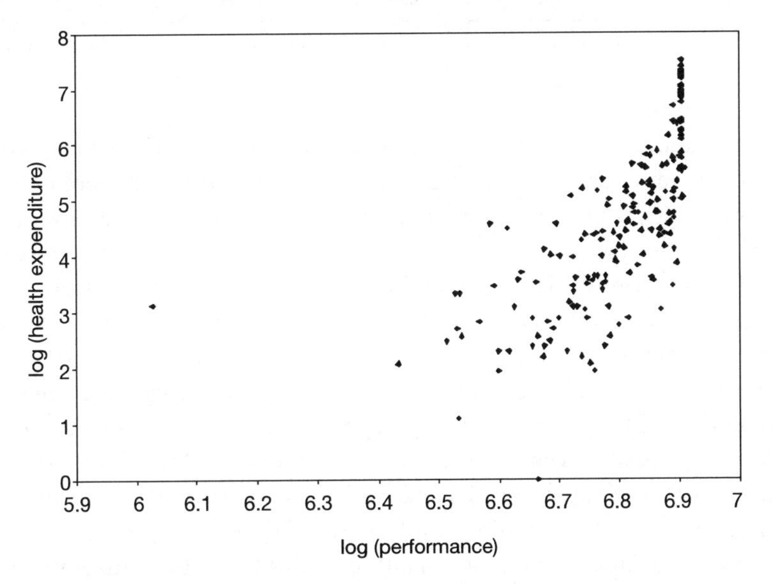

FIGURE 5.3 National health system performance on responsiveness compared with health expenditure in international dollars (WHO, 2000a)

However, health system performance is not always related straightforwardly to health service expenditure. For example, data shown below suggest that the USA and Switzerland, spending over 1600 international dollars per annum on health care, had

continued ⇒

Box 5.1 Continued

indices of responsiveness and fairness which were no better than the UK, Spain, or even Jordan, which were spending much less. India appeared to achieve relatively high levels of performance in terms of fairness, although expenditure was less than 100th of that in the western countries shown here. This illustrates the importance of considering various aspects of service performance when assessing international geographies of health care provision and consumption. It is not sufficient to consider only levels of expenditure.

Country	Public expenditure on health	Fairness index	Responsiveness index
Ethiopia	7	0.906	0.733
India	11	0.962	0.876
Jordan	119	0.958	0.981
Spain	855	0.971	0.995
United Kingdom	1156	0.977	0.995
United States of America	1643	0.954	0.995
Switzerland	1833	0.964	0.995

rational and well informed choices about the kinds of services they wish to consume. Their preferences are expressed through democratic processes and voters in different areas may choose varying strategies for provision of local services. This 'public choice model' has been associated with 'new right' ideologies of health care. For example, Tiebout (1956) suggested that voters in an area who disagree with the majority view have three options: *exit* (to move elsewhere), *voice* (to campaign for change through the democratic process) or *loyalty* (to accept the view of the majority). The public choice perspective is problematic as a basis for justifying local variation in health care provision for various reasons, as Mohan (1995) explained. For example, not all individuals in the population are equally articulate and influential in the democratic process. Those who most need to use health care are likely to be disempowered or even disenfranchised (e.g. children, people who are physically frail or mentally infirm). By contrast, certain actors, including members of the medical profession, have much more influence. Also, in some countries, like Britain, local administration of health care is not controlled by local government (although social and environmental health services are responsibilities of local authorities), so it is unclear how local democratic processes can influence national health service policy. Furthermore, some elements of health service infrastructure, such as hospitals and clinics, are relatively difficult to change quickly in response to public opinion.

PHYSICAL INFRASTRUCTURE AND SPATIAL ACCESS

Castells (1977) also pointed out that the city in the capitalist system provides a physical infrastructure for reproduction of labour power in the form of facilities for consumption of essential services at the local level. The physical elements of service infrastructure are relatively fixed in space (for example, school buildings, hospitals, railways) and they are often most accessible to populations living closest to them.

Health geography has devoted considerable attention to the spatial distribution of health facilities as a factor influencing heath care use. This has been researched over a long period. For example, Hunter and Shannon (1985) have considered the 19th century work of Jarvis, who identified a *distance decay* effect in the use of hospitals, showing that rates of use of hospital facilities for geographically defined populations tend to decline as distance from the hospital increases. The importance of distance as a variable in access to health care has continued to be of concern to geographers up to the present day. Joseph and Phillips (1984) and Thouez (1987) discussed research in this field in the 1980s, and Gatrell (2002: 158–61) has reviewed several more recent examples of studies of use of secondary and tertiary health care services that demonstrate distance decay effects.

Spatial organization of health care can be considered in the light of broader geographical perspectives on the catchment areas of service facilities. These have included central place theory, which invokes the idea of a hierarchy of service centres, such that lower order centres offer services to a local catchment, while higher order centres provide a larger range of services over a wider area (Carter, 1981). The concept of the 'gravity model' extends these ideas about the geographical range of catchments. This model specifies a distance decay effect in use of a service facility. It estimates utilization as inversely proportional to the squared value of distance between the consumer and the facility. The distance travelled (or time taken to travel) to hospitals has been used in the estimation of likely use of hospital facilities by different communities and to assess how the reorganization of hospital services changes access to services. For example, Congdon (1996c; 2000) used gravity modelling to assess the regional impact of hospital closures in north-east London on access to accident and emergency services. Lowe (1996) examined the impact on access to hospitals that might result from the introduction of universal coverage in a hospital system in the USA and also the potential impact from hospital closures. In this study, the changes were expected to be most significant for poor neighbourhoods.

Proximity to health care facilities is most likely to affect access in sparsely populated areas, where distances to be travelled are greater. In low income countries, problems of access to health care in rural areas are especially severe due to the generally low levels of health care provision, the more extreme remoteness of rural settlements, the poverty of the population and the lack of access to transport (e.g. Phillips, 1990; Phillips and Verhasslet, 1994; Curtis and Taket, 1996).

Research on access to health care in rural parts of developed countries also demonstrates variation in health care use and health outcomes associated with distance to be travelled to health facilities. Reviews include Haynes and Bentham (1979), Fielder (1981), Gesler and Ricketts (1992), Watt and Sheldon (1993), Higgs (1999), *Health Services Research* (1989). Compared with urban populations, rural residents may have relatively low rates of health service use, especially those who are have limited mobility due to lack of transport or infirmity. Evidence is reviewed by Haynes (1991), Bentham and Haynes (1992), Bentham et al. (1995), Greenburg (1991), Gesler et al., (1999), Nemet and Bailey (2000). For example, Haynes et al. (1999) recently reported differences in hospital admission rates between urban and rural populations in East Anglia, UK, which related to distance from hospitals. After controlling for population need and supply of services, populations living furthest from hospitals had acute hospital admission rates 17% lower than in the most accessible areas. Use of emergency care, geriatric hospitals and psychiatric hospitals declined even more markedly in proportion to distance to hospital (see Figure 5.4). A study of asthma mortality rates in the same region showed that, after controlling for local social conditions affecting asthma risk, mortality rates were 31% higher in populations living more than 30 minutes travel time from a hospital, compared with those living up to 10 minutes away from the nearest hospital (Jones et al., 1999).

The spatial distribution of health facilities and problems of physical accessibility therefore contribute to landscapes of consumption relevant to health. These factors are especially important in remote rural areas, for populations with restricted physical mobility (such as elderly people and those with physical disabilities), and where travel time to hospital is crucial to the outcome of treatment (as for accident and emergency services). Populations in inaccessible places may not use health care services they need, and this has implications for equity in access to health care. However, spatial accessibility interacts with other structural factors

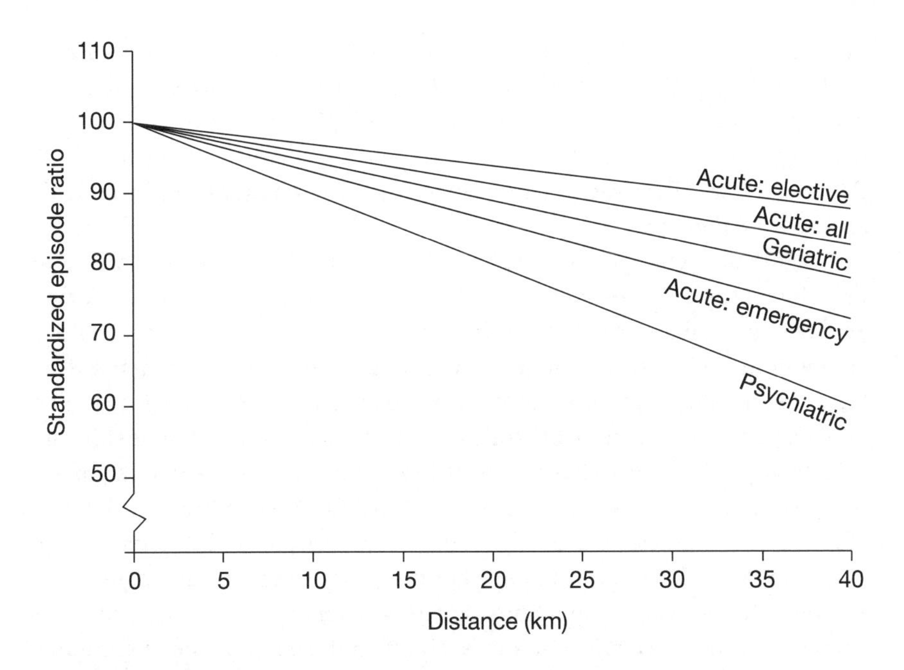

FIGURE 5.4 Distance decay in use of hospitals in East Anglia, controlling for population need and provision of services (reproduced from Haynes et al., 1999: Figure 1, with permission from Elsevier)

and with individual characteristics, and it is often not the dominant factor influencing patterns of use.

PROCESSES OF CHANGE IN LANDSCAPES OF CONSUMPTION

The factors influencing landscapes of consumption, summarized above, are subject to changing conditions. Forces for change include three processes given particular attention here: health system reforms; growth of consumerism and commodification of health and health care; and technological development. Figure 5.5 summarizes the following discussion, which illustrates some impacts on administrative structures, social milieux and spatial structures associated with these three processes of change.

HEALTH SERVICE REFORMS AND THEIR IMPACT ON LANDSCAPES OF CONSUMPTION

In many countries, reforms of health care systems have reorganized health services to operate as *managed markets* for health care. This means that government plays a more important role in funding and regulating health services than in delivery of health services. These trends have been discussed in several reviews and typologies of health care systems (e.g. Curtis and Taket, 1996; Altenstetter and Björkman, 1997; Ham, 1997; Ranade, 1998). This section describes resulting changes outlined in the first row of Figure 5.5.

Significant changes have taken place to introduce health care 'markets' in some countries. For example, Britain and countries of the former Soviet Union, like Russia, have introduced reforms during the 1990s to create internal health care markets within state health care systems. They have established administrative divisions within the health system between agencies responsible for funding and commissioning health care, and the organizations delivering health services to users. Separate parts of the system are supposed to compete with each other to provide the most efficient and effective use of health care resources. The health care available to a local population therefore depends on the contractual agreements between the commissioning agencies and the providers. These reforms also allow a more significant role for the commercial

	Structures in landscapes of consumption		
Processes of change	Administrative and political structures	Social milieux	Spatial structures
Health system reform: the shift to managed markets	Changing role of state and independent sectors in health service provision	Impact of contracting on informal and voluntary sectors	Potential for increased differentiation in service infrastructure between areas
Consumerism and commodification	Emphasis on consumerism Regulation of markets for health care and the healthy body	Smart consumers and health services as 'lifestyle' products	Themed sites for health care delivery
Technological development	Issues of resource allocation for high tech vs basic medicine Health information systems	Cyberspace Utopias or dystopias	Telemedicine

FIGURE 5.5 Processes of change and their impacts on landscapes of consumption

independent health care sector. Russia adopted a style of administration involving commercial agencies managing contracts for state funded health care (Curtis et al., 1995; 1997). In Britain, private as well as public providers compete for state funded contracts to supply some types of health services (Mohan, 1995).

In the USA, where the health system already operated as a 'market', government-led reforms have been directed towards greater regulation and control of health service provision and costs. These reforms are influenced by the principle of *managed care*. Health care is funded differently for different groups of users: by government on behalf of citizens; by employers on behalf of their employees; and by some individuals through private insurance. Managed care organizations (MCOs) compete in a regulated health care market to manage these funds, using them to pay health care providers to deliver health services to specified groups of users. The health services available to a user will depend on the health care plan of the MCO with which he or she is enrolled. Gold (1998) argued that access to health care in the USA is influenced by structural and financial characteristics of health care markets, the features of different health care plans and personal attributes of users. MCOs are expected to provide health care of a given standard at a competitive price, providing good value for money and containing health care costs. However, Grembowski et al. (2002) argued that MCOs are tending to merge together and become more powerful in the health care market, which may limit competition. MCOs are exercising growing control over the care that doctors provide to their patients and this may make doctors less responsive to patients' individual requirements.

The shift to managed markets has also changed the pattern of involvement of communities in informal and voluntary provision of health and social care. Milligan (2001: Chapter 6) demonstrated that state funding was important for the operation of voluntary services. She reported a varying emphasis on contracts in the state and the voluntary sectors. In some areas, local governments were contracting out much of their social care activity to the voluntary sector. In this situation, large voluntary organizations with well developed administrative resources were able to compete for funds more successfully than smaller organizations. Voluntary organizations were also less likely to be awarded funding if they were serving groups of users that were not priorities for local government support. In other areas, local government continued to directly provide a larger proportion of care for the elderly. Milligan suggests that these processes were contributing to growing geographical inequalities in provision.

Changes to the role of the state and private sectors are also important for the spatial pattern of service infrastructure. For example, Mohan (2002) discusses the impact of the Private Finance Initiative in Britain, by which independent organizations are involved in the building of new NHS hospitals on a commercial basis. This has resulted in several new hospital developments in different parts of the country. However, the commercial cost and viability of these schemes depends on a range of factors that vary from one area to another. These include the type and scale of hospital workload, the support provided by local community services to care for patients as they are discharged, the cost of land and construction, the cost of labour to staff the non-medical functions of these hospitals and the potential for income generation. Mohan suggests that financial criteria were at least as important as population need in determining the decisions about where these new hospitals would be built. He questions whether this provides a sound basis for strategic hospital planning.

Although managed markets are regulated by government, they are intended to encourage flexibility and responsiveness to local consumers' requirements. This can easily result in growing inequality of provision between areas. In the case of changes to the British National Health Service, Exworthy et al. (1999) described an increasingly complex system, with elements of quasi-markets, quasi-hierarchies and quasi-networks coexisting. The British health service reforms introduced by the Conservative government in 1990 were deemed by the Labour administration in 1997 to have produced excessive fragmentation and lack of coordination between different parts of the system and to have resulted in unacceptable inequities in provision (Great Britain Parliament, 1997). Mohan (1998a; 1998b) emphasized the significance of local political mobilization and participation for diversity in health care. Similar observations have been made about the changes introduced to the Russian health care system, discussed by Curtis et al. (1995; 1997). Reforms were implemented in different ways in different local areas, resulting in greater geographical variability of health care.

If managed markets encourage greater geographical diversity in health care provision between geographical areas, this raises issues of territorial justice. An equitable pattern of health care, according to conventional models of collective consumption, is one that distributes health care resources in proportion to socially agreed criteria. Normally these are criteria such as medical need, merit or common good, rather than ability or willingness to pay for services or social position. For example, Davies (1968) developed a model of *territorial justice* in the geographical distribution of welfare that advocates a regional allocation of resources in proportion to local needs of the population. According to this type of criterion, diversity of provision in managed health care markets is not of itself inequitable, provided that all communities have equality of access to health care which effectively meets their needs. However, according to many critics of the managed market model, uneven development is a natural feature of market driven systems (as discussed in Chapter 4). Furthermore, as discussed above, critics of public choice theory suggest it is difficult to make local health services democratically accountable and responsive to the needs of all users. It is therefore difficult to ensure that managed markets meet requirements for social and territorial justice in the provision of care.

COMMODIFICATION OF HEALTH AND HEALTH CARE

A growing number of studies in geography and other disciplines has explored the commodification of health and health care. This is linked to consumerism and the use of commercial strategies for marketing of health care and the healthy body. (See the second row in Figure 5.5.)

Research by Kearns and Barnett (1999), considered in Chapter 2, discussed efforts by private health care companies to make hospitals attractive to consumers by making them almost like theme parks. Kearns and Barnett (1997) have also examined other aspects of the marketing strategies employed. For example, they consider the development in New Zealand of accident and medical centres (AMCs). These have generated greater competition in primary care markets. Legislation allowing more freedom to health care providers to advertise their services led to primary care publicity campaigns. A range of messages and symbols was employed in advertising by these private companies, and by other primary care providers, as they all competed to provide for the same local populations. These advertisements included emphasis on *competitive prices* for health care (including 'special offers' in the form of free health checks, for example). The advertising also stressed the *range of services* available and presented these facilities as *comfortable and friendly* to users. An example of a publicity flyer, for a group of GPs, offered the sense of personal rapport and proximity which patients are likely to be seeking with a medical adviser; they publicize their practice as 'your own local GP patient service' and 'the doctor that knows you best' (1997: 177). These trends are changing the character of health facilities and the ways they are represented to users.

Similar trends are also noted with respect to complementary and alternative medicines, which are increasingly being marketed as 'lifestyle products'. For example, from a sociological perspective Eastwood (2000), in a study of the deployment of alternative therapies by general practitioners in Brisbane, Australia, has discussed the commodification of nature, spirituality and alternative health practices. Wiles and Rosenburg (2001), Kelner and Wellman (1997) and Sharma (1992) review evidence concerning the factors which influence varying rates of use of alternative therapies. These include indicators of social position such as gender, age, education and income, as well as health status. Communities which differ in terms of these factors will also differ in their propensity to use alternative care. Although commodification of health and health care has been documented in greatest detail in western countries, similar processes are also evident in other parts of the world. For example, Smith (1993) comments on the commodification of traditional Chinese medicine which has accompanied the restoration of forms of capitalism to rural areas of China.

Wiles and Rosenburg (2001) argued that consumers of these alternative health care products are acting as 'smart consumers' making independent, informed choices about a range of possible health care options. Kelner and Wellman (1997) suggested that this attitude is associated with social values in western societies which empower (and expect) individuals to take responsibility for their health. The decision to use a form of alternative therapy may be a way of asserting oneself as an individual who can act independently of institutional structures such as government funded medical care. Dissatisfaction with biomedicine is often a factor in the use of alternative care. Use of health care may also express the desire to identify with (or distance oneself from) particular social groups or types of lifestyle. Wiles and Rosenburg (2001: 221) for example, suggested that greater use of alternative medicines in British Columbia, compared with other parts of Canada, may be an expression of a regional identity and lifestyle which favour the rejection of conventions and the adoption of more 'natural' and ecologically sustainable lifestyles.

This trend towards commodification raises interesting ethical and philosophical issues about the role of the state in regulating markets for health related products. There is international and local variation in the extent to which government administrations aim to regulate alternative medicines and control commercial strategies associated with commodification. Pellegrino (2001) saw commodification of health care as a central feature of the health care system in the United States and argued that this detracts from the ethics of patient care. Ashcroft (1999) argued that the combination of increased commodification and marketing of herbal medicines, in the absence of information on the efficacy of such alternative therapies, raises interesting issues about the benefits to consumers. On the one hand, consumers should be protected against the marketing of alternative therapies that are potentially harmful, or which do not provide the health benefits claimed for them. On the other hand, a strategy aiming to license all herbal health products may result in rising prices and more unequal access for consumers with different income levels. This might not benefit consumers much either, particularly in the case of products that have been shown to be safe in a long history of use, and which people will want to continue using, even if scientific evidence of their efficacy is equivocal.

Issues of ethics and regulation also extend to questions concerning marketing of the human body. For example, Glasner and Rothman (2001) comment on the ethical issues raised by the commercial application of genomics. Hanson (1999) argues that patenting of genomic material has benefits in terms of development of medicine which outweigh, for the moment at least, the ethical disadvantages. Palsson and Rabinow (2001) discuss the controversy over issues of property, ownership and access to human genomic material raised by the way that the genomic information on the Icelandic population was made available for research.

TECHNOLOGICAL DEVELOPMENTS: CYBERSPACE UTOPIAS OR DYSTOPIAS?

Continuing technological advances are producing effects summarized in the third row of Figure 5.5. These cause dilemmas for administrators of health systems over whether to invest in advanced technologies for health care and information management, or to concentrate resources on more conventional, basic health provision. In poor countries with very limited health care expenditures (see Box 5.1) these problems are particularly difficult.

One example of modern technology, which has important implications for the geography of health care, is the development of telecommunication and information systems. In some places where physical access to care is especially limited, new telecommunication technologies are used to try to break down the barriers of distance. Several authors have commented optimistically about the potential for profound and beneficial changes in access to health care in what Marsh (1998) has termed the 'global telemedical information society'. Nesbitt et al. (1999) and Picot (1998) discussed how telemedicine may enhance health care in underserved rural and inner city areas of the USA and Canada. Lovell and Celler (1999) described the potential for telemedicine primary health care networks in Australia, linking primary, secondary and tertiary services to improve access to health care for patients, irrespective of their distance from urban areas. Whitten and Cook (1999) described the use of telemedicine to improve health care provision through schools in poor inner city areas in Kansas, USA, and Woods et al. (2000) discussed the application of telemedicine to extend outreach services for sickle cell disease to rural areas of the USA. Bowater (2001) reported the use of video conferencing to link primary medicine in remote Australian mining towns with urban medical centres, which reduced the need for transfers of patients for specialist treatment. Similarly, Tually et al. (2001) describe the development of nuclear medicine teleradiology in Australia, which, while more expensive than conventional radiology, may save costs of transfers to cities for unnecessary examinations.

These examples suggest scenarios for the geography of health care in the future corresponding to what Graham (1997) describes as 'cyberspace utopias'. New technology will liberate societies from the need for high density urban living, and they will be more equitable, democratic and environmentally sustainable and less socially divided than at present. Castells (1989; 1996) has presented a vision of the *informational city* where geographically defined places would lose their meaning, in terms of sociogeographical power relationships, and be replaced by 'spaces of flows'. Mitchell (1995) has described the *city of bits*, no longer rooted to particular places, with an expanding range of services delivered at home. At an international conference in medical geography, several presentations considered arguments that telemedicine would offer a way to tackle hitherto intractable problems of intra- and international inequality in health care provision (Cutchin, 2000; Löytonen, 2000; Mayer, 2000; Shannon, 2000).

However, it is also possible to take a more pessimistic perspective on the impact of telemedicine. For example, its rather slow and uneven development has been noted in the USA (LaMay, 1997; Huston, 1999), Australia (Mitchell, 1999) and South Africa (Gulube and Wynchank, 2001). Obstacles include technical, organizational and attitudinal problems. LaMay (1997) suggested that the private sector in the US has been limited by technical and regulatory constraints, and commented that, if telemedicine generated a surge in demand for health care, this might be resisted because of rising health care costs. Swanson (1999) pointed out that development of telemedicine in underserved areas may do little to help integrate groups such as the mentally ill or the illiterate into their local communities. Telemedicine might also result in patients bypassing local services, with the possibility that this might lead to further reductions in local provision in poorly served areas. Some practitioners may find it difficult to acquire the necessary IT skills to use telemedicine effectively. The quality of care possible through telemedicine may also be different than in face-to-face consultation. Miller (2001) reviews several studies of doctor–patient communication using telemedicine. These were generally positive, but drawbacks included the lack of touch and of observation of non-verbal behaviour. Mayer (2000) voiced ethical concerns that telemedicine might be detrimental to aspects of the doctor–patient relationship depending on contact in physical rather than virtual space.

Thus telemedicine also has the potential to create what Graham (1997) has described as 'urban dystopias'. These are associated with: increasing social polarization between those with access to new technology and those without; desertion of inner cities by those who can leave, resulting in urban decay and degeneration of key services like health care; growing power of global industries controlling new technologies which will be detrimental to national democracy; increasing withdrawal from public spaces into highly protected and isolated domestic environments; lack of physical mobility and associated loss of physical fitness. Shannon (2000) suggested that geographical variation in medical care, rather than being eliminated, will become more complex as telemedicine creates new health care landscapes. Cutchin (2000) suggested that telemedicine is creating virtual regions of care, over which certain health care providers have increasing power. He argued that health geography should conceptualize this 'new frontier' of telemedicine and focus on the implications for accessibility and equality of care.

GEOGRAPHIES OF PRIMARY CARE USE

The perspectives discussed above are important for landscapes of consumption of primary health care. The World Health Organization declaration, agreed in Alma Ata in 1978, identified primary health care as of crucial importance for health for all people, in low income and high income countries. The WHO definition of primary health care is very broad. It extends beyond medical services and includes the following elements:

- education concerning prevailing health problems and the methods of preventing and controlling them;
- promotion of food supply and proper nutrition;
- an adequate supply of safe water and basic sanitation;
- maternal and child health care, including family planning;
- immunization against the major infectious diseases;

- prevention and control of locally endemic diseases;
- appropriate treatment of common diseases and injuries;
- provision of essential drugs.

Primary *medical* care comprises ambulatory health services delivered in community settings such as local clinics and doctors' surgeries and in patients' homes. They are distinguished from secondary and tertiary health care, provided in hospital settings through specialist medical outpatient clinics or as inpatient treatment of patients occupying a hospital bed. Primary care also involves a range of other services provided by other personnel in primary care teams, including nursing care, health promotion advice, occupational and physical therapy and chiropody. Primary *medical* care includes community based general medical services by doctors or other medically trained personnel, and these are given particular attention here.

Although societies invest large resources in the hospital care sector, the primary care sector is the most essential part of any health care system, as it offers care to the largest proportion of the population, and is an important source of prevention of illness, as well as treatment of disease. Even people with severe illnesses often receive a significant part of their care through primary services, although they may also be treated in hospital. This is reflected in Figure 5.6, showing primary care as the

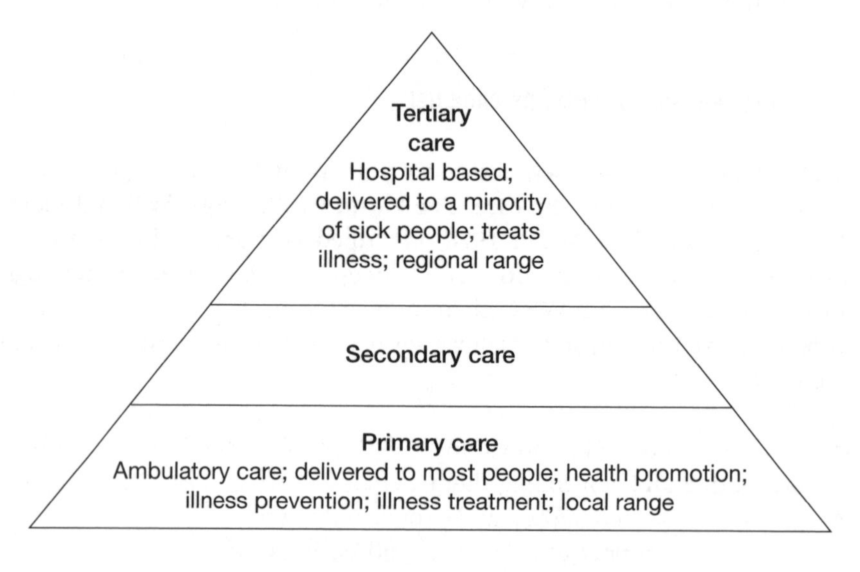

FIGURE 5.6 The health care system hierarchy: primary care as the foundation

'foundation' for a health care system, supporting the other elements. Primary care needs to be delivered locally, as many people need to use these health services relatively frequently, and it is uneconomic for users or carers to travel large distances in the process. The catchment areas of primary care facilities are therefore relatively limited in their geographical range and the distance decay effect on use is quite strong.

At the international scale, health systems vary in the organization of primary health care and in the proportion of their population who have access to these services. Wealthy countries with well developed biomedical health systems have the capacity to deliver such services to all of their population (although as discussed below, they do not always achieve this). Some poor countries are unable to provide primary care to all of their populations, especially those living in very remote rural areas or informal urban settlements. This may be partly because much of the limited resources available for health care are spent on care in hospital facilities. There are also logistical difficulties in supplying services to very sparsely distributed rural populations where transportation systems are poorly developed, and settlements lack basic infrastructure.

Figure 5.7 illustrates primary medical care coverage in terms of maternity care in the countries of the Pan-American Health Organization around 1990. The proportion of births attended by a professional in the US and Canada was over 90% but in poorer countries, such as Guatemala, Ecuador, Paraguay and Bolivia, the proportion was as low as a quarter or a third. This is important since the risk of women dying from pregnancy related complications in Latin America and the Caribbean is 1 in 130, compared with only 1 in 3700 in North America.

In a different part of the world, the health impact of poor access to maternity care was studied by Midhet et al. (1998), who demonstrated greater risks of maternal mortality, for mothers having their first child, in more remote areas of rural Pakistan. They found that access to local primary health facilities, rather than the distribution of specialist hospital maternity units, was important. Having a local source of contact with medical services, even if it was not a specialist maternity unit, would encourage use of health care in the event of complications at birth.

Wealthy countries also vary in the organization of primary care and in the eligibility of different population groups to access the system. Thus for example Box 5.2 summarizes, very schematically, some features of the systems of access to primary care in Britain, France and the USA, which differ in terms of funding and organization of health systems. Because of these differences there are variations in the factors that influence access to care.

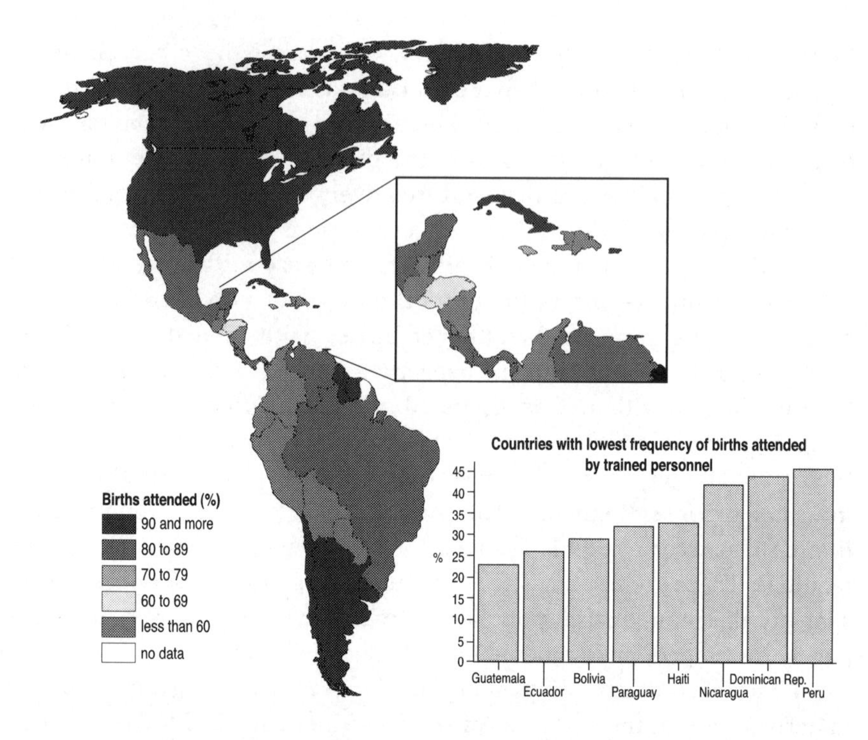

FIGURE 5.7 Frequency of births attended by personnel by country in the region of the PanAmerican Health Organization, 1990 (PAHO, 1999)

In the US, access to medical services depends to a large degree on health insurance status. A substantial majority (over 16% of the population under 65 years of age) are uninsured and dependent on 'safety net' health care provision by the state through Medicaid (Carrasquillo et al., 1999). Ayanian et al. (2000) reported that in 1998 about 33 million people aged 18–64 in the US were uninsured. Insurance cover often depends on ability to pay for private insurance (either directly or through an employer's scheme), and it is a significant factor for accessibility of general medical care in this type of system. Ayanian et al. (2000) reported a survey showing that over half of uninsured people in poor health said they were unable to see a physician when they needed to because of cost. Physical access as well as cost may affect accessibility. Health care systems where the private sector plays an important role, such as the USA, often have large geographical variations in the provision of medical care, including primary medical services (Shannon and Dever, 1974; Knapp and Hardwick, 2000; Carrasquillo et al., 1999). Rural areas and poor urban areas tend to be relatively underserved. Commentators such as Whiteis (1998) see this as a result of the tendency for uneven

Box 5.2 Organization of access to ambulatory general medical services: differences between national health systems in Britain, France and the USA

The following table summarizes some key features of general medical services in three different countries with different systems of finance and eligibility for care: Britain, France and the USA.

The National Health Service (NHS) in Britain is resourced by central government budgets, largely funded through general taxation, and the whole population is eligible to use this system. Patients register with one general practitioner (GP) who contracts with the NHS to provide the patient with access at all times to comprehensive general medical services. Payment of GPs is determined by the terms of their contract and is based on a complex set of criteria including 'capitation' payments for each patient registered on the GP's NHS list. Patients can choose which local GP they register with and can change their doctor. However, the choice may be limited if there are few GPs practising locally, and GPs may refuse to register patients if their lists are fully subscribed or the patient lives outside the catchment area of the practice. GPs most often operate in group practices with several practitioners sharing premises and treating patients registered with the group as a whole, giving patients access to more than one doctor within a practice. Practices determine their own catchment areas, though these are monitored by the NHS to ensure all areas of the country are covered. In some areas, deemed to be well served by GPs, there are NHS restrictions on the establishment of new practices. There are also incentives to GPs to set up practices in areas which are less well served. Consultation with practitioners is free of charge but patients who are not exempt pay a contribution to the costs of medicines the GP prescribes. GPs can also refer patients to receive hospital care under the NHS, and, apart from trauma services, there is no self-referral to the hospital sector for NHS patients. This gives the GP a pivotal role in the NHS, controlling access to secondary and tertiary services. Patients who are not registered with a GP (particularly those who are highly mobile or homeless, for example) have more limited access to the health care system. Practices are variable in the way that they are organized, and in the range of services and treatment which they provide for patients. For those that are registered with a GP, issues therefore arise about variability in primary care in the NHS.

France operates a health care system funded through a number of social health insurance schemes which cover almost all the population. Over 80% of the population is covered by the dominant, state regulated insurer. Ambulatory general medical services are provided by generalist physicians who function as independent practitioners. A large proportion of the costs of consultation and prescriptions is reimbursed through the medical insurance system and the cost of consultations is regulated. Many patients have complementary independent insurance cover for that part of the cost of general medical care not covered by social insurance. Recent health reforms have introduced a system of universal health coverage (*couverture maladie universelle*, CMU). For those on low incomes it has the effect of providing more complete cover for costs of medical care. There are no restrictions on where doctors may set up their practices and patients can make use of more than one general physician. Access to

continued ⇒

Box 5.2 Continued

hospital care does not depend on referral by one particular physician or general practice. For some types of ambulatory medicine (e.g. gynaecology) patients may alternatively use specialists who practise in community based settings. Often groups of specialists will share premises to provide ambulatory access to a range of medical specialities at one location.

While in both Britain and France a minority of patients choose to consult a GP outside the national health care system, in the USA the majority (about 60% of patients) are covered for general medical services through independent medical insurance plans. These are often paid, at least partly, by employers. Managed care organizations (MCOs) have become a very widespread method of organizing funding of privately insured care through contractual agreements. MCOs contract to manage an insurance fund for provision of health care to a group of enrolled patients, covered by an insurance plan (see main text). The patient will normally use the general physician, or specialists, to whom the health insurance plan gives access. Insurance companies and MCOs may be selective over which groups of patients they will accept. The types

Country (type of funding)	Type of professional	Type of coverage	Eligibility	Relation to hospital sector
Britain				
Central budget funding; quasi-market structure	General practitioners (independent contractors to NHS)	Comprehensive general medicine (24 hour access to GPs under the NHS); no fees for consultation but some patients pay for prescriptions	Patients registered with only one GP (have choice of GP under NHS)	Referral to most hospital care via GPs: 'gatekeeper' role for GPs
France				
National Health Insurance	Generalist physicians or specialists in community clinics	Government regulated insurance schemes; comprehensive cover	Patients can consult more than one generalist/ specialist under the health insurance system	Referral to hospital via generalists or specialists
USA				
Regulated private insurance with public safety net; MCO model is commonly used to manage care	General physicians or specialists	Coverage and cost depend on health insurance; private insurance plan or 'safety net' public cover, mainly through emergency rooms	Determined by insurance coverage and ability to pay	Access to hospital depends on insurance plan and MCO or Medicaid/ Medicare eligibility

continued ⇒

Box 5.2 Continued

of medical services covered are determined by the terms of the insurance plan. Many patients are not covered by medical insurance, and a state 'safety net' scheme (*Medicaid*) provides access to basic medical care. Many other patients receive care which is funded through *Medicare*, a second publicly funded scheme for older patients who often have complex and expensive health care needs which are difficult and costly to cover through private insurance. Access to hospital care depends largely on the coverage and service arrangements specified in the patient's insurance plan (or by *Medicare/Medicaid*).

All of these systems have potential for inequality of access to primary care. In Britain and France these arise mainly from variation in the services provided by GPs and uneven geographical distribution of GPs. These national systems are intended to provide cover for the whole population, though in practice some marginalized groups are not well served. In the USA there are greater inequalities in entitlement to care and its cost, due to differences in insurance plans covering different groups of patients. MCOs have a significant influence over access to primary care.

development in market driven primary care that parallels uneven development in capitalist systems generally (discussed in Chapter 4).

However, even countries with more integrated national health systems find that it is difficult to maintain an even distribution of primary care over the national space in order to promote equality of access. For example, in France, Lucas-Gabrielli et al. (1998) and Vigneron (1997) report marked differences in the numbers of general physicians relative to population for different areas. For labour market areas in France, the number of general practitioners per 100,000 people varies from 55 to 142 (Lucas-Gabrielli et al., 1998: 58). There is much more generous provision in the south of the country and in Paris than in other regions, particularly in old industrial regions in the north where health is relatively poor (Lacoste, 1997). There is a greater concentration of general practitioners in the centres of cities, and the workload of doctors in working class areas is higher than in wealthier areas.

Taylor (1998) compared the problem of uneven distribution of GPs in the US and the UK. While the issue has been of concern in both countries, responses in the US have been relatively piecemeal because there is no national mandate or coherent strategy to address the problem. Variation in access to primary care provided by the British NHS, on the other hand, has been a major policy issue because of the emphasis on territorial justice and universal access to the NHS and the importance of registration with a general practitioner as the principal gatekeeper to the health care system.

In the 1970s there was already concern that the inverse care law applied in general practice in Britain. Tudor-Hart's (1971) concept of the 'inverse care law' has often been used to summarize an inequitable situation in which the relative level of provision of care for local communities is inversely proportional to their relative need for health care. This is the reverse of the objective of equity, or territorial justice. New concerns about equity have arisen in the 1990s because health service reforms introduced by the Conservative government created more diversity in general practice. So called 'fundholding' practices were introduced in 1990, which could exercise greater control over the spending of NHS funds for their registered patients. This was intended to encourage innovation in provision of health care and to make use of resources more responsive to patients' needs. It was in line with policies to introduce managed markets for health care. However, it also meant that the primary and hospital care provided for patients became more variable, depending on which practice they belonged to, and there was concern that an inequitable, two-tier system of health care might be created as a result. Labour government reforms in 1997 moved away from fundholding models. This led to the creation of primary care trusts (PCTs) to oversee general practice throughout a PCT area, and to ensure that NHS funds were used equitably and efficiently to meet all patients' needs for primary and secondary care (Great Britain Parliament, 1997).

Geographical research has contributed to the debate about equity of access to general medical services in Britain. Various authors have reviewed evidence of limited spatial access to primary care in rural areas of Britain (e.g. Haynes and Bentham, 1979; Joseph and Phillips, 1984; Watt and Sheldon, 1993; Watt et al., 1993; Higgs, 1999; Moon and North, 2000). Doctors' surgeries are concentrated in larger settlements serving rural areas. While this makes for better access to care for the majority, the range and quality of care are more limited for a minority of less mobile, isolated rural residents (Brown, 1999). Phillips (1979) examined access to primary health care in rural Wales and found that more privileged socio-economic classes, with better mobility, were more likely to travel further to use primary care facilities offering a wider range of primary care. Less privileged groups tended to use more limited local facilities. Curtis (1982) showed that the centralization of primary care in towns in semi-rural areas of Kent resulted in wide variation in travel distances for elderly populations and that the distances involved could be problematic for frail elderly people with limited mobility. Asthana et al. (1999) reported that in south-west England there was an urban bias in the establishment of fundholding practices, so that this type of organization of primary care would not have been an accessible option for some rural populations. A stream of work from geographers at the University of East Anglia has established links between access to primary care and use of services in the relatively rural English county of Norfolk. Haynes and Bentham (1982) found associations between accessibility and use of general practice by rural populations. Fearn et al. (1984) showed that older residents in a rural village, which previously had no GP surgery, were more likely to consult a doctor when a new mobile surgery was introduced. Bentham et al. (1995) showed that spatial accessibility was one of several factors influencing uptake of cervical cancer screening. Jones et al. (1998; 1999) found that, where distance to the surgery was longer, asthmatics were less likely to consult with GPs, and less likely to be referred to hospital.

A considerable body of evidence also indicates differences in primary care provision in urban areas associated with poverty and wealth. In poorer parts of British cities primary care has often been shown to be more limited in quantity or quality than in richer areas. For example, Knox (1978) calculated access opportunity to GPs in Edinburgh using information on the spatial distribution of population and GP surgeries as well as data on car ownership and the relative speed of public and private transport. He demonstrated lower access opportunity in poorer parts of the city. Jarman (1981) pointed to differences in the pattern of provision of primary care between inner London and outer London, and Powell (1990) underlined the point that quality of care is at least as important as the quantity of provision. Jenkins and Campbell (1996), Blatchford et al. (1999) and Naish et al. (2000) have demonstrated that within the same urban areas, patterns of service use in different practices vary in ways which are difficult to explain in terms of patient needs, and seem more likely to be due to differences in clinical practice between different practitioners. Gillam (1992) identified 'another example of the inverse care law' in health promotion activities of general practitioners in Bedfordshire, including the town of Luton. Deprived populations with high mortality levels have the greatest need for health promotion. However, Gillam showed that practices with relatively high mortality in their practice populations, and those receiving extra payments for patients from deprived areas, were less likely to offer health promotion than practices serving less deprived populations with below average mortality (Figure 5.8).

These studies suggest that a number of organizational aspects of primary care contribute to differences in access to general medical services in the British NHS, such as proximity to surgeries, uneven provision between rich and poor areas, and the quantity and quality of

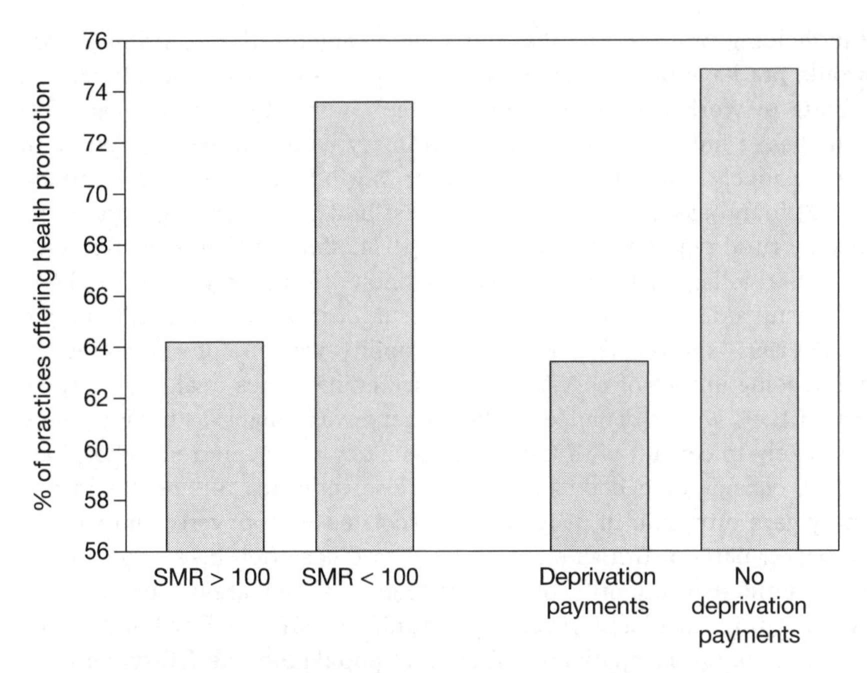

FIGURE 5.8 Proportion of practices offering health promotion clinics in Bedfordshire, UK (data from Gillam, 1992: Table 2)

provision in different practices. Additionally, aspects of the interaction between GPs and their patients are important for access. Eyles and Donovan (1990) reported that patients often feel unable to communicate well with their doctor, or do not understand the advice and treatment they are offered. Patients may be uncomfortable about the surgery facilities, the limited time available for consultations, or not being able to consult a GP of the same sex or ethnic group as themselves. In urban areas with large minority ethnic populations there may be special problems of communication and the acceptability and appropriateness of care (e.g. Donovan, 1986a; Kelleher and Islam, 1996). Campbell et al. (2001) report evidence of variability among ethnic groups in patients' perceptions of their general practitioner.

In spite of all this evidence, it is difficult to tackle these issues of inequality in primary care in Britain because of the status of general practitioners as independent contractors to the NHS. Although geographical allocation of resources for hospital care has been determined with reference to indicators of population need since the 1970s, action to make local variation in resources for general practice equitable has been more limited (Bloor and Maynard, 1995). However, some financial incentives are offered to GPs establishing practices in relatively underserved parts of the country, those who practise in rural areas with many patients living at a distance from the surgery, or those who draw patients from localities which are deemed to be relatively 'deprived' in terms of a range of socio-economic indicators (Senior, 1991; Curtis and Taket, 1996). In London special initiatives to improve inner city general practice were funded under the London Initiative Zone (LIZ) programme (Moon and North, 2000: 122–4). A national programme of primary medical service pilot schemes has encouraged innovative strategies which will make primary care more accessible to vulnerable groups at special risk of poor access to care. Schemes included salaried GPs, outreach services and better joint working between different professionals involved in primary care (Carter et al., 2000). A telephone information system called NHS Direct has been set up to offer basic medical advice to people who find it difficult to contact their general practitioner for help, especially outside office hours (Payne and Jessop, 2001). The primary care trusts established following the 1997 NHS reforms (Great Britain Parliament, 1997) are beginning to develop ways of monitoring variation between GP practices to address problems of poor performance or inequity of provision for some patients. Forde-Roberts (1999) discussed the emphasis in British health care reforms in the 1990s on consumer participation in the NHS, which might make primary care more responsive to patients' needs. However, her research showed that participation was in fact limited and there was inequality in the extent to which different groups of users felt able to influence the health care system. All of these initiatives show how national and local policies in Britain are attempting to tackle geographical and social inequalities in access to primary care.

'FOOD DESERTS': THE LINKS BETWEEN FOOD RETAILING AND POPULATION HEALTH

Food consumption illustrates how consumption and commodification outside the medical care sector can be important for health and also exemplifies the significance for health of private (as opposed to collective) consumption. Food security and food poverty are major issues in poor countries. Jenkins and Scanlan (2001) estimate that more than half of the world's poorest countries have significant problems of food availability for their populations. As discussed in the previous chapter, this is an important contributor to the health differences between rich and poor countries. Access to food supplies is insufficient for some groups because of profound socio-economic inequality, militarism, violence and inadequate food distribution networks. Argenti (2000) commented on urban food distribution networks in the cities of developing countries, which often rely largely on traditional markets rather than modern supermarket retail facilities. These are of inadequate capacity and standard to feed the rapidly growing population in many cases. Poverty and inadequacy of supply cause large scale food insecurity in some of these cities. The International Food Policy Research Institute (IFPRI, 1998) reported, for example, that 24% of households in Accra, Ghana were food insecure. Even where food supplies are more assured, changing patterns of food consumption have potential to change health. Popkin (2000) comments that urban populations in many developing countries are undergoing the 'food transition', a change in dietary habits towards an increased intake of fat and sugars which results in growing public health problems of obesity.

Some of the same problems apply, albeit to a lesser degree, in wealthy countries where the food supply should be ample to provide a healthy diet to the whole population. Dowler and Raats (1998) have suggested that countries like Britain need to learn from the experience of poor countries since there are population groups in Britain for whom food poverty is significant. Poor and disadvantaged groups show evidence of poorer quality of diet and their nutritional status has declined in recent years particularly in terms of the nutritional quality of the food consumed (Dowler, 1997; Dowler and Calvert, 1995; Dowler and McConnell, 1993; Leather and Dowler, 1997; Craig and Dowler, 1997).

There is evidence that in deprived areas, diet tends to be poorer, and this is not entirely explained by the income of individual households. In Clydeside, Scotland, results from the Twenty-07 Study have shown that on average diet was worse in more deprived areas of Glasgow (Ellaway and Macintyre, 1996; Forsyth et al., 1994). Ecob and Macintyre (2000) reported that some of this difference seemed to relate to the area of residence and not just to individual purchasing power. People in higher income households tended to have a poorer diet if they lived in a deprived area. People in low income households tended to have a poor diet regardless of the level of deprivation in the area where they lived. Diez-Roux et al. (1999) examined risks for artherosclerosis, including poor diet, in four differing communities in the USA. They found that unhealthy eating was associated both with low individual income and with low average area income.

Nutritional deficiency in countries like Britain arises to a large extent from low income, but organization of, and access to, food retail outlets also contribute to the problem. This has given rise to discussion about the existence of 'food deserts' in the UK in areas that are poorly served by shops selling nutritious food at affordable prices. 'Food deserts' result partly from trends in the geography of retailing, which is increasingly dominated by supermarkets operated by major retail chains, especially in grocery shopping, and decentralized to suburban or out-of-town locations. For example, Jarvis et al. (2001) discussed these changes as they have affected London, and commented that the trends may disadvantage poorer and less mobile consumers in inner city areas. Jarvis et al. (2001) reported on the hierarchy of retail outlets in the city in 1994. The largest, metropolitan scale centres were all located in the more suburban and affluent outer London boroughs. Other boroughs were almost all served by at least one shopping centre in the next largest category, 'major centres'. The exceptions were the two poorest boroughs, Newham and Tower Hamlets. These two areas have, until recently, been largely dependent on small independent shops and street markets for local grocery supplies. Jarvis et al. (2001) also comment on a new trend, since the early 1990s, for branches of major food retail chains to open in the city centre. Some of these, however, are specializing in the 'high end' of grocery marketing, selling expensive, ready-made meals which are useful for relatively well paid workers in small households, but less so for larger families on low income.

Cummins and Macintyre (2002) report that in Glasgow, Scotland, the cheapest foods available locally in poor areas were of poor nutritional value. Dowler (1997), Dowler and Raats (1998), Dowler and Calvert (1995), Diez-Roux et al. (1999), and Sooman et al. (1993) have also reported evidence, from both Britain and the USA, that access to affordable and convenient sources of healthy food is relatively poor in deprived urban areas, and that this may contribute to unhealthy eating on the part of the local population. Problems of poor access to retail facilities are also likely to be important for poor populations in remote rural areas of Britain where local shops have been closing at a rapid rate in recent years (Great Britain Parliament, 2000; Guy, 1991).

The local organization of food retailing could influence public health in quite complex ways. Accessibility of shops and cost of food for the consumer may be just one part of the problem. Public health may also be affected by the quality of the products and the standards of hygiene associated with their storage and sale, which may vary between different retailers. Consumer information (especially on nutritional value of food) and the range of products available may be important. For example, in ethnically diverse inner cities, consumers may require specific types of products determined by ethnic customs and religious requirements, such as halal or kosher foods or certain imported foodstuffs from particular countries. Small retailers in places like London's East End may cater more specifically for these requirements than large chain stores. Food retailing spaces may also have other functions relevant to public health. Shops are significant as sources of employment for local people and the patterns of recruitment and working conditions will influence health (see Chapter 4). Also, local retail venues can provide centres for social interaction, as discussed below, which may contribute to the social capital of a local community.

The processes influencing local patterns of food marketing and food security also operate at a much wider geographical scale. Lang (1999) argues that we need to consider food economies in the world today as 'hypermarket' economies, with a relatively small number of multinational food distributors exercising considerable power over the global system of production and consumption of food. McKee and Lang (1997) have argued that it is necessary to separate commercial interests in the food industry from responsibility for ensuring health of consumers. Kearns and Barnett (2000) illustrate the association between a highly commercialized international food retailing company and the health service sector, in their analysis of the controversial location of a McDonald's restaurant in the foyer of the 'Starship' children's hospital in New Zealand (see Chapter 2).

Differences in diet may also depend on 'food cultures' and the ways in which food and eating relate to lifestyle and sense of identity. Bourdieu (2000: 186) develops the idea of the 'food space' in which different social groups (defined in terms of cultural and economic capital) adopt patterns of food consumption reflecting patterns of taste that reinforce the lifestyles associated with their position. From a cultural geographical perspective, Bell and Valentine (1997) have also discussed the social and geographical factors which influence patterns of food consumption. Customs and practices relating to preparation and consumption of food are strongly linked to social, cultural and ethnic traditions and so patterns of eating are expressions of cultural identity.

Bourdieu's discussion of tastes in food (first put forward in 1979) makes what today we might consider excessively essentialist distinctions between men and women and between social classes in the matter of eating habits. He suggests, for example, that 'It is in fact through preferences with regard to food . . . that the class distribution of bodily properties is determined', and that

> in the working classes, fish tends to be regarded as an unsuitable food for men, not only because it is a light food, insufficiently 'filling' which would only be cooked for health reasons, i.e. for invalids and children, but also because,

> like fruit (except bananas) it is one of the 'fiddly' things which a man's hands cannot cope with and which make him childlike. (2000: 190)

Bell and Valentine (1997) demonstrate how certain places derive their identity from certain types of food and drink. Diet may be represented as part of one's geographically defined identity. Stereotypical views of people belonging to certain places may include expectations of the types of food they consume. (Stereotypes of beer drinkers in northern England might be one example). Fashions in eating are partly determined by the marketing strategies of the food industry, but also reflect lifestyle aspirations and the type of self-image which the consumer desires. Being seen in locations reserved for eating or purchasing expensive or unusual food can demonstrate that one belongs to an exclusive group. Bell and Valentine (1997) also describe food shopping as a social event and suggest that some retailers play on this behaviour to enhance their sales. Demonstrating knowledge of sought-after forms of food and drink (for example, as a connoisseur of fine wines) can convey aspects of social standing as well as reflecting certain patterns of consumption. May (1996) discusses how some individuals adopt a form of cultural sophistication, using sources of food which they see as 'authentic', such as street markets, rather than ordinary shops.

Some commentators suggest that we are seeing the globalization of culture, and that this is reflected in the widening range of different types of national cuisine available in global cities, and the ways that these are blended to create new, cosmopolitan eating styles. The consumption of foods seen to be characteristic of minority ethnic populations can be associated with globalization of culture in large cosmopolitan cities. However, where consumers seek crude forms of foreign cuisine consumed in poorly furnished settings, it may also reflect racist stereotypes depicting different ethnic groups, and the food they eat, as exotic and primitive (see May, 1996: 59).

The social and geographical dimensions of food consumption are important for health because they help us to understand the dietary behaviour of individuals (Murcott, 2000). Although eating habits are partly a matter of personal choice, these choices are also affected by conditions in the area and community in which individuals live. Furthermore, widely held popular images of food consumption typical of particular places may serve to reinforce local dietary patterns. These relationships are not, of course, limited to consumption of food; similar arguments and observations can be made with respect to other types of health related activity such as smoking (e.g. Brown and Duncan, 2000).

Ideas drawn from social and geographical theory can inform efforts to tackle such inequalities in private consumption which impact on health. Williams (1995) and Cockerham et al. (1997) have pointed out that these interpretations are important because they highlight the constraints placed on individual agency and choice in decisions about consumption. Individual choices are limited by the structures within which actors have to make these choices, so healthy consumption is not solely a matter of individual responsibility. If society wishes all its members to make 'responsible' choices', in matters such as healthy diet, it is important to try to create conditions which make these choices feasible, desirable and acceptable for everyone. This involves some regulation of the geographical distribution and accessibility of retail outlets in rich and poor areas, control over the costs of healthy products and control of marketing strategies, to advertise healthy products as associated with desirable lifestyles, and to clearly label unhealthy products as undesirable.

A MODEL OF LANDSCAPES OF CONSUMPTION AND HEALTH

Geographies of consumption are important to health variation in many ways. Figure 5.9 summarizes some of the relationships described in this

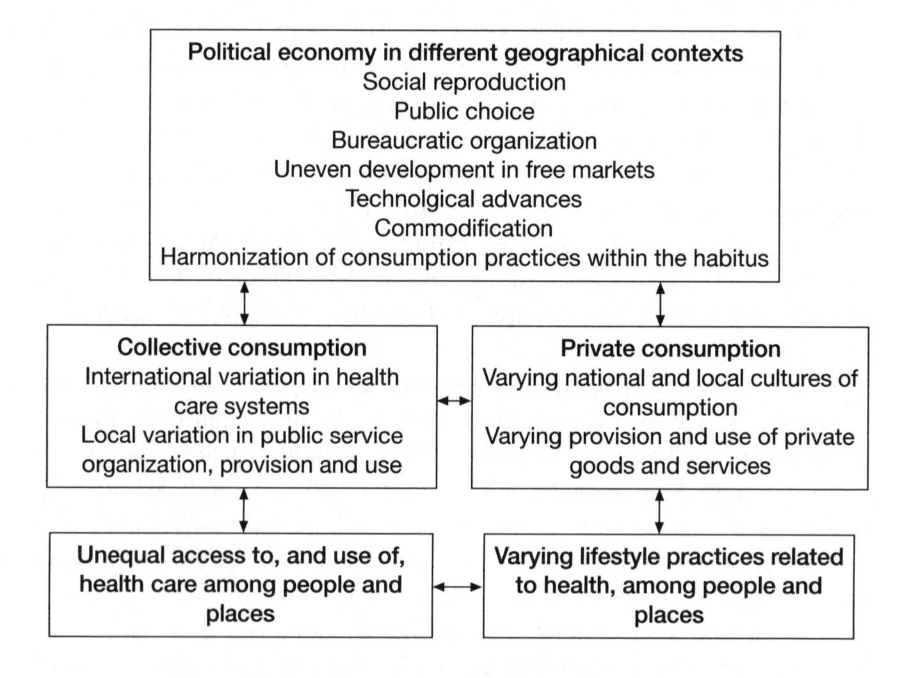

FIGURE 5.9 Geographies of consumption and health inequality

chapter. Conditions in the political economy influence variability in key processes discussed here. They include processes of social reproduction; public choice; bureaucratic organization; uneven market development; technological advances; commodification; harmonization of consumption practices within societies and communities. These result in differences of consumption of goods and services that are important for health. They may be provided collectively, such as state funded health care, or privately, such as food. There are variations between countries, and more locally between areas within countries.

Some geographical variability in consumption relevant to health is desirable, since populations vary in their needs. However, variation that does not correspond with need is likely to perpetuate health inequalities. Therefore studies of geographical differences in patterns of use of health care are often used to assess the performance of health care systems and target health care resources to populations more equitably in proportion to their needs. Geographers are also increasingly interested in theories of consumption which emphasize the social and geographical context of consumption. These show how characteristics of places and communities are reflected in, and reinforced by, individual consumption practices affecting health.

FURTHER READING

A fundamental theme in this chapter is varying access to health services, and national differences in health care systems. For an up-to-date introduction students may also wish to refer to a good review of research on varying access to health services by: Gatrell (2002), Chapter 5, 'Inequalities in the provision and utilization of services', 135–65.

Students may also be interested to see a number of earlier texts that have set out this agenda from a health geography perspective, and to note how geographical interpretations of the issues have evolved over time: Shannon and Dever (1974) Joseph and Phillips (1984); Thouez (1987); Jones and Moon (1987).

For those who are interested in primary care in Britain, a comprehensive discussion is provided in: Moon and North (2000).

For a more detailed discussion of the evolution of the British National Health Service, and a geographical critique of the role of markets in health care, I would recommend: Mohan (1995); Mohan (2002).

Other excellent reviews, providing a theoretical framework for the geography of health and welfare services, are provided by: Pinch (1985); Pinch (1997)

OBJECTIVES AND QUESTIONS

Students should find this chapter useful in learning about:

- conceptual models of social and geographical variation in access to and use of health services;
- the significance of collective consumption and private consumption for geographies of health care;
- the factors which differentiate health systems in different countries;
- the geographical relationships between service provision, accessibility and use and the trends which are bringing about change in these relationships.

Students may like to use this knowledge to answer the following questions:

1 With reference to two specific countries, compare and contrast the main factors which produce social and geographical variations in access to primary health care.
2 What are the implications, for the geography of health care, of the introduction of managed markets into the health care system?
3 To what extent, and why, does what you eat depend on where you live?
4 How will the advent of telemedicine affect the geography of health care?

6 Ecological Landscapes: populations, air, water and ground

This chapter is particularly concerned with risk factors for disease in the physical environment and the biological and spatial attributes of human populations that influence these risks. The relationships between these are presented here as key dimensions of 'ecological landscapes'.

These 'ecological' perspectives are especially focused on biological and chemical pathways of risk for specific, medically recognized diseases. The research reviewed in this chapter particularly concerns variations in space and time in exposure of human populations to these risk factors. Exposure depends partly on spatial proximity to hazards in the environment, so geography as a spatial science contributes to knowledge in this area by describing and analysing the uneven spatial distribution of different human populations in relation to variable environmental risks.

This chapter starts with an explanation of human disease ecology. This leads to a discussion of biological and spatial attributes of human populations that may influence variability in human susceptibility and exposure to biological and chemical risks. Some geographical studies use data on spatial distributions of human populations to model the diffusion of infectious diseases. This approach is reviewed here, using examples of geographical research on HIV/AIDS.

This is followed by a consideration of perspectives from landscape epidemiology, which combine information on human populations with data on natural attributes of the environment. Attention focuses on illustrations from studies of the varying prevalence of malaria.

Next there is discussion of the potential and the limitations of strategies from environmental epidemiology for measuring varying human exposure to chemical pollution. Studies of air pollution and health are used as examples.

The chapter concludes with a discussion of illustrations of the literature concerning environmental justice. These studies are of special relevance to the theme of this book, since they attempt to establish whether unequal exposures to environmental risks for different social and economic groups contribute to the socio-economic health inequalities reviewed in earlier chapters.

ECOLOGICAL LANDSCAPES: CLASSICAL ROOTS AND CONTEMPORARY RELEVANCE

This chapter considers the legacy of some of the earliest research in the geography of health, concerning the spatial pattern of diseases in human populations. This area of the geography of health is at the core of 'classical' medical geography, employing strategies such as analysis of human disease ecology; disease diffusion; landscape epidemiology; environmental epidemiology. These medical geographical perspectives continue to be important in the geography of health and they have more recently been developed using contemporary techniques for mapping and modelling the pattern of diseases in space. More comprehensive discussions have been provided by other authors such as Learmonth (1988), Meade and Earickson (2000), Gatrell (2002) and Cromley and McLafferty (2002). This chapter focuses on selected illustrations, designed to demonstrate the types of concepts and methods employed and their relevance for the debate about health variations.

The theoretical paradigms and methods reviewed here are based on natural and computational sciences and medicine. This is in contrast with perspectives discussed in the previous chapters, which draw on other understandings of health and illness from the social sciences and humanities. The emphasis here on biological attributes of human populations and their spatial distribution therefore contrasts with conceptual frameworks discussed so far in this book, which emphasized social, economic and political pathways and took a broader view of health and illness.

That said, it is not possible to completely divorce these different perspectives, and this chapter illustrates how an understanding of the social and behavioural factors reviewed earlier in this book can help to understand the patterns of ecological environmental risk discussed here. The examples show how evidence from research of this sort may be used to inform public health strategies, with the aim of reducing health inequalities, although in some cases the nature of the evidence makes it difficult to interpret into policy and action.

PERSPECTIVES FROM HUMAN DISEASE ECOLOGY

Conceptual frameworks such as human disease ecology and landscape epidemiology consider the biological and physical aspects of environment and how these interact with human activity to produce variable risk of diseases. Barrett (2000), Meade and Earickson (2000), Howe (1997) and Foster (1992) provide thorough reviews of these, and make the point that this type of approach has a very long pedigree. Meade and Earickson cite Hippocrates, writing about 400 BC who states that:

> when one comes into a city . . . he ought to consider its situation, how it lies as to the winds and the rising of the sun . . . the waters which the inhabitants use, whether they be marshy and soft or hard and running from elevated and rocky situations, and then if saltish and unfit for cooking; and the ground, whether it be naked and deficient in water or wooded and well watered and whether it lies in a hollow, confined situation, or is elevated and cold; and the mode in which the inhabitants live. (2000: x)

Meade and Earickson (2000: 25–39) present the conceptual framework for human disease ecology in the form of a 'triangle' of three interrelated groups of factors, comprising population, habitat and behaviour. A slightly revised version of their model is presented in Figure 6.1. This includes *population* factors comprising biological characteristics of age and sex, and genetic makeup. The physical space–time distribution of human populations is also important in spatial analysis of disease and environmental risk, since uneven spatial concentration and movement in space and time of human populations influence the level and duration of exposure to environmental risks and the probability of disease transmission between people. *Habitat* includes the physical dimensions of the natural and built environment as well as the social environment. *Behavioural* factors in Meade and Earickson's model include social organization, beliefs and technological resources of societies. Foster (1992) also discusses a similar formulation of what he calls the *health field concept*, which considers the interaction of four sets of factors: human biology, environment, lifestyle and health care systems. Two of these groups of factors, human biology and environment, are given special attention in this chapter.

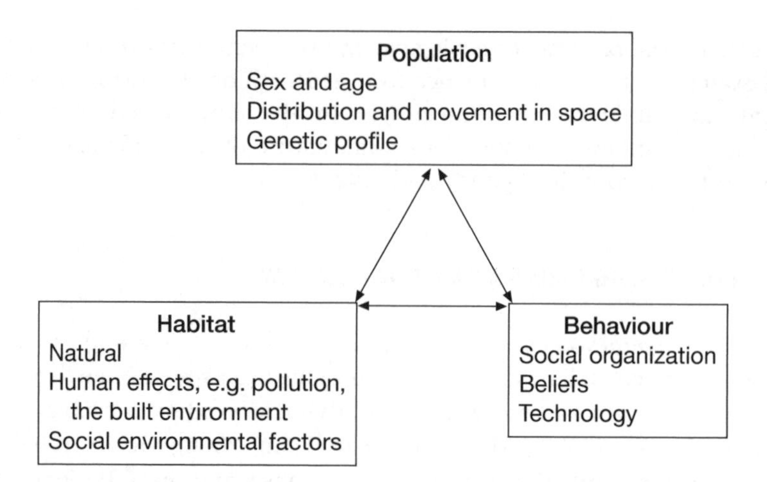

FIGURE 6.1 The triangle of human disease ecology (derived and adapted from Meade and Earickson, 2000: 25)

While groups of factors in human disease ecology are shown here in separate categories, this does not mean they are independent, and this is emphasized by the connections between them, indicated in Figure 6.1. For example, to fully understand the demographic and genetic composition of populations and their spatial distribution it is necessary to consider the social and behavioural factors that help to explain these attributes. Similarly, varying chemical pollution of the environment is closely linked to human organization and activity. The social and behavioural aspects of disease risk, including lifestyle and health care, have been considered in earlier chapters of this book, using conceptual models more suited to understanding these social processes. Many of the studies reviewed in this chapter pay less attention to explaining these social and behavioural factors, and may seek to 'control' for social conditions as 'confounding factors'. In this way, attention is focused mainly on modelling varying exposures to physical environmental risk and biological susceptibility, and the spatial patterns of diseases in populations that result.

BIOLOGICAL AND SPATIAL ATTRIBUTES OF HUMAN POPULATIONS

This chapter focuses on humans as biological organisms, unevenly distributed in space (in contrast with the emphasis in earlier chapters on people as social beings). Let us first consider the biological and spatial attributes of human populations that are often included in research on ecological landscapes. Figure 6.1 summarizes a number of these: demographic composition, genetic attributes and the changing spatial distribution of human populations.

Demographic composition of human populations

Age and sex are important demographic factors for risk of disease and death. In a geographical area, age and sex composition of the population are the product of the key demographic processes of fertility, mortality and migration. Age composition, in particular, shows considerable variability between areas. Most statistical geographical work on health variation seeks to 'control' for these factors, using measures that standardize for differences in age and sex composition of the population, because susceptibility to different diseases varies with age and sex, and mortality is naturally greater in older age groups. Without controlling for variations in demographic composition it would therefore be difficult to establish whether environmental differences between areas are significant for variability in morbidity and mortality.

Age composition is related in quite complex ways to the health status of the population, as shown in the earlier discussion of the epidemiological transition (see Chapter 1). For example, in some parts of the world, mortality is high, often due to low standards of living and poorly controlled infectious diseases. Fewer people reach advanced ages, and fertility rates are high, so populations tend to be more youthful. In countries where infectious diseases are better controlled, populations are likely to be older and to show higher rates of morbidity and mortality due to degenerative diseases.

Genetic attributes of human populations

The genetic composition of the population represents a theme of growing importance in our conceptualization of the biological factors in health geography. Iannoccone (2001) has emphasized that the environmental risks to health considered in this chapter have variable impacts on individuals depending partly on their genetically determined susceptibility. McGuffin et al. (2001) suggest that the main role of environment in varying patterns of disease is in differentiating genetically related people. Thus the combination of genetic and environmental risks will generate greater diversity in health outcomes than would be produced by these effects separately. Many physical diseases involve an inherited aspect of risk, including some that affect large numbers of people, such as heart disease, cancers and diabetes. McGuffin et al. (2001) also draw attention to the wide range of mental diseases and behavioural traits which show distinct genetic variation, including: Huntington's disease, Alzheimer's disease, schizophrenia. They suggest that nearly all behaviours show moderate to high heritability. This means that individual behavioural risk factors for disease may to some extent be genetically determined rather than depending solely on environment and learnt behaviour.

The genetic profiles of people who make up the populations of different areas, and the interaction of these genetic characteristics with environmental factors, therefore have considerable potential significance for the geography of health, though to date the contribution of geography in this field has been quite limited. A body of research has built up which examines the gene geography of populations and concentrates especially on differences in the genetic makeup of populations that have historically been separated, spatially and socially. For example, Howe (1997) discusses the geographical distribution in Britain of populations with different blood groups and links this to the ancient history of occupation of the country by different groups such as Celts, Romans, Anglo-Saxons and Scandinavians. He summarizes evidence suggesting that the susceptibility to certain diseases may vary according to blood group. Work by Schneider et al. (1995) demonstrated that populations in central Europe separated by the Ural mountain ranges have distinct blood groups indicative of a long history of separation and lack of opportunity for genetic mixing.

Thus, in the past, physical distance and topographical barriers effectively separated human populations and created distinct gene pools. These genetically different groups may vary in their longevity and their susceptibility to health problems and environmental risks to health. Growing regional and international movements of populations may affect the pattern of geographical diversity of genetic risks for populations. These processes are particularly evident in urban areas that are frequently the destinations for human migration.

Knowledge of genomics is changing social understandings of identity and risks to health (Kerr and Cunningham-Burley, 2000). For example, advances in genomics have the potential to transform the ways that people conceptualize their ethnic and family identities and their response to health risks. For ethnically (and genetically) diverse populations of major cities the implications of research in genetics therefore have special significance. Race, though culturally important, determines few continuous traits in genetic terms, and genetic makeup is not strongly related to physical appearance, so there is little scientific basis for racist applications of genomics (Paabo, 2001).

Nevertheless certain genetically inherited characteristics, affecting susceptibility to particular diseases, are unevenly distributed among racial groups. Sickle cell disease is often cited as an example of an inherited condition that affects certain racial groups, particularly populations originating from certain African, African Caribbean and Mediterranean regions. It is of particular concern in areas of Britain and the US where there are large concentrations of people in ethnic groups with these origins (Donovan, 1984). While disease is caused by genetic and biological factors, its impact on the person with the disease and their family will depend on social factors including the response of the health services and the social construction of the disorder (see Chapter 3).

Patterns of marriage within relatively small communities can also be relevant for genetic risk factors and this sometimes is evident in minority ethnic populations which show strong social segregation from the wider society in which they live. Overall and Nichols (2001), for example, have examined the impact on health related genetic makeup of close patterns of intermarriage in two Asian communities living in the UK. These populations showed a high degree of intermarriage within a limited pool, defined by their family and social group, through large numbers of marriages between cousins, or by marriage within the same caste. This increased the probability of genetic disorders associated with 'recessive' genes. These disorders occur in individuals who have inherited from both parents the genetic characteristic associated with the disease. The risks in this instance are genetic, and they may not be well understood in the populations concerned. However, the patterns of intermarriage which influence these genetic risks for health are the result of social processes such as those reviewed in Chapter 3, including the social cohesion of the Asian communities concerned and their experience of segregation and exclusion from the majority British population. Thus the biological dimensions of genomics are related in complex ways to the social processes discussed in earlier chapters of this book.

Data on the genetic profile of human populations may provide valuable information about risk of disease, and there may be benefits to be derived from medical or scientific interventions to help reduce the risks associated with genetic attributes. However, there are also some dangers in using genetic risk assessment or manipulating genetic material. One problem is the potential for 'genetic hypochondria' causing people to spend their lives waiting for a disease which never occurs. Another potential difficulty is the possibility of discrimination against people with an above average risk of certain types of disease (Horton, 1996). There is also a risk that genetic components of health risk will begin to disproportionately overshadow attention to environmental factors (Paabo, 2001), so that we will pay insufficient attention to aspects of the social and physical environment that are important for human health and health inequality.

Spatial attributes of human populations

It is necessary to take a dynamic view of human populations in space and time, particularly by considering the effects of migration on biological profiles of human populations and the length of exposure they have to environmental risks in particular places.

Several studies have compared the disease patterns of immigrant populations with those of the host population in the destination country and with the population in the country of origin. These studies aim to explore how far international differences in patterns of human disease may result from environmental factors, as opposed to genetic risk. Some of these studies concluded that populations of recent immigrants have risks for disease that reflect those in their country of origin and differ from those in the host population. However, some studies have also suggested that, over time, the epidemiological profile of these immigrant populations becomes more similar to the host population, and that, in particular, the 'second generation' (children of immigrants) seem to have epidemiological characteristics closer to those of the host population than to their parents. These findings are taken to show that environmental factors are relatively important, since genetic factors do not explain the way that risk of disease seems to change for migrants moving between different environments in different countries.

Balzi et al. (1995) studied Italian migrants to Canada, showing that they had lower risks of certain cancers (of colon, breast and lung) than the host population, but higher risks of stomach cancer. Their children showed risks for these diseases that were more similar to the average Canadian population. McCredie et al. (1999) compared immigrants to Australia from East and South East Asia with Australian born populations. Among recent migrants, resident less than 10 years, nasopharyngeal, stomach and liver cancer rates were higher, but colorectal, breast and prostate cancers were less common, than in the Australian population generally. Immigrants with 30 years of residence in Australia had rates of these cancers that were more similar to the native Australian population.

Moradi et al. (1998) compared Scandinavian immigrants to the USA with US born and European Scandinavian populations. Scandinavian populations living in the US and in Europe had lower risks of lung cancer and higher risks of stomach cancer than US born white populations. Risks for colorectal, breast and prostate cancers approached those of the US population. It was suggested that risks for lung cancer and stomach cancer such as smoking and *Helicobacter pylori* infection might be retained after migration, but that the other types of cancer might have a stronger relationship with environmental factors in the new country of residence.

Stellman and Wang (1994) compared proportional cancer mortality rates among Chinese immigrant populations in New York City with US born whites in the city and populations in Tianjin, China. For these three groups, the study showed the relative contribution of different types of cancer to overall cancer mortality. The pattern for Chinese migrants was intermediate between those for the native US and the Chinese population for several types of cancer.

Tsugane et al. (1999) studied Japanese immigrants in São Paulo, Brazil and compared their cancer rates with those of the local native Brazilian population and rates in Japan. One of their findings was that rates of colorectal cancer were not significantly greater in the migrant Japanese population than in populations living in Japan. Other studies had shown that Japanese migrants to the US had higher rates of colorectal cancer than populations in their country of origin. This might be further evidence of differential environmental effects in different countries producing variable disease risk in genetically similar populations.

Other research on migration has helped to clarify its importance for varying length of exposure to environmental factors. Rogerson and Han (2002) and Boyle et al. (1999) have provided useful discussions of the impacts of internal (within country) migration on environmental exposure and health variation.

Rogerson and Han (2002) argued that in the United States, contrary to common perception, populations were not becoming much more residentially mobile, and that most residential migration takes place over fairly short distances (median distance of 5–10 miles). Also longer distance moves were concentrated especially in certain states that see high rates of both inward and outward migration. Those with especially high migration rates had unusual characteristics, such as Alaska or Hawaii, the District of Columbia, or small western states with small population bases. Populations in such states are quite likely to have experienced environmental health risks in other regions of residence in the past, making it more difficult to judge how their health status relates to their present residential environment. However, in many other states the population has been more 'stable' (comprising long term residents in the region) than might have been supposed. Therefore they will have been exposed to the health risks prevailing in the region over a long period. Rogerson and Han calculated adjusted lung cancer risks for counties of New York State, controlling for levels of population migration. Adjusted and unadjusted risks were different in Orange and Bronx counties, since a large proportion of residents in these counties had lived there less than 20 years. In contrast, Fulton and Queens counties, with more stable populations, had rather similar adjusted and unadjusted risks. The variable impact of migration affected the comparability of the risk of lung cancer for county populations.

Boyle et al. (1999) examined variation in standardized morbidity ratios (based on information on long term illness reported in the census). They studied morbidity differences in small areas of England and Wales in relation to deprivation and population migration, finding that short term population migration (measured in terms of proportions moving in the year preceding the census) was associated with morbidity rates as well as deprivation. They point out that some of the most depressed areas, with high levels of morbidity, also had high levels of stability. They suggest that analyses of local variation in health may need to pay more attention to differences in rates of population migration as well as socio-economic conditions.

These two studies suggest that populations in depressed industrial areas of the US and England and Wales tend to have relatively low levels of population turnover and inflow. These areas also often have rather poor socio-economic status and are disadvantaged in terms of health status. If environmental factors in these areas are also detrimental to health, the populations living there are likely to have been exposed to these factors for a significant period of their lives. This would increase the chances that physical environmental risks may contribute to the poorer health of populations in poor areas. This possibility is explored in more detail below, in the light of literature concerning environmental justice.

THE SPREAD OF INFECTIONS IN SPACE AND TIME: SPATIAL DIFFUSION MODELLING AND THE EXAMPLE OF HIV/AIDS

For some communicable diseases, contact between people is of primary importance in the risk of contracting the disease. Geographical proximity of infected and non-infected populations can be powerful predictors of the chances of infection. Spatial modelling of disease diffusion is especially relevant for the study of such diseases. These approaches are based on a long established stream of work in geography, concerned to observe and to simulate the temporal and spatial spread through societies of human phenomena such as innovations, ideas or contagious diseases

(Hägerstrand, 1952). Models that accurately simulate the geographical spread of diseases in the past can provide the basis for predictions about how new infections will spread in future. This may be useful for planning interventions in public health and medicine to control communicable diseases (Haggett, 1994). However, these models cannot predict risks of infection for particular people because this depends on individual characteristics and behaviour.

These models often employ basic information about the spatial distribution and mobility of human populations. They may also incorporate other information that estimates the susceptibility of the population to infection. Predictions are made using mathematical expressions of spatial diffusion processes that are typical of infectious diseases. Diseases tend to spread in time and space, outwards from their source, in ways that can be predicted to a large degree from the spatial distribution and movement of human populations. Communication is more likely between people who are geographically close together than between those who are separated by large distances. This relationship is expressed as some mathematical function of the distance between places, such as the squared value of distance, to reflect the increasing impact of distance decay effects in human contact as distance increases.

The total numbers of contacts between persons are greater among dense concentrations of people in big cities than among sparsely distributed populations in small settlements. However, there may be barriers that inhibit communication in some directions more than others. These can include topographical features such as water masses or mountain ranges that are difficult to cross. This means that the 'friction of distance' effect can be irregular in space, depending on the distribution of such physical barriers. (Also, social barriers to communication, discussed in earlier chapters of this book, may inhibit the spread of diseases, even over relatively short distances, but these are usually more difficult to represent in spatial diffusion modelling).

Human disease diffusion in relation to distance is complex, since patterns of diffusion are hierarchical through city systems. Rates of communication are sometimes more rapid between large cities in different regions than among smaller settlements within the same region. New infections are very often introduced to a region via the 'highest order' city, which has the largest population, the greatest rates of contact with areas outside the region and a dominant position within the region. The discussion of globalization in Chapter 1 considered how space-time compression is making movement between major cities in very distant places increasingly common. The growing human traffic between world cities is important for the spread of diseases, as it is for other human phenomena.

There are many examples of geographical approaches to ecological landscapes employing spatial analysis and disease diffusion modelling (see also Meade and Earickson, 2000; Cromley and McLafferty, 2002). Such diffusion models have been used to study the spread of various infectious diseases, including influenza and measles, which are readily communicated between people (Cliff and Haggett, 1988; Haggett, 1976; Pyle, 1969; 1986). More recently, similar techniques have been used to describe and model the global and regional spread of the HIV (human immunodeficiency virus) and the resulting forms of morbidity referred to as AIDS (acquired immunodeficiency syndrome). This is a suitable candidate for diffusion modelling because the AIDS syndrome is caused by a virus which is passed directly between humans as a result of physical exchange of semen or blood (mainly through unprotected sex, shared use of hypodermic needles, medical transfusion of blood which has not been screened for HIV, or transmission from mother to foetus in the womb). Disease diffusion studies do not provide direct information about the underlying *reasons* for the geographical pattern of infection, but careful analysis of diffusion patterns can lead to hypotheses about the behavioural, social, political and economic factors likely to explain the development of the epidemic. The models sometimes include indicators representing social and behavioural factors that may affect susceptibility to infection, drawing upon other research of the type already reviewed in earlier chapters.

Spatial diffusion studies of HIV illustrate the tendency for disease to spread outwards from a core area, subsequently reaching other nearby populations. Large, 'high order' cities, and especially global cities, which dominate the national urban system are often the 'gateways' through which infectious diseases are introduced to a region and are the sites of the 'core areas'.

Smallman-Raynor et al. (1992) have recorded the global diffusion of HIV in detail, showing that it seems to reflect the geography of lines of international communications between global cities. The history of the worldwide spread of the disease showed that the first cases of infection in a country were often reported in its main urban centres and could sometimes be traced to infected individuals moving between countries. Gould's (1993) maps of the geographical spread of HIV infection and AIDS in the USA in the 1980s showed that the areas of initial infection were in large, widely separated cities in eastern and western parts of the country, and that infection spread into more rural regions from these 'core' areas of infection. Other sexually transmitted diseases also show this pattern. For example, at the scale of a single city, Becker et al. (1998) used geographic information systems to examine the pattern of gonorrhea in Baltimore, USA. The study showed a concentration around a 'core' zone of 13 spatially contiguous areas with the highest rates. As radial distance from this core increased, the rates declined.

While spatial diffusion modelling describes these patterns of diffusion, it does not fully explain the pathways of diffusion involved. Other research has helped to elucidate the processes producing these spatial diffusion patterns. The geographical pattern of AIDS may increase partly because inward migration of already infected people increases the total level of prevalence of infection in receiving areas. For example, Cohn et al. (1994) found that most HIV positive patients in North Carolina lived in rural communities and a substantial proportion had moved into the state after being infected with HIV elsewhere. The most commonly cited reason for moving into North Carolina was that migrants were seeking social support. Gatrell (2002: 187–8) reviews other evidence from research by Ellis and Muschkin (1996) and Graham et al. (1995) which also suggests that people with HIV/AIDS may move into rural areas seeking social support, especially from their family.

The movement of infected persons into areas previously free of infection also gives rise to new cases through contagion, and there is often debate about how social and behavioural factors may influence this. For example, Gould's (1993) discussion of the diffusion of AIDS in sub-Saharan Africa emphasized contagion along the trucking routes linking East and West Africa as result of sexual contacts between truck drivers and people living in the communities along the main east to west transport routes. However, Hunt (1996) pointed out that there is actually little hard evidence to support this thesis, and cited Smallman-Raynor et al. (1991) who favour an alternative explanation based on the role of movements of military troops. Hunt (1996) also discusses a third theory, concerning the breakdown of traditional family relationships in rural communities due to separation of married people under systems of migrant labour. This is associated with changes in sexual behaviour and an increase in the number of people having sexual relationships with multiple partners.

Movement of populations associated with economic activity may also be indirectly linked to spread of AIDS in other countries. In the USA, Wallace (1990) and Wallace (1991; 1993) has examined the more local spread of the infection in New York City and the surrounding area (see also Chapter 3). They showed that prevalence of AIDS was highest in poor inner city areas of New York City. In suburban areas, prevalence was greatest where rates of commuting to New York City were highest.

Obviously commuting to work would not in itself constitute a risk for HIV infection. However, the increased rates of human contact between the central urban area and the commuter suburbs appeared to be associated in some way with a greater chance of spread of the infection. For example, there might be an increased probability that commuter populations will engage in risky behaviours that cause infection to spread from the core area, or people infected with HIV may migrate to commuter suburbs.

Information about the core areas of infection, the spatial distribution of susceptible populations and the rates of population mobility between areas has been also applied in models to predict the future spread of HIV. For example, Löytenen and Arbona (1996) reported on a model to predict the spread of HIV and AIDS in municipalities in Puerto Rico. Their model incorporated information about the population size of municipalities and also data on population mobility. They also used data on income per capita, since they considered that poor populations were more susceptible to social factors exacerbating risk of infection. The level of infection with syphilis was included as an indicator of behavioural risk of sexually transmitted infections. These variables were used to statistically model the varying risk of HIV infection in different municipalities, illustrated in Figure 6.2. Using this model of risk of infection, the rate of spread of HIV was predicted over the period 1993–7. The fastest rate of infection initially was in the larger towns, demonstrating the importance of the

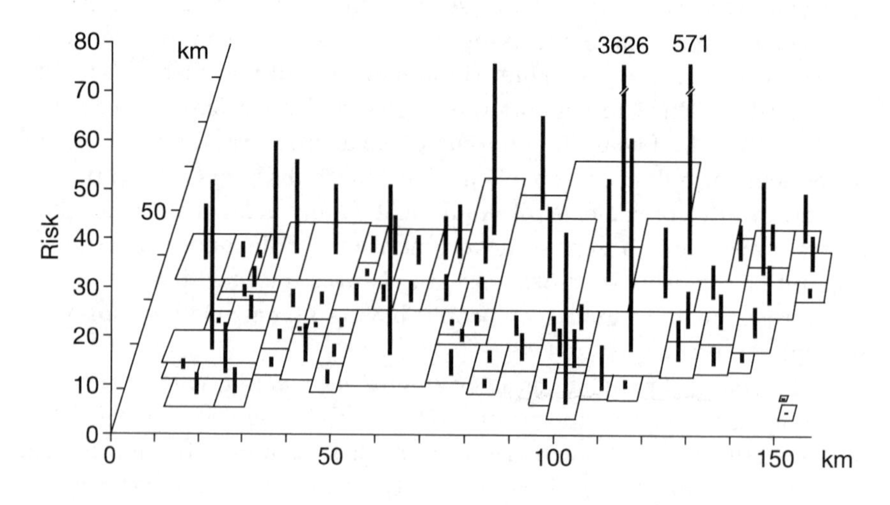

FIGURE 6.2 Varying risk of contracting HIV and AIDS in municipalities in Puerto Rico (reproduced from from Löytenen and Arbona, 1996: Figure 5, with permission from Elsevier)

urban hierarchy in the diffusion process. Subsequently the infection was estimated to have spread out to other areas. The model suggested that diffusion of the HIV infection to more rural areas was slower, and communities on small islands off the coast of Puerto Rico were 'protected' by virtue of their physical separation from the mainland, which limited communication.

These examples illustrate the significance of population distribution and population migration for our understanding of the geography of health. To some extent it is possible to summarize the likelihood of diffusion of diseases mathematically, on the basis of quantitative information on population distribution and likely migration patterns. This kind of approach draws upon 'social physics' traditions in geography, which have modelled human behaviour almost as though it was a natural phenomenon similar to gravitational forces, flows of water or atmospheric circulation. These perspectives provide insights into the population factors in disease ecology which are important in defining spaces of varying risk.

These examples of HIV diffusion also show the connections between spatial analysis and research on social, political and economic processes already considered in earlier chapters. Such processes depend on political ecology, political economy, social relations and individual human behaviours. Even the most sophisticated spatial modelling does not fully represent these processes, so it can only approximate the actual spread of disease at a general level. In order to fully explain disease diffusion it is also necessary to understand the variability of social factors influencing human behaviour in space and time, using research strategies discussed in earlier chapters of this book.

THE IMPACT OF THE NATURAL ENVIRONMENT: LANDSCAPE EPIDEMIOLOGY AND THE EXAMPLE OF MALARIA

The spatial analysis of HIV discussed above mainly concentrates on attributes of human populations that predict risks of disease transmission. Other research on human disease ecology needs to focus strongly on non-human attributes of the environment as well as human factors. This is illustrated by the ecological landscapes associated with malaria. A useful conceptual framework is provided by landscape epidemiology. This is concerned with the environmental aspects of places that are significant for varying disease risk. It links classical ideas from geography, about what makes places distinctive, with ideas from epidemiology (the science concerned with the 'natural history' of diseases and their pattern in populations). Landscape epidemiology has provided a way of classifying habitats in terms of disease risk. Meade and Earickson (2000) attribute the original idea to a Russian, Eugene Pavlovsky, who used it to explain the geographical pattern of diseases. According to Meade and Earickson (2000), the conceptual framework involves ideas of *biomes*, *realms of evolution*, *realms of culture* and the natural *nidus* for a disease. The biomes are biotic regions comprising certain topography, fauna and flora, which result from particular combinations of latitude, altitude, climate and geology. These need to be seen as dynamic and changing because of processes such as emergence of new species of animals and plants and migration of species. The landscape epidemiology approach tends to emphasize biotic and topographical features of the natural environment. However, it also involves concepts of culturally distinct human populations living in different parts of the world with beliefs and behaviours which influence population health. A nidus or focus for a disease is an area where the disease is constantly present because of the local conditions which favour its maintenance. The nidus provides a source of the disease, from which it may spread to other areas.

The thinking of Jacques May (1958) has also influenced studies in landscape epidemiology. May proposed a model (see Figure 6.3) which classified the different elements involved in transmissible diseases. These included the *agent* which causes the disease (often a micro-organism such as a virus or bacterium), the human *host* who is made ill by the disease and involved in transmission of the disease to other people, and *vectors* which carry the disease agent to the human host (these include animals and insects which may act as intermediate hosts, harbouring the agent during part of its life cycle). Other factors in May's model include *geogens* (aspects of the environment which create the conditions for disease transmission) such as water, vegetation and climatic factors.

The transmission of malaria may be interpreted in terms of Jacques May's model. Learmonth (1988) presented the complex set of factors which come into play, including the agent, the *Plasmodium* protozoa (single cell parasite), acting on the human host, who develops malaria as a result. The *Plasmodium* parasite is transmitted between people when the vector, a female mosquito (especially of the *Anopheles* variety), takes a blood meal from an infected person, and in the process intakes the *Plasmodium*, which can then be injected into the bloodstream of an uninfected human host which the mosquito bites later. This complex set of processes is summarized in Figure 6.4. Learmonth's original versions of this diagram showed that some factors involved would be variable

Agent
Causes the disease,
e.g. *Plasmodium* protozoa causing malaria

Vector
Transmits the agent to the host,
e.g. *Anopheles* mosquito in malaria

Geogens
Physical environment factors,
e.g. humidity, temperature, vegetation for malaria

Host
Human made ill by communicable disease,
e.g. malaria

FIGURE 6.3 May's model of the elements in transmissible disease epidemiology, using the example of malaria (based on May, 1958; Learmonth, 1988)

between countries, particularly the types of intervention which were undertaken to try to control malaria.

Symptoms of malaria in the human host can include fever, shivering, pain in the joints, headache, vomiting, convulsions and coma, and the disease exacerbates the effects of anaemia. If left untreated, the condition can progress to severe forms and cause death, especially in the case of malaria caused by one species of the parasite (*P. falciparum*).

The geographical extent of areas affected by malaria has been reduced since the 1950s, but the disease continues to be of serious concern internationally because it causes large scale morbidity and mortality. About 40% of the world's population live in areas where malaria presents a public health problem (in about 90 countries). There are between 300 million and 500 million cases of the disease each year (of which over 90% are in sub-Saharan Africa) resulting in an estimated 1 million deaths annually, the majority of which are among young children. About 3000 children under five die each day due to this disease. The social and economic toll is considerable, estimated at around 1–5% of gross domestic product in Africa, for example (WHO, 1998a).

A range of factors in the natural environment influence the risk of transmission, such as climate, topography, vegetation and the availability of stagnant water providing breeding grounds for the mosquito. Also, factors determined by human behaviour are crucial, including personal protection by prophylactic medical treatment, or use of measures such as

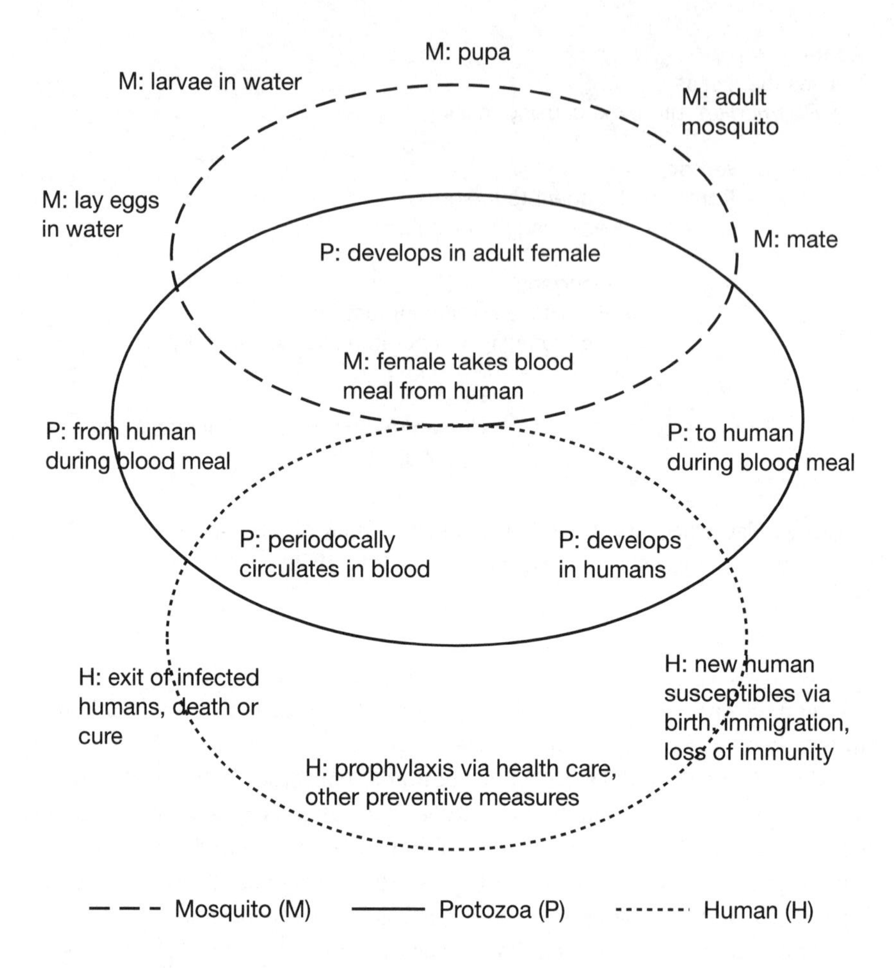

FIGURE 6.4 Interlocking life cycles involved in transmission of malaria (after Learmonth, 1988: Figure 10.6)

mosquito nets, repellents and protective clothing. Human genetic characteristics influence susceptibility as well. One reason for the persistence of the gene causing sickle cell disease in African and Mediterranean populations may be because it provides some protection against malaria. Environmental management also plays a role, including use of insecticides and removal of uncovered stagnant water sources.

There is evidence that the geographical range of the disease may be extending back into areas where it had previously been controlled. Factors in this re-emergence include environmental change due to human activities like forest clearance, mining and agricultural irrigation, as well as the impact of global warming. Unplanned developments in urban areas of low income tropical countries are also areas where the risk of the disease is increasing. Other factors include: the development of more drug resistant strains of the disease; failure of health services to provide comprehensive prophylaxis and treatment; and movements of populations.

Human migration can also be significant for changing patterns of malaria. International movements, such as those arising from globalization of business and leisure activities, can also contribute to the pattern of malaria. Britain saw 2364 cases of 'airport' malaria imported by travellers in 1997, for example. Population movements within countries affected by malaria also influence the spread of the disease. For example, in some low income tropical countries it has been noted that migrant city workers returning for weekends to their rural family homes are at risk of contracting the disease and potentially infecting others on their return to the city (WHO, 1998a; Crail, 1999).

There is some debate over whether forced migration, caused for example by famine or war, has had a significant impact on the distribution of malaria. Kazmi and Pandit (2001) reported that the influx of Afghan refugees between 1972 and 1997 into the North West Frontier Province of Pakistan has been associated with a shift in the broad scale spatial pattern of malaria in the region, with prevalence rising in the areas receiving refugees. However, Rowland et al. (2002) have questioned this finding because, they argued, the data used by Kazmi and Pandit were not sufficiently comprehensive to record all cases of malaria. Rowland et al. (2002) suggested that additional risks for refugees showed a more complex pattern, depending on the local risks of the epidemiological landscape where groups of refugees were staying, such as local climate and proximity to water providing breeding grounds for mosquitos.

Understanding these essentially geographical aspects of malaria can contribute to efforts to combat the disease. The Global Malaria Control Strategy promoted by WHO (1998a) involves four basic elements:

1 provision of early diagnosis and prompt treatment for people with the disease;
2 planning and implementing selective and sustainable preventive measures (including control of the vector mosquitos);
3 early detection of major outbreaks of the disease to prevent or contain epidemics;
4 strengthening of local research capacities to promote regular assessment of the malaria situation in countries that are affected.

WHO (1998b) advocates a range of measures to control malaria in countries like Africa. These include upgrading health delivery systems;

Box 6.1 *Modern approaches to epidemiological landscapes: using GIS methods to estimate geographical variation in risk of malaria in Kenya*

Snow et al. (1998) describe an example from Kenya of the use of geographical information systems (GIS) to estimate the likely prevalence of malaria, and its impact in terms of illness, death and use of hospital care, in different parts of the country. The work draws on epidemiological theories and evidence which provide an understanding of the factors in the natural environment which influence transmission of malaria.

Transmission requires the survival of the vector mosquito between at least two contacts with human hosts. This depends on climate and other aspects of the natural environment, and survival is reduced in areas with low rainfall or low temperatures. Rainfall and temperature data over 60 years were collated from records of weather stations. Vegetation levels were also measured using satellite sensor data. A GIS was used to bring these sources of data together to describe varying physical environments (epidemiological landscapes) across the whole country.

Estimates of persistent levels of prevalence of malaria in human populations were produced for populations of some areas of Kenya where epidemiological surveys had been carried out. The relationship between varying geographical conditions and malaria prevalence in the surveyed areas was established. On the basis of this relationship, and using the country-wide data on physical environment, a GIS was then applied by the researchers to estimate likely levels of transmission of malaria in different geographical areas over the whole of Kenya. Population data were used to calculate the distribution of population in relation to areas with different estimated prevalence. High transmission areas were those with highest risks of contracting malaria. In these areas young children were especially susceptible.

continued ⇒

increased use of bed netting (nets coated with insecticide); developing new drugs for victims already infected with malaria and coordinating the development and testing of new malaria drugs and vaccines; developing methods to address malaria in emergencies (e.g. refugee and post-war situations). Because the disease is so strongly influenced by the interaction between people and their natural environment, geographical perspectives are very relevant to research and strategies aiming to control the disease. The WHO campaign to 'roll back malaria' also includes the use of geographical mapping of the disease to assist prevention and control measures (WHO, 1998b).

Geographical methods for mapping and modelling malaria show some parallels with disease diffusion modelling discussed above. However, studies of malaria need to incorporate more detailed data on the natural environment, as well as information on human populations. Authors such as Bergquist (2001), Coetzee et al. (2000), Snow et al. (1998), Manguin and Boussinesq (1999) and Thompson et al. (1996) describe how

Box 6.1 Continued

Using these results, together with information on levels of morbidity, mortality and hospital use typically associated with malaria in young children, it was also possible to estimate the likely burden of disease and death and the use of hospital resources. High and low stable transmission areas contrasted as follows:

Burden of illness and use of health care due to malaria in areas of Kenya with different levels of transmission of the disease

	Level of transmission	
	Stable and high	Stable and low
Mortality per 1000 children aged 0–4 years	9.77	0.55
Hospital admission rates per 1000 children aged 0–4 years	32.30	21.38

The estimates could not be expected to be completely accurate, partly because the models used rely on statistical estimates, not direct measurement of prevalence of malaria in all parts of the country. Furthermore, the models used do not allow for geographical variation in human behaviour. However, this example does illustrate how information on physical attributes of the 'epidemiological landscape' can be applied using modern computing methods to assess areas most likely to be at risk of malaria and to estimate the likely impact of the disease and the cost of providing treatment of malaria.

data acquired by remote sensing, combined with geographical information systems (GIS), make it possible to generate very accurate geographical information on key environmental factors such as temperature, rainfall and vegetation cover. This information on the natural environment, combined with data on distributions of human populations, can help to identify the areas where diseases like malaria are likely to be endemic.

Snow et al. (1998) used a combination of epidemiological survey data, meteorological records and satellite information on vegetation cover, brought together using GISs, to estimate the likely levels of malaria prevalence in different parts of Kenya. They were also able to estimate the likely burden of disease among young children, who are particularly susceptible to malaria, and the level of hospital use which was likely to result (shown in Box 6.1). In a similar example, Booman et al. (2000) discuss the use of GIS to plan a malaria control programme in South Africa. By mapping the notified cases of malaria, the authors were able to show the varying spatial pattern of the disease and identify spatial clusters in certain areas. There was a conspicuous east-to-west gradient, with cases increasing towards the border with Mozambique. This information made it possible to target malaria prevention strategies more effectively.

Carter et al. (2000) reviewed a number of other examples of the use of geographical information to assess the risk of malaria in populations and target resources to fight the disease more effectively. Beck et al. (2000) commented that new generations of remote sensor systems combined with increased capacity of widely available GIS could mean that in future this type of modelling could extend beyond the research community and become part of routine disease surveillance and control systems. However, Thompson et al. (1996) pointed out that, while this type of mapping method can identify areas where transmission of the disease is most likely, it is still necessary to have information on the detailed local ecology, and the species of mosquito likely to be involved, in order to have an accurate picture.

Furthermore, the natural environmental factors which are important for malaria are not static but are changing. Martin and Lefebvre (1995) commented on how malaria has changed its geographic range significantly in the past in response to climatic change, and that therefore global warming is likely to lead to spatial shifts in the pattern of the disease. They discussed the use of modelling techniques, based on information about different scenarios for climatic change, which suggest that there will be a growth in seasonal malaria and that this will affect populations which have not previously been exposed to the disease and have low resistance. It may become more common in wealthier countries in northern regions of the globe as well as in the poorer southern countries affected at present.

Human behavioural factors important for malaria transmission may be more difficult to map effectively. For example, Carter et al. (2000: 1406) cited the following domestic factors having a significant bearing on malaria risk in Ethiopia:

- type of roof and eaves of dwelling;
- whether the dwelling has windows;
- number of people sleeping in a room;
- whether there is a separate kitchen;
- whether there are animals sleeping in the house;
- use of irrigated land.

Since behavioural factors are likely to vary in association with economic and social position, they may contribute to socio-economic inequalities in risk of infectious diseases like malaria. Bastos et al. (1999) discussed the links between HIV/AIDS and malaria distribution in Brazil. These diseases were both seen to be expanding in certain areas and populations. Malaria has recurred in Rio de Janeiro, which had been free of the disease for 20 years. Outbreaks of both diseases occurred in the industrialized area of São Paulo, apparently associated with needle sharing among injecting drug users. The authors suggested that risks of malaria are exacerbated in 'bridging areas' which connect undeveloped forest zones of Brazil, where malaria is endemic, with modernizing urban areas where injecting drug users are more concentrated.

Winch (1998) argued that some of the lessons learned from the HIV epidemic also apply to changing prevalence of malaria. The populations lacking basic human rights, such as control over their land, access to water and sanitation and political empowerment, are least able to control risk factors in their environment or to protect themselves against these risks. They are therefore most likely to contract HIV or vector borne diseases like malaria.

Thus malaria provides a powerful illustration of the ways that the biological and topographical environment, interacting with human activity, can shape the geographical distribution of a disease, and how a landscape perspective can contribute to measures of disease surveillance and control. It exemplifies the interplay of fauna and flora of a region with climate and topography, as well as human effects on the local ecology and the distribution and movement of human populations. An important aspect of the research reviewed here is the potential for identifying *areas* (as opposed to individuals or social groups) toward which efforts to combat malaria can be targeted to make more effective use of resources. The geographical modelling described above provides a scientific and quantitative way to identify *spaces of risk* for the disease.

The geographical perspectives on malaria mentioned here also illustrate how the classic landscape epidemiology approach has evolved with new technology and methods. Geographical approaches are useful in the fight against malaria because the disease is quite well understood from a medical and epidemiological point of view, giving a clear idea of the physical environmental factors which are known to be important. At the same time, most commentators highlight the need to take into account social and behavioural factors, which have been discussed in more detail in earlier chapters of this book. It is the interactions between individual people, societies and their environment which are crucial for the geography of this disease.

THE IMPACT OF THE POLLUTION SYNDROME: ENVIRONMENTAL EPIDEMIOLOGY OF AIR POLLUTION AND RESPIRATORY HEALTH

Some research on the environmental factors important for disease focuses on problems which are less well understood than the situation for malaria. This type of research often falls into the category of *environmental epidemiology*, defined in a report cited by Needleman as 'the study of the effect on human health of physical, biologic and chemical factors in the external environment, broadly conceived' (1997: 263). The example considered below is research on the impact of air pollution on respiratory health. Researchers in environmental epidemiology aim to discover the risks to health associated with aspects of the physical and biological environment. Geographical perspectives and methods are often employed. However, studies in environmental epidemiology are fraught with complex problems from the scientific point of view. Needleman (1997) suggested these include the following questions: 'What is considered environmental?', 'What counts as credible research?', 'What does applied epidemiology mean in the context of environmental health?'

Problems of how to define environmental conditions arise because they can potentially be very broad in nature and in scale. Since interactions between the physical and the social dimensions of environment are likely to be important, a concentration on the physical and biological dimensions alone will only provide partial explanations of health variation. Furthermore, environmental factors may operate at the micro level (e.g. within the home) or at the macro level (e.g. over whole regions or even at the global scale). It is possible to study these effects using analyses at the level of whole populations and the broader ecosystem level, or by means of studies of individuals, or even at the molecular level to examine physiological processes. Pekkanen and Pearce (2001) recommended making studies at all these levels, starting from the population/ecosystem level in order to make sure that the different aspects of environmental risk are kept in perspective. This implies that a geographical framework taking a 'whole system' perspective, including these broad effects (as discussed in Chapter 1), is a suitable starting point for research in environmental epidemiology.

Further difficulties occur because of controversial questions about how to interpret the findings of scientific research. Chapter 1 considered issues of the ecological fallacy and the atomistic fallacy, which are relevant for research in environmental epidemiology. Studies in environmental epidemiology often search for environmental effects on health by using areas as the units of analysis and testing the ecological correlation between local disease rates and pollution levels. However, neither exposures nor health status are necessarily the same for all individual residents of an area. Even if, hypothetically, exposures were the same for all members of a local population, individuals would be likely to have differential susceptibility; not all of them would react to environmental hazards in the same way or to the same degree. Thus, for example, children are often considered to be more susceptible than adults to environmental hazards. People with different genetic predispositions may be differently affected by environmental factors.

Furthermore, populations are exposed to a range of possible health hazards in the environment. It is important to identify and quantify the whole history of exposures. It may not be easy to attribute health variation to a single type of hazard. The exact physiological processes by which a hazard may affect human health are not always very well understood. Thus the associations found between environmental factors and health often provide limited understanding about *why* these associations occur. This situation is referred to as 'black box epidemiology' (see Figure 6.5). Needleman (1997) cited Susser (1996) who called for a different way of thinking about causation, represented by the metaphor of 'Chinese boxes' with different theories of causation nested inside one another.

Pekkanen and Pearce (2001) also argued that environmental epidemiology is complicated because of the need to study different and interrelated exposures to potentially harmful environmental factors which often appear to carry relatively small health risks. Many of the associations being tested in environmental epidemiology are relatively weak in statistical terms (Hemon, 1995). Hemon distinguished between inherently weak associations and cases of 'diluted risk', which can in fact be important, but appear small because the design of epidemiological survey methods does not allow them to be detected.

The relationship between air quality and respiratory health is a good illustration of these complex problems addressed by environmental epidemiology. This issue exemplifies what Meade and Earickson (2000)

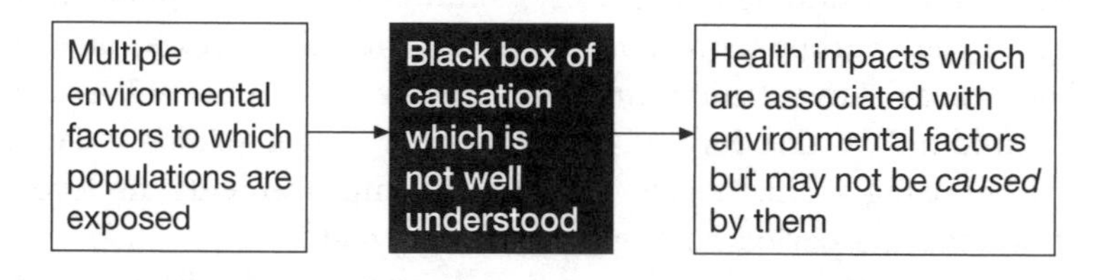

FIGURE 6.5 Black box epidemiology

referred to as the 'pollution syndrome' – the combination of environmental changes which arise from human development. The pollution syndrome is often most severe in urbanized and industrial areas, so that it has a special relevance for health in cities.

The problem of air quality and respiratory health is an important issue given that in many parts of the world air quality now falls short of the standards which are thought to be acceptable for health. Developed industrial nations face significant problems in this respect. For example, Iannoccone suggested that 'arguably the most important environmental challenge facing the American population in the next 25 years is air pollution' (2001: A10). This may even be an issue in areas which appear to meet accepted standards for maximum levels of pollution. Fairley (1999) reported a study on daily trends in mortality due to respiratory and cardiovascular disease in relation to estimated levels of air pollution in Santa Clara County, California, USA. The study showed that respiratory mortality was higher during periods of greater pollution with sulphates and carbon monoxide, even though average levels of these pollutants are lower than in other parts of the USA and fall within accepted air quality standards. It was suggested that data on average levels of pollution masked the variability in the pollution level in Santa Clara County, and failed to detect short periods of high pollution levels which were significant for health.

The question of air quality is even more pressing in parts of the world where rapid and large scale industrialization in recent decades has been realized with little in the way of environmental management or regulation. For example, Jedrychowski (1995) and Krzyzanowski (1997) drew attention to the high levels of environmental pollution in Eastern Europe, arising from a history of environmental mismanagement over several decades. These resulted in relatively hazardous levels of various forms of pollution, including air pollution due to sulphur dioxide and particulates. In many parts of Eastern Europe air pollution levels exceed levels considered safe for health, and are worse than in Western Europe, even though economic restructuring and related changes in industrial activity have produced some improvements in air quality. Drgona and Turnock (2002) examined levels of pollution due to sulphur dioxide, nitrogen oxides, carbon monoxide and particulates in Slovakia since 1985. Although these showed some improvement by 1997, they still exceeded levels found in Western Europe. In the case of nitrogen oxides and carbon monoxide, rising traffic levels meant that improvements in levels of pollution over the period had been relatively small.

Establishing the impact of air pollution on human health is scientifically challenging. Jedrychowski (1995) suggested that while there is some evidence that short term high levels of these pollutants can have an impact on the health of vulnerable populations, the poor scientific standard of the studies makes interpretation difficult. Duhme et al. (1998) pointed to the increasing trend in the prevalence of asthma over recent decades in several countries. They argue that this could result from a number of different factors, including changing levels of outdoor pollution. However, they saw little clear evidence from epidemiological studies for a causal connection between asthma and levels of pollutants such as sulphur dioxide or particulates in the atmosphere, so the question of whether this pollution is associated with asthma is unresolved scientifically. They argued that we should also be considering the effects of indoor as well as outdoor pollutants, and of other lifestyle factors.

Authors such as Krzyzanowski (1997), Gatrell (2002) and Cromley and McLafferty (2002) have discussed systematic strategies to assess the extent of exposures and their effects on health. These involve: measuring the type and amount of the pollutant released; assessing the likely exposure of the population; assessing the ways this may affect health; and assessing the risk of damage to health. This approach would help to identify which are the most damaging pollutants from the health point of view, and which are the most vulnerable populations, so that action can be especially targeted towards these.

Geographers have contributed to the research in this area. Dunn and Kingham (1996) discussed research relating to air pollution levels and health in northern England. Their discussion highlights several of the issues already considered above. One of these is presented in Box 6.2, illustrating how studies of air pollution and respiratory health can be difficult to design in ways that provide a complete and accurate view of the factors involved. They also commented that a geographical perspective can provide a useful framework for considering the interaction of the factors which are important for respiratory health within a particular place. This helps to integrate and interpret the results derived using approaches employed by toxicologists (focusing primarily on varying levels of pollution) and epidemiologists (focusing primarily on variations in health and their relation to risk factors, especially at the individual level). Dunn et al. (2001) also reported on the use of different analytical methods to test relationships between respiratory health and air pollution. Epidemiological comparison of sites, GIS modelling and raised incidence modelling gave different impressions of the risk to health from proximity to a factory producing volatile organic compounds. They recommended using a combination of methods in this type of study, to derive a more complete picture.

Box 6.2 An example of environmental epidemiological strategies for studying the relationship of asthma to air pollution

Dunn and Kingham (1996) reported on a study to examine whether there was a relationship between emissions from a hospital incinerator in Preston, England, and the respiratory health of children living in the area. Measures from the stack of the incinerator provided information on emissions of chemicals such as sulphur dioxide. A mathematical model was used to estimate the dispersion of these pollutants in the air around the stack. This model used information on the stack and on the meteorological conditions in the area. It produced pollution 'contours', indicating varying levels of exposure to pollution for populations in different positions in relation to the stack. GIS methods were used to relate this information on exposure to data on the respiratory health of children in the area, derived from a health questionnaire survey. Figures 6.6 and 6.7 show the residential distributions of children with asthma and of a 'control' group of children without asthma.

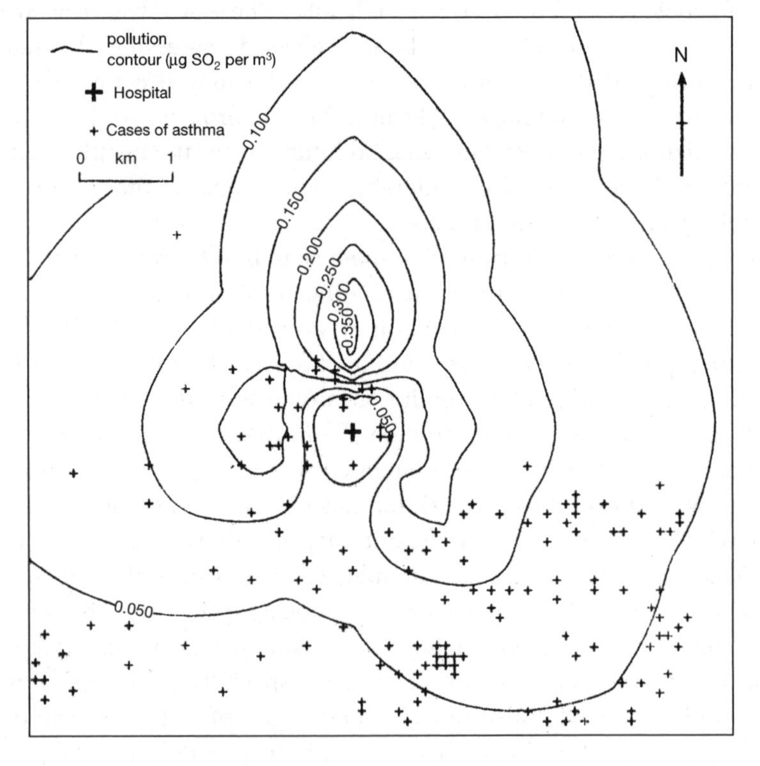

FIGURE 6.6 Residential distribution of sample of children with asthma, in relation to level of pollution from hospital chimney stack
(reproduced from Dunn and Kingham, 1996: Figure 1, with permission from Elsevier)

continued ⇒

Box 6.2 continued

Using the odds ratio of risk of asthma for children inside and outside the areas of heaviest pollution, the researchers tested whether children inside the more polluted area had a greater risk of asthma. However, the odds ratio was 0.62, with a confidence interval of 0.33–1.16, so there was no evidence of a statistically significant association.

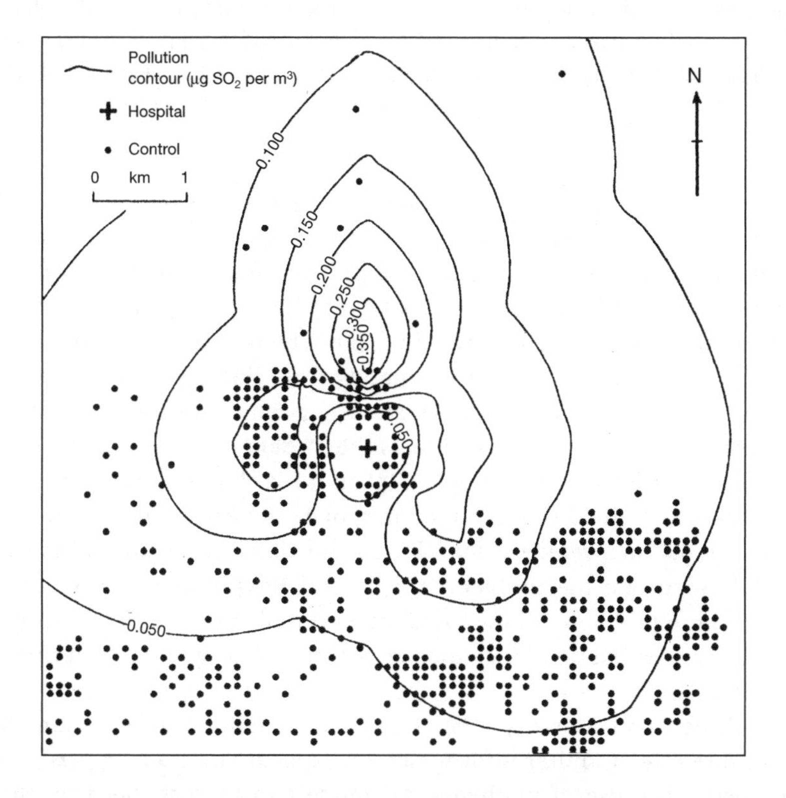

FIGURE 6.7 Residential distribution of 'control' sample of children without asthma, in relation to level of pollution from hospital chimney stack (after Dunn and Kingham, 1996: Figure 2)

The information on varying levels of pollution in the children's residential area was also added to the data from the questionnaire survey, to test whether estimated air pollution was statistically associated with risk of asthma. After controlling for other risk factors such as smoking or alcohol consumption in the family, occupational group or housing conditions, there was no evidence that air pollution data helped to explain the pattern of asthma in the survey population when these other factors were taken into account.

continued ⇒

Box 6.2 continued

The authors commented on the challenges involved in this research. The accuracy of the model of dispersion of pollution relies on the quality and comprehensiveness of the emissions and meteorological data. Also the model may not accurately reflect minor variations in topography which can affect pollution levels. There may also be other emissions, from sources such as road traffic, which are independent of that coming from the hospital stack. There may also be difficulties in detecting health problems such as asthma and in measuring some of the intervening variables such as health related behaviours and other risk factors within the home.

Various geographical techniques have been developed in environmental epidemiology to tackle the challenges faced in research on air pollution and health. These have also been described by Gatrell (2002) and Cromley and McLafferty (2002). Gatrell (2002) explains that different approaches are needed, depending on whether the pollution is widespread over a whole area, or emanating from linear sources like roads or point sources like factories. Several authors have discussed the difficulty in testing for environmental effects when data on exposure to air pollution are derived from samples taken at a limited number of fixed locations. A method is needed to interpolate the level of exposure likely to occur in the intervening spaces. Beyea and Hatch (1999) discussed the use of GIS to tackle the problem. Diggle and Rowlingson, (1994) also considered the problem of how to examine spatial variation in risk from data on the residential point locations of samples of people representing cases and controls.

Esmen and Marsh (1996) stated that making rigorous estimates of exposure often requires information on exposure over a long period. They discussed the use of mathematical models to predict the dispersion of pollutants from the sources of emission, using information on a range of factors. These included: the position and height of the stack; the concentration and rate of emissions; the variability of the resulting plume of pollution; the average wind speeds and directions over time; the altitude of each site and its distance from the source of pollution; and the hilly or flat character of the terrain. They showed that it was possible to use the model to make reasonably accurate estimates of exposure to pollutants from a stack over time at a site in Tucson, Arizona (Figure 6.8). The model results were more accurate than estimates produced by simply averaging data from nearby monitoring stations. It is not always possible to obtain such detailed information over long time periods. Even where good data

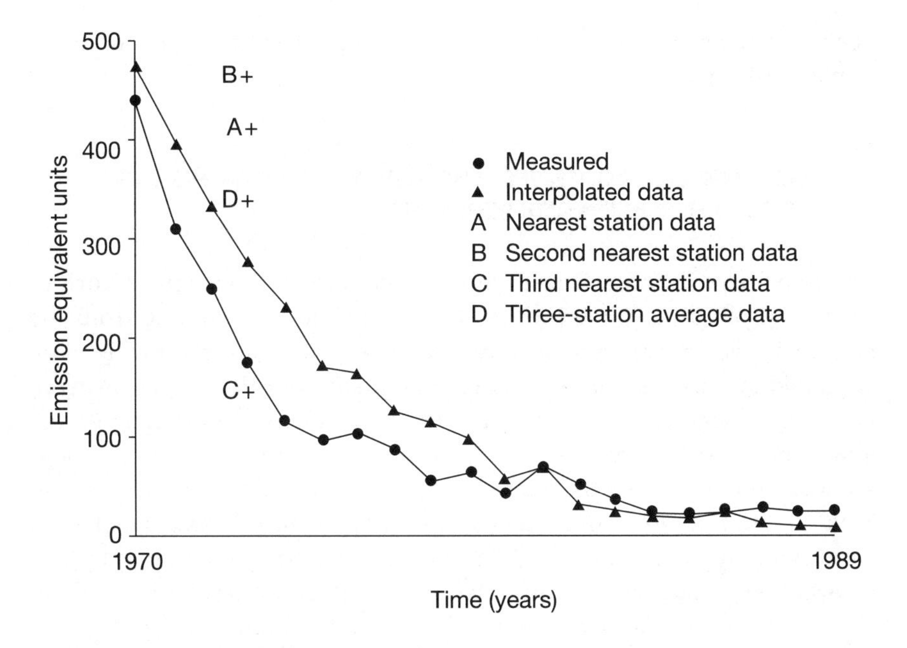

FIGURE 6.8 Comparison of measured and predicted pollution exposure levels (from Esmen and Marsh, 1996: Figure 6)

are available, there are limitations to the use of information on estimated air quality at a site in terms of likely exposure of individuals. Assumptions have to be made about the amount of time people spend out of doors, the amount of time they spend in their home location, and seasonal and age dependent changes in lifestyle.

Bellander et al. (2001) reported an attempt to incorporate data on individual residential histories into a model of exposure to pollution from traffic and domestic heating. This study compared two samples of people in Stockholm: one comprising patients with lung cancer and the other a 'control' group who did not have cancer. Their residential addresses at different points in the past were mapped. Reconstructed data on emissions of nitrogen oxides and sulphur dioxide were used to estimate levels of pollution in different areas of the city at three historical time points, in order to estimate lifetime exposure to air pollution for the samples of people in the study. It was found that there was likely to have been a good deal of variation in individual exposure.

The research in environmental epidemiology reviewed here therefore demonstrates the challenges involved in trying to establish clear scientific evidence about the importance of environmental pollution for health. In the conclusion to this chapter we consider the implications for public health policy.

ENVIRONMENTAL JUSTICE: THE CONTRIBUTION OF PHYSICAL HAZARDS TO SOCIO-ECONOMIC HEALTH INEQUALITIES

Work on air pollution and respiratory health also demonstrates a further major complication in environmental epidemiology, stemming from the fact that poorer populations often seem to be relatively highly concentrated in areas of poor environmental quality. For those working from an epidemiological standpoint, this makes it difficult to disentangle the effects of hazards in the physical environment from the impact of poor socio-economic conditions. Studies in environmental epidemiology often seek to 'control' for 'confounding' variation in health associated with socio-economic conditions, in order to establish whether physical environmental factors have an independent effect on health (e.g. Dolk et al., 1995). Sieswerda et al. (2001), for example, examined geographical indicators of 'environmental disintegrity' in relation to health and included data on economic indicators in the analysis. They concluded that, after controlling for GDP, in large scale analyses there was no association between trends in pollution and population health.

From the perspective presented in this book, however, these relationships between socio-economic conditions and physical hazard are not merely 'confounding' factors, but are central to the issue of whether certain social groups face systematic health disadvantages. Authors such as Coughlin (1996) have argued from this perspective that environmental epidemiology has a role to play in debates about environmental justice. A number of authors have argued that exposure to environmental hazards such as air pollution is likely to be worse in areas occupied by disadvantaged communities than in wealthy areas. This would be a situation of environmental injustice, since populations that are already poor and socially excluded are likely to be further disadvantaged by environmental hazards. They are least likely to be able to protect themselves against the possible health effects of pollution and they are also at greater risk of health problems for other reasons. If evidence concerning environmental justice is to be really useful in policy making and action on public health, it is important to be clear about the findings and what action is likely to be effective to redress environmental injustice (Cutter, 1995).

Theoretically there are a number of reasons why environmental injustice might occur. Boyce et al. (1999) compared the 50 states of the USA in terms of voter participation, fairness of taxation, educational attainment and Medicaid entitlement, to assess the distribution of power. They postulated that where the distribution of power between social groups was more uneven, environmental protection policies tended to be weaker and the environment is more polluted. Helfand and Peyton (1999) considered different groups of actors whose activities influence the siting of potentially noxious facilities: facility operators; community planners; and residents of affected areas. Residents in disadvantaged communities with low incomes and large proportions of minority ethnic groups, and with relatively little influence over social, economic and political processes, are likely to be relatively disempowered in decisions over the location of sources of pollution and the operation of the housing market. In contrast, wealthy, influential communities would be less likely to tolerate high levels of pollution in their area.

Williams (1999) argued that the geographical scale at which political debate is focused is important for environmental justice, and that the scale at which a problem is generated may not match the scale at which it is resolved politically. For example, a local decision might be made in one community about location of a facility with potential to pollute a much wider area. Williams (1999) also stressed that it is important to understand the *processes* which generate unequal risks, as well as to demonstrate variations in environmental risks.

Jerrett et al. (1997; 2001) discussed conceptual models of the processes which might produce environmental inequalities and associated health effects. They considered the possibility that populations living in more polluted areas are 'compensated' through lower cost housing and higher wages. They also conceptualized processes generating differences in environmental exposures (see Figure 6.9). The key processes in the political economy are: the impact of market economies in terms of polluting effects and uneven development; differences in political power between groups; institutional structures which affect decisions about the distribution of land use. These simultaneously influence the geographical siting of polluting facilities as well as generating socio-economic segregation in space. The model also shows the 'mediating' factors that can complicate the relationship between exposure to pollution and health effects, such as behavioural factors, occupational risks and genetic predisposition.

However, although environmental justice seems theoretically likely to occur, it is not always demonstrated empirically. Bowen (2002) reviewed studies of environmental justice and concluded that we lack sufficient empirical foundations to be certain how far the environmental injustice scenario really applies. Jerrett et al. (1997) also discussed the

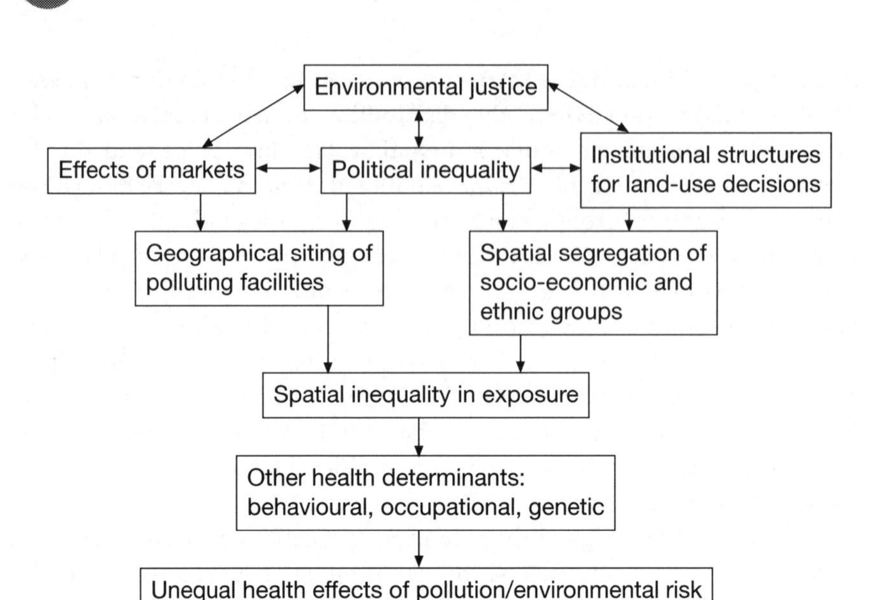

FIGURE 6.9 Conceptual model of environmental justice (after Jerrett et al., 2001: 957)

methodological issues which complicate the discussion of environmental justice.

Numerous studies in the USA and Canada have examined data on air pollution risks in relation to the socio-economic and ethnic profile of the population, of which some examples are considered here. It is difficult to draw clear conclusions for a number of reasons. These studies seem to provide equivocal evidence as to whether or not environmental justice applies. However it may not be reasonable to consider them as a coherent body of evidence. First, they differ in the indicators of risk of pollution employed: for example, proximity to potential industrial pollution sources (e.g. using the Toxics Release Inventory; US Environmental Protection Agency, 2003), or emissions over the long term, or short, extreme pollution events. Most of these studies have examined industrial pollution from factories, and they did not include other sources such as traffic. Also a variety of measures of socio-economic position have been used. Associations between pollution, race and class may be influenced by 'confounding' variables such as population density and employment structures, which were often not controlled for in the analysis. Furthermore, the studies are from different regions of North America, where varying conditions may apply. In addition, it is clear from some of the studies that even within the same area, and when employing similar methods of analysis, results at different geographical scales are not consistent, so that scale is an important aspect of the problem.

Perlin et al. (2001) found from a study of three regions in the USA (parts of West Virginia, Louisiana and the Baltimore area) that African Americans were more likely than whites to live in households with incomes below the household poverty line, to live closer to the nearest industrial emissions source, and to live within 2 miles of multiple industrial emission sources. African American households were also more likely to have children under five. The authors concluded that African American children are more likely to live close to sources of industrial air pollution.

Cutter and Solecki (1996) examined patterns of exposure to toxic airborne pollutants in the south-eastern USA. They found that chronic (small scale, long term) exposures were rather evenly distributed, but that acute (large scale, short term) exposures were more clustered geographically. The analysis suggested that there was no clear geographical association with ethnicity profile of the population. There was a positive association between wealth and exposure. They suggest that there are complex processes involved in varying exposures.

Jerrett et al. (1997) examined the socio-economic variables associated with varying levels of pollution in the 49 counties of Ontario, Canada. They found an association between pollution and four variables representing socio-economic status of the population: manufacturing employment, urbanization variables, dwelling value, and household income. It may be that their results were affected by the difficulties of modelling with a number of inter-correlated variables. Some aspects of their results ran counter to the idea of environmental injustice. The association between income and pollution was positive, that is pollution was greater in higher income areas. At a finer geographical scale, Jerrett et al. (2001) also showed an association between value of dwellings in small areas of Hamilton and their exposure to air pollution from particulates, such that lower cost housing was in areas with greater exposures. Low income and unemployment were also associated with exposure risk. However, the significance of the variables depended on the type of model used.

Cutter et al. (2001) examined the relationship between the location of environmental risks and federally assisted public housing in cities in the US. Associations between risk potential from hazardous facilities did not show a clear association with income or race, but there was some evidence that areas with higher risks were becoming poorer in some cities.

Some studies have examined variation in exposure at different geographical scales. Cutter et al. (1996) showed that in South Carolina, USA, at the county scale, wealthier white metropolitan communities were more likely to live close to potential sources of pollution. They did not find strong evidence for an association between community income and ethnic profile and risk of proximity to pollution sources at the more local level. Tiefenbacher and Hagelman (1999) found that at county level in Texas, areas with large proportions of black and other minority populations tended to have higher numbers of short term releases of toxic air pollutants, but at census tract level within counties, there was little evidence that neighbourhoods with black and minority populations were more at risk. Downey (1998) found that in the state of Michigan the distribution of racial minority groups was associated with pollution measured by the Toxics Releases Inventory. However, within urban areas, race was not so strongly associated with pollution. A complex relationship between race and income was noted.

Thus, while several studies do suggest that poor and/or black populations experience relative environmental disadvantage, these results do not support the idea of a straightforward or consistent situation of environmental territorial injustice, such that poor, black communities are systematically and unjustly exposed to industrial air pollution in every case. Instead, there are complex, differing patterns at various geographical scales.

ECOLOGICAL LANDSCAPES AND HEALTH DIFFERENCE: SCIENCE AND HEALTH POLICY

Research illustrated above on HIV infection, malaria prevalence and health effects of air pollutants exemplifies recent development in methods for geographical information systems, using approaches which take account of variation in both space and time. This dynamic view of ecological landscapes and the relationships between human health and environment is an important aspect of contemporary perspectives in health geography. Other fundamental issues in geography have also been raised here, including: questions of the geographical scale at which key processes operate and can best be studied; the relationship between risks operating at the individual and area levels; problems of the ecological fallacy and the atomistic fallacy.

The examples discussed above also highlight the importance of a rigorous scientific approach, but reveal the limitations to the extent that natural and medical sciences can inform human action in relation to problems like the risks of infectious diseases or the pollution syndrome. It is often difficult to offer concrete and unambiguous evidence about questions concerning the health risks of physical environmental factors.

There is a danger, therefore, that in the absence of clear evidence there will be insufficient attention paid to the potential health hazards in the environment, or else that public fears will be aroused without justification. This is especially important for disadvantaged groups in society, who may be affected by environmental injustice.

Botti et al. (1996) discussed the epistemological debate which is necessary to help societies interpret epidemiological research on ecological landscapes. Such interpretation may depend on essentially subjective judgements and values which are not scientifically based. These include, for example, the precautionary principle, which emphasizes avoidance of possible harm. Needleman (1997: 269) advocates a 'weight of evidence' response to the uncertainty involved in this area of science, whereby a health risk is assumed when there is scientific evidence suggesting a causal link, even though the causal pathways may not be fully understood and proven. Policy makers and interest groups concerned about issues such as environmental justice need to consider evidence presented in terms of probabilities rather than certainties, and base their decisions partly on values, rather than solely on science.

Soskolne and Light (1996) and Frankel (1996) have discussed ethical guidelines concerning how, and in what circumstances, environmental risk should be researched. On the one hand, communities with concerns over pollution may want the risks investigated, even if research is unlikely to produce conclusive evidence. On the other hand, epidemiological investigations may raise fears in communities that are not justified by the low level of risk that the research may reveal.

Grandjean and Sorsa (1996) also argued for ethical guidelines concerning the approach to information on how individuals' genetic makeup may affect their susceptibility to environmental risks. Major ethical issues are raised by genomics in relation to human health. The question of who should be provided with information on genetic risks, and in what circumstances, is a difficult one, both ethically and legally. Genomics education is essential for the general population, health providers and health policy makers, but it may be difficult to provide information when genetic risks have to be expressed in terms of probabilities rather than certainties, and when the relative risks of illness associated with particular combinations of genetic and environmental factors are not very well understood (Armstrong et al., 1998; McConnell et al., 1998). Jeffords and Daschle (2001) argue for a proper balance between privacy concerns and fair use of genetic information. These issues are likely to have particular implications for people who are socially excluded, to whom health information has often not been provided appropriately or effectively in the past, or who face particular risks of discrimination because of their race.

We can conclude from this chapter that physical environmental and biological factors are important for health geography and for inequalities in health. Spaces of risk for human health are partly defined in terms of these factors. Theoretically and empirically the geography of health needs to be informed by a scientific understanding of physical environmental risks to health, the distribution of human populations in relation to these risks and their biological characteristics (including their genetic profiles). Furthermore, physical environmental processes interact with social, economic, political and ethical factors.

Figure 6.10 presents a conceptual framework for socially unequal environmental risk. In this diagram, the key elements of habitat, behaviour and population characteristics from Meade and Earickson's model in Figure 6.1 are combined with ideas about social and political processes of environmental justice proposed by Jerrett et al. (2001) from Figure 6.9. These may lead to socially and spatially unequal exposures to physical risks to health. These varying exposures, operating together with other health determinants, generate unequal health effects for different population groups in different areas. It is important to pay attention to these interactions if we are to have a sound evidence base for action in public health. However, the examples considered here also show that some of the risks and processes involved are so complex that scientific knowledge

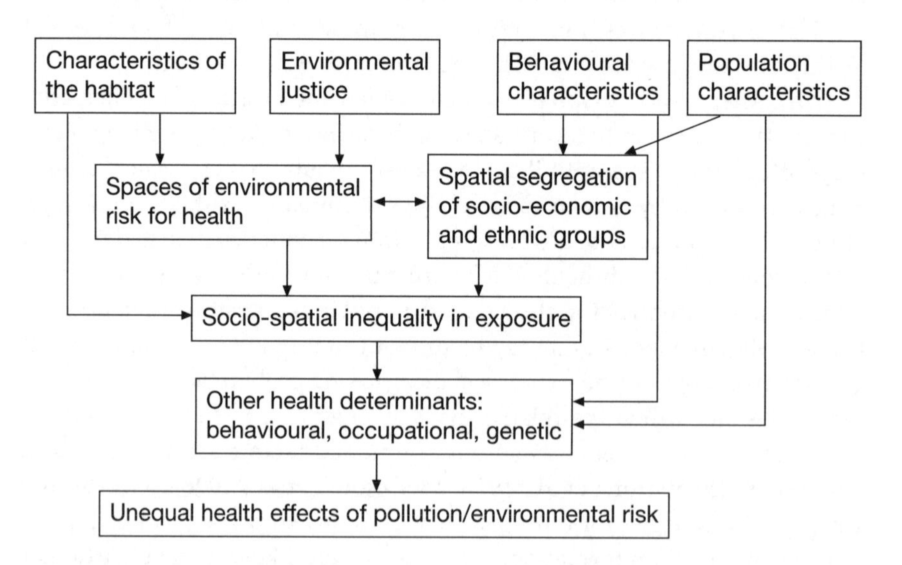

FIGURE 6.10 Conceptual framework for socially unequal environmental risk (adapted from Jerrett et al., 2001: 957; Meade and Earickson, 2000)

and methods are not yet sufficiently developed for them to be completely understood.

FURTHER READING

Useful and more comprehensive discussions of the issues raised in this chapter are to be found in: Gatrell (2002); Meade and Earickson (2000); Howe (1997).

A good introduction to the use of GIS in public health is: Cromley and McLafferty (2002).

Students seeking more information on particular diseases may find useful material on the WHO website: http://www.who.int/en.

OBJECTIVES AND QUESTIONS

The material covered in this chapter should help students to:

- understand some principles of epidemiological landscape perspectives in health geography;
- understand the relevance of geography for diseases like malaria and be familiar with examples of use of GIS in malaria prevention programmes;
- know about the methods that have been used in environmental epidemiology;
- know about some of the difficulties which researchers face when they try to distinguish effects of air pollution on health;
- consider the ways that results of research on epidemiological landscapes may be used to inform health policy, and understand the difficulties of interpreting scientific evidence which is complex and sometimes equivocal;
- understand the significance of the interaction between biological and physical environmental risks to health and socio-economic factors.

To help to consolidate this knowledge, students may wish to try answering the following questions:

1 Discuss the potential and the limitations of geographical information systems as tools to assist in programmes to control malaria.
2 With reference to specific research findings, critically discuss the assertion that 'Varying exposure to air pollution in North America contributes to environmental injustice between social and ethnic groups in the population.'
3 Discuss critically the assertion that 'Environmental epidemiology has made very little useful contribution to our knowledge about whether air pollution damages respiratory health.'
4 What do you understand by gene geography, and how important is it for the geography of health?

SECTION II

Integrating landscape perspectives: Aspects of health in urban settings

7 The Geography of Mental Health in Cities

Mental health and mental illness are of enormous human significance. This is underlined by the emphasis placed on mental health in a recent *World Health Report* by the World Health Organization, which encouraged all nations to give priority to it (WHO, 2001a). The recommendations included improving education and awareness about mental health. This chapter considers geographical perspectives that have contributed to knowledge about mental health and mental health care.

To start the following review, the nature of mental health and illness and the evolution of mental health care are summarized. The following discussion concentrates especially on examples of studies of mental health variation in urban settings in high income countries, with a particular focus on London, UK, and New York City, USA.

Some useful conceptual models are introduced, concerning the interaction of 'breeder' and 'drift' effects and the idea of the 'service dependent ghetto'. The discussion also shows how the five types of 'landscape', reviewed in the previous chapters, help to explain geographical variations and inequalities in mental heath and illness. Examples reviewed here relate to the impact on mental health of: physical and biological risk; material wealth and consumption; collective consumption (especially of mental health services); social and political control; and therapeutic aspects of the physical and social landscape.

The research reviewed in this chapter illustrates geographical inequality in mental health. It demonstrates that there are geographically unequal risks of developing mental illness. Socio-geographical factors also contribute to disadvantage experienced by people who already have mental illness.

MENTAL HEALTH AND ILLNESS

The US Surgeon General's report on mental health (US Department of Health and Human Services, 1999) reviewed concepts of mental health, arguing that mental and physical health are not separate dimensions, but are strongly linked. However, a distinction was made between mental health, which relates to successful performance of mental functions, and somatic health, relating to bodily functions. Mental disorders are those in which alterations in mental, as opposed to bodily, functions are paramount. Understandings of mental health are based on value judgements and are socially and culturally constructed. For example, Prior discussed the social representation of mental illness and argued that it can be understood as the result of products of thought and social practices which include 'physical structures where mental illness is observed and treated; the organizational arrangements through which it is analysed and monitored; and the forms of professional practice which surround the . . . objects of psychiatric disorder' (1993: 6). (These points are also reviewed by Pilgrim and Rogers, 1999: 18–19.) This means that the idea of mental health is not fixed but variable between societies, cultural and social groups or individual people, creating potential for geographical variation in perceived mental health between spatially separate communities.

Furthermore, mental functioning is affected in many different ways by the interaction between people and their social and physical environment, which means that geographical factors influence mental health and wellbeing. The attributes of places, as well as those of individual people, are therefore important for mental health in the population.

The geography of mental health is often concerned with geographical variation in measures of psychosocial health, wellbeing and quality of life collected in surveys of the population. A range of instruments is used to produce these measures, and Bowling (1997) has compiled a comprehensive review. They include measures of life satisfaction, happiness or general psychological wellbeing, such as the *Affect-Balance Scale*, described by Bradburn (1969). This instrument includes positive and negative descriptions of psychological health. Examples (cited from Bowling, 1997: 117) are as follows:

Things going your way.
Excited, interested in something.
Pleased about having accomplished something.

Broad indicators of 'psychosocial' health are also increasingly being included in surveys and population censuses. For example, in Britain in 2001, questions were asked about long term illness and general health. These do not relate exclusively to mental health, but they are, at least in part, indicators of general psychosocial health or malaise. A question which was asked in several British sample surveys, and included in the 2001 population census in Britain, was:

Over the last twelve months, would you say your health has on the whole been:
Good?
Fairly Good?
Not Good?
(Office of National Statistics, 2000: 24–5)

Survey instruments which are more focused on psychological illness include the General Health Questionnaire (GHQ) (Goldberg and Williams, 1988) which is widely used in the UK and is employed in the *Health Survey for England* (Joint Health Surveys Unit, 1997). Another instrument which has been used to measure levels of psychosocial malaise or distress in population surveys is the Nottingham Health Profile (Hunt et al., 1986) which covers six health dimensions included psychosocial aspects such as social isolation, lack of energy, emotional reactions and sleep disturbance. This instrument requests respondents to state whether or not descriptions of malaise apply to them. There are 38 items including, for example:

I'm tired all the time.
Things are getting me down.
I take tablets to help me sleep.

Figure 7.1 shows results from surveys of inner and outer city populations in London and Manchester in the early 1980s (Curtis, 1987). Inner city populations tended to show worse health and the inner/outer city contrast was more striking for London than for Manchester. These results suggested relatively poor health in poor, highly urbanized areas, but they also showed that patterns of illness reporting are variable in different urban contexts.

According to more medical definitions, mental illness can be interpreted as 'diagnosable mental disorders' (US Department of Health and Human Services, 1999). Medically diagnosed mental illness takes a number of forms, as indicated in Box 7.1, which can affect people in the

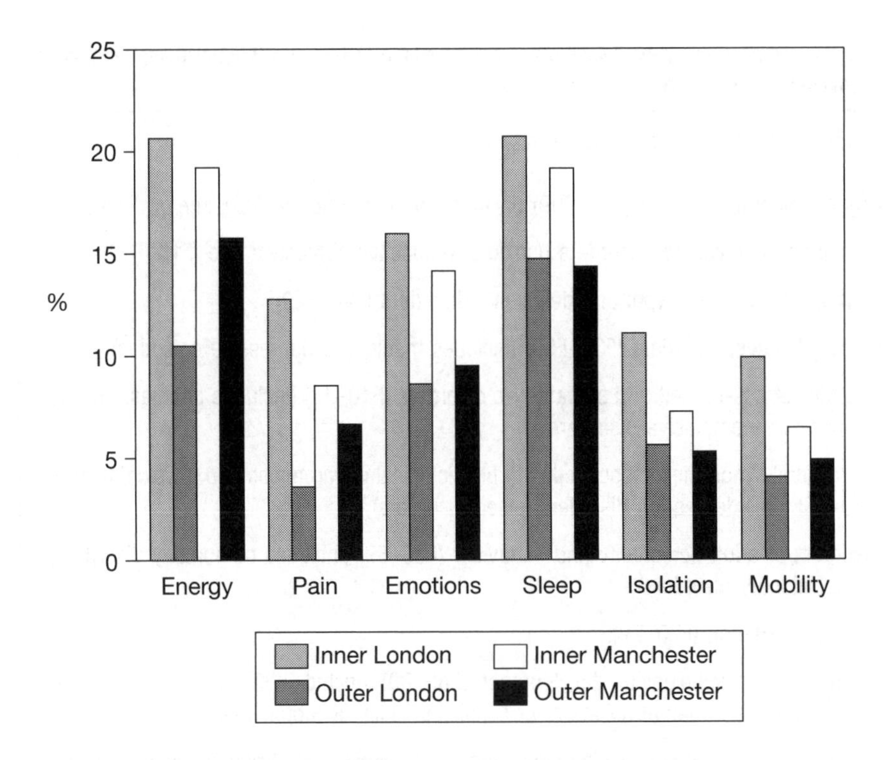

FIGURE 7.1 Scores on Nottingham Health Profile: inner and outer city populations aged 45–64 (Curtis 1987)

population at different stages of their lives. Many of these conditions are chronic, enduring, health problems, but some may involve shorter episodes of illness from which the person recovers. Diseases classified under mental and behavioural disorders include, for example, organic disorders such as dementia produced by Alzheimer's disease (a chronic, degenerative condition, which most often affects elderly people). At the other extreme of the life cycle, problems in childhood can include hyperactivity, conduct disorders and difficulties with scholastic development or motor skills. Schizophrenia usually starts to manifest in late adolescence or early adulthood, often resulting in serious long term illness. Affective or mood disorders such as depression, or neurotic conditions such as anxiety disorders, can affect people of all ages and may be either short or long term health problems. In severe cases, mental illness can be associated with mortality and, for example, levels of suicide in populations are often taken as indicators of outcomes of severe mental illness.

These mental disorders and outcomes of mental illness also show variations in geographically defined populations, partly because of the

Box 7.1 Mental and behavioural disorders as coded in the* International Classification of Diseases, *version 10

Chapter V: Mental and behavioural disorder

Organic mental disorders (F00–F09) (includes dementia due to Alzheimer's disease);

Mental and behavioural disorders due to psychoactive substance use (F10–F19);

Schizophrenia, Schizotypal and delusional disorders (F20–F29);

Mood (affective) disorders (F30–F39) (includes manic and depressive episodes);

Neurotic, stress-related and somatoform disorders (F40–48) (includes phobias, anxiety or obsessive-compulsive disorders)

Behavioural syndromes associated with physiological disturbances and physical factors (F50–59) (includes eating disorders, sleep disorders);

Disorders of adult personality and behaviour (F60–69) (includes personality disorders, gender identity disorders);

Mental retardation (F70–79);

Disorders of psychological development (F80–89) (includes speech and language disorders, developmental disorders of scholastic skills or motor function);

Behavioural and emotional disorders with onset usually occurring in childhood and adolescence (F90–98) (includes hyperkinetic, conduct or emotional disorders disorders).

Source: WHO, 1992

distribution of populations with different risks of mental disorder, but also because attributes of places can impact on the risk of mental illness. Given the many different dimensions of mental health, there is no single geography of mental illness, and this discussion cannot encompass all the different geographical factors relating to every different type of disorder. However, the following discussion considers examples that illustrate the importance for the geography of mental health of the 'landscapes' reviewed earlier in this book.

THE EVOLUTION OF MENTAL HEALTH CARE

Mental illness in populations is often assessed in terms of use of mental health care, especially for more severe mental illness requiring significant levels of medical treatment and social support services. Since not everyone with mental illness will be known to health services, or receiving care, this approach only measures part of the total burden of mental illness in communities. However, in countries with fairly advanced medical information systems, it has the advantage of being based on routinely collected data which are less costly to obtain than information gathered by special surveys in the community. Those using mental health services will usually have been diagnosed as having a clinically recognized mental illness. People with similar mental illnesses do not always receive the same care, however. Their treatment varies according to the model of mental health care adopted in different places and at different points in time.

Various authors have discussed how mental health care has developed historically, in different countries, reflecting changing views of the nature of mental illness and the appropriate therapies. Scull (1979), Rogers and Pilgrim (1996), Edgington (1997) and Milligan (2000b), for example, gave accounts of these trends in the UK, while Dear and Wolch (1987) and Elpers and Levin (1996) described changes in the USA which have been broadly similar. These authors summarize changing constructions of mental illness since the 18th century, associated with evolving models of care and a changing role for the voluntary sector. They have described the development of the psychiatric asylum as the site for provision of mental health care, where people with mental illness were separated from the wider community and experienced institutional regimes of care and treatment.

At the start of this period, the dominant view of mental illness was as a subhuman, deviant condition requiring 'incarceration' in institutions. Conditions in such places were often particularly poor and they resembled prisons, shutting mentally ill people away from general society rather than offering effective therapy. By the late 18th century, psychological disorder began to be seen as a condition which could be treated through more humane and philanthropic 'moral treatment', combining psychological and social interventions delivered in institutional settings. This was associated, during the 19th century, with efforts (including legislative measures) to improve conditions in mental asylums, to make them more humane and therapeutic. In the 20th century, notions of mental illness as biological dysfunction became more dominant in psychiatric medicine, leading to emphasis on a medical approach and administration of drug therapies.

In addition to developments in psychiatric medicine, other processes influenced the way that mental health care developed. Rogers and Pilgrim

(1996) reviewed other explanations for the rise of asylum modes of care in the late 19th and early 20th centuries, which included the increasing rates of mental disorder recognized in British society; developing ideas of humanitarianism and benevolence towards people affected by mental illness; changes in the 'Poor Law' system for administration of state welfare; and the breakdown of traditional community support due to growing urbanization and industrialization. Scull (1979) put forward an argument that the development of institutional care models has been associated with the increasingly dominating role of a capitalist political economy.

In the latter half of the 20th century, commentators described a trend away from institutional care of mental disorders (Dear and Wolch, 1987; Elpers and Levin, 1996; Rogers and Pilgrim, 1996; Milligan, 2000b). In Britain, for example, this was associated in the 1960s with more critical views of the benefits of institutional care, and concerns over the quality of life and health outcomes of institutions. This was followed in the 1970s by a fiscal crisis of the state which led to a far-reaching reappraisal of state funding for health and welfare, and raised questions about how feasible it was for the state to fund institutional mental health care. The view emerged that mental health care resources could be used more effectively and efficiently in non-institutional settings, and the process of deinstitutionalization of mental health care began. Thus, for example, several of the major mental health institutions surrounding London started to close in the 1980s.

By the end of the 20th century, associated with this trend towards care in the community, more emphasis was being placed on the social environmental factors which are important for mental health, and which are discussed in more detail below (Laws and Dear, 1988; Milligan, 2000b). This also led to a greater emphasis on community based interventions to support mentally ill people living outside institutions (Department of Health, 1996). There has been a growing recognition that different care 'sectors' provide important support to people with mental illness in the community. The Surgeon General's Report (US Department of Health and Human Services, 1999) refers to four key 'sectors': psychiatric services; general health services; human services (e.g. education and welfare); care provided by voluntary organizations and non-professional, informal carers, such as relatives and friends of people with mental illness.

Most recently there appears to have been a further shift in the public debate about mental illness and the appropriate social and medical response. This places more emphasis on the *hazards* of mental illness, both for the person with the illness and for those around them. Commentators have begun to question how far it is possible to shift the balance of care of mental illness out of institutions and towards the community. Particularly in Britain, measures to introduce greater control over people with mental illness living in the community are therefore also a feature of the development of community based mental health care (Department of Health, 1999b, 1999c). Moon (2000) explored how themes of *confinement* have recently re-emerged in policy literature and popular discourse. He noted that community health services are presented as having to control danger to society posed by some mentally ill people. These geographical perspectives connect with discussion in other disciplines. For example, from an anthropological perspective, Mossman (1997) considered the view that the growing numbers of mentally ill, homeless people encountered in public spaces in the US result from abandonment of mentally ill people by the psychiatric profession through the process of deinstitutionalization. Mossman interpreted this widely held notion of 'abandonment' as a 'myth'. In anthropological terms a myth is a narrative embodying a collective view held in a society. Ducq et al. (1997) also discussed competing explanatory hypotheses of socio-economic background or deinstitutionalization as causes of homelessness.

Moon (2000: 248) commented that these developments reflect variable social understandings of risk associated with mental illness. Also, there is a varying balance in societies between, on the one hand, values of individual aspiration and freedom, and on the other hand, the importance of social capital and shared responsibility to the community. The recurring theme of confinement of people with mental illness highlights one reason why socio-geographical theories of control and surveillance (discussed in Chapter 3) are pertinent to an understanding of the geography of mental health.

From a geographical perspective, therefore, we can observe evolving geographies of mental health care, at least in high income countries, with both institutional and community based settings now identified as important spaces of care and treatment for mental illness (Philo, 1997; Wolch and Philo, 2000). Research on the nature of these spaces of care, and what takes place within them, reveals how mental illness is understood by medical professionals, by researchers in social science disciplines, and by society more widely. It also contributes to knowledge of what affects mental health. Later in this chapter the discussion returns to the concept of 'therapeutic landscapes' introduced in Chapter 2, to consider the value of this perspective for the study of settings intended to be beneficial to mental health.

MENTAL HEALTH IN URBAN SETTINGS

Two well known models are often used to conceptualize the complex processes involved in the geography of mental health and illness. They involve ideas about 'breeder' and 'drift' effects and the development of the service dependent ghetto. These are discussed below. This chapter then goes on to argue that the various 'landscape' perspectives reviewed in the first section of this book can be used to extend our understanding of geographical variations of mental health.

VARIATION IN URBAN MENTAL HEALTH: BREEDER AND DRIFT HYPOTHESES

There has been considerable discussion about the reasons for geographical variation in mental health among urban populations and the debate has often centred on the relative importance of 'breeder' and 'drift' effects. Jones and Moon (1987) trace the origin of ideas about these effects from literature on urban sociology and urban mental health published in the 1930s and 1940s. The breeder thesis postulates that certain conditions (particularly those of socio-economic deprivation and social malaise) increase the risk for the residents in disadvantaged areas of developing mental health and behavioural problems. The drift hypothesis emphasizes processes by which people with mental health problems or anti-social behaviour tend to move towards relatively disadvantaged areas of a city. As was suggested by Giggs (1975), and by other authors subsequently, research often indicates that in fact these are not so much competing hypotheses, but complementary explanations of complex processes influencing the geography of mental illness.

THE SERVICE DEPENDENT GHETTO

In their analysis of deinstitutionalized mental health care in North American cities, Dear and Wolch (1987) put forward a particularly coherent and comprehensive model of the geography of mental health. The strength of this model lay in the way that they brought together ideas about three sets of factors relating to: urban development and the political economy of cities; the roles of stakeholders involved in decisions about mental health care policy and its implementation; and change in the policies and ideologies of care determining the organization of mental health services (see Figure 7.2). They argued that deinstitutionalization has resulted in a concentration of mentally ill people in poor inner city

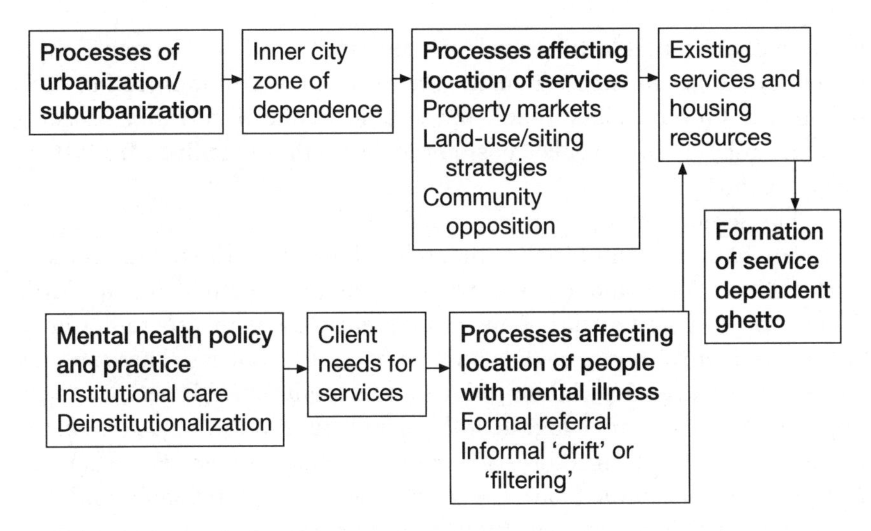

FIGURE 7.2 The service dependent ghetto (derived and adapted from Dear and Wolch, 1987: 12)

areas where they rely on a range of service provision. This contrasted with the semi-rural location of 19th century asylum institutions, designed to separate people with mental illness from wider society and to offer what was considered a therapeutic (natural and peaceful) landscape for care.

Dear and Wolch suggested that certain conditions in poor urban areas may affect the distribution of people with mental illness and the use of mental health care. Lower property prices in poor areas make it easier for mentally ill persons to find affordable housing so that there are 'drift' effects producing greater concentrations of mentally ill people in poor areas. Greater levels of need in the population encourage location of mental health services in poorer areas. In order to meet these needs, mental health care professionals tend to group together to form a 'critical mass' of specialized personnel. This concentration of care attracts even more clients dependent on these services, thus contributing further to the geographical concentration of persons with mental illness: the 'service dependent ghetto' effect.

Dear and Wolch also pointed to the NIMBY ('not in my back yard') syndrome in negotiation over the location of facilities for people who are mentally ill and who are perceived as likely to behave in ways which are not socially acceptable to the wider community. Affluent communities, which are influential and articulate, are most successful in rejecting such facilities if they consider them undesirable, while disadvantaged communities may be in a weaker position to reject these facilities from their neighbourhood.

Thus a number of reinforcing effects are hypothesized to lead to relatively high concentrations of mentally ill people in the poorer parts of inner cities. These models were developed principally from a North American perspective, and some subsequent studies in geography have examined their relevance to the situation in other countries. These studies suggest international variability of trends in mental health policy and its implementation and they highlight the significance of the local context for the development of structures for care of mental illness. For example, Milligan (1996) examined patterns of provision in Scotland and found that Dear and Wolch's model was reflected in the continued concentration of some non-residential facilities for mentally ill people in larger towns. However, residential provision was less concentrated in urban areas. She attributes this partly to the more powerful position of statutory planning agencies in the UK compared with the US.

Milligan's study highlighted an increasing spatial separation and dislocation of residential accommodation from other support services for mentally ill people, which has also been noted in studies elsewhere. For example, Kearns and Joseph (2000: 167) reported similar findings in their analysis of the spatial organization of residential and health care opportunities for mentally ill people in Auckland, New Zealand, which is influenced by geographies of consumption (especially in the housing market and in the increasingly privatized market of mental health care).

Jones (2000) compared the policies for the deinstitutionalization of mental health care in Sheffield, UK, and Verona, Italy. Jones commented that the situation in Sheffield did not accord with the service dependent ghetto model proposed by Dear and Wolch, since the geographical patterning of the housing stock in Sheffield led to concentrations of mental care provision in some of the more affluent areas of the city rather than in deprived inner city areas. Jones suggested that differences in the organization of mental health care in Sheffield and Verona were due to: the historical legacy of provision in each city; the degree of decentralization of health policy decisions; the role of health professionals in championing reforms and of planners in exercising control over new developments; and the patterns of land use and the housing markets. This study therefore highlighted again the variable development of health care internationally and the ways that the geographical context of provision and consumption of care can influence service provision.

While the service dependent ghetto model is not equally applicable in all settings, it is nevertheless a very useful starting point for analysis of the geography of mental health in the city. It also shows a number of connections with the conceptual frameworks described in the earlier chapters of this book. For example, it emphasizes the complex links between geography of health and the spatial unevenness of capitalist economies; theories of control by powerful social groups, bureaucracies and urban 'gatekeepers'; and geographies of consumption. There are particularly interesting parallels with the writings of Wallace (1990) and Wallace and Wallace (1991; 1993) on the 'desertification' of inner cities and its relevance for health, discussed in Chapter 3.

URBAN LANDSCAPES AND MENTAL HEALTH

There is a large research literature on the processes contributing to variation in mental health, and the following review explores some of this from the perspective of different geographical landscapes. The focus is particularly (but not exclusively) on what is known about the situation in cities in Britain and the US, such as London and New York City. These cities are interesting to consider because they share certain characteristics which seem likely to generate the kinds of effects predicted by the 'service dependent ghetto' model. They are global cities with large populations (just over 8 million people in New York City in 2000, and 7 million in London in 1995). There is a good deal of internal diversity in these urban populations, in terms of both social and economic inequality and ethnic diversity (Curtis, 2001; and see Chapters 1 and 4). These places represent highly urbanized environments, where the impacts of urban living on mental health are most likely to be felt. They are also very dynamic cities, at the forefront of processes of urban change and regeneration. An international comparative perspective also allows consideration of how different policy contexts influence patterns of mental health care in urban settings. London and New York, for example, are located in countries where mental health policy has for some time been moving towards deinstitutionalization of mental health care, but with important national differences in the policy process and the national health systems, as discussed in Chapter 5.

LANDSCAPES OF PHYSICAL AND BIOLOGICAL RISK

Most of the urban hazards for mental health relate to social conditions and material poverty and are considered in the following sections on material wealth and consumption. However, there are some physical and biological factors in the wider urban environment which have salience for mental health. These include physical environmental conditions generating stress and hazards to mental health.

Biological and physiological pathways are likely to be important in linking psychosocial stress to risks of both infectious and non-communicable diseases, so that we can see mental health variation as part of a wider pattern of varying health in the population. Pollard (1997) cited studies from low income countries which show that physiological indicators of stress are higher amongst urban dwellers than rural populations. He also reviews research demonstrating physiological reactions to stressful urban environments, associated with changes in hormonal levels. Blood levels of adrenaline and cortisol increase as the individual experiences psychological stress. These in turn raise blood pressure and serum cholesterol, increasing risks of cardiovascular disease. Also immune function is reduced, leading to greater risks of succumbing to infectious diseases or cancers. This provides a hypothesized physiological link between mental state and risks for physical health.

One aspect of urban living which may generate stress and influence mental health is the intense level of movement between different environments experienced as part of the typical time–space path of the urban resident. As noted in Chapters 1 and 6, time–space compression and the spatial organization of workplaces, residential areas and urban amenities in cities have led to high levels of short term movement between and within cities, at scales ranging from international migration and travel to daily travel to work patterns inside the city. Pollard (1997) reviews evidence from a number of studies on how travel in the city may raise levels of stress. These suggest that stress levels increase with the distance travelled and time taken, and use of crowded public transport may raise stress levels for urban dwellers. It is postulated that stress arises at least partly because of a sense of lack of control over one's environment. James (1991) showed that adrenaline levels and blood pressure are typically higher for people living in New York when they are at work and lower at home. Thus the environmental stressors for individuals appear to vary as they move through their daily action spaces. The time–space geographies of individual life paths (e.g. discussed in Chapter 3) are therefore important to our understanding of risks to mental health.

Another set of factors, which may have relevance for variation in mental health at the individual level, are differences in the biological makeup of the population (considered in Chapter 6). Some mental illnesses, such as depression, have a complex pattern of causation, and although genetic makeup does not fully explain the pattern of these diseases, it may be a contributory factor associated with increased risks for some individuals. For example, genetic factors are related to risks of bipolar (manic depressive) disorders (Wilson, 1998). Also, Gurling et al. (2001) have identified the genetic markers for schizophrenia. Whether genetic factors would account for variation in mental health of populations in different parts of a city is less clear, however.

Various aspects of physical environment may have significance for mental health in major cities. Some of these arise from environmental pollution of various kinds, such as noise and chemical pollutants. Meteorological conditions and daylight exposure are other examples of physical conditions with relevance for mental health (Meade and Earickson, 2000), though these are less specific to cities.

One aspect of environmental pollution of possible relevance for mental health is ambient noise. Stansfeld et al. (2000; 2001) have reviewed the evidence for the impact of ambient noise on health. Several studies in London have examined the impact of aircraft noise on mental health, especially in the vicinity of Heathrow airport, on the western edge of the city. A survey by Tarnopolsky et al. (1980) revealed greater levels of acute symptoms such as depression, irritability, difficulty getting to sleep and night waking. Jenkins et al. (1981) and Kryter (1990) showed that locally higher levels of aircraft noise were associated with higher admission rates to local psychiatric hospitals. However other factors were more strongly related to admissions, and, as discussed above, hospital use is only an approximate indicator of mental illness in the population. It is possible that these types of relationships arose because of the greater average social disadvantage of populations living closer to the noise nuisance of the airport. Watkins et al. (1981) did not find a link between aircraft noise and use of primary care or rates of prescription of medicines. The evidence for effects of noise on mental health of the general population is not, therefore, consistently clear, although it does seem that people with mental illnesses such as depression or anxiety are particularly likely to be annoyed and disturbed by noise (Stansfeld et al., 2001).

Several studies have considered noise impact on mental health and cognitive abilities of school children, including some studies in London and New York (Evans and Maxwell, 1997; Maxwell and Evans, 2000; Stansfeld et al., 2000; Haines et al., 2001). In areas where aircraft noise is high, children tend to have poorer measured levels of cognitive ability, attention, concentration, auditory discrimination and speech perception, memory and reading ability. However, children living close to large airports are also likely to come from relatively disadvantaged families, and disentangling the effects of noise pollution from socio-economic impacts requires careful analysis. It appears that exposure to noise pollution may be compounding other patterns of educational disadvantage for poor children. Therefore these studies seem to exemplify problems of environmental epidemiology and environmental justice discussed in Chapter 6.

Some chemical pollutants can be important for human mental health and development. Schneider and Freeman (2000), for example, summarized findings from epidemiological studies of the impact of lead absorption on children, which can have effects on physical health as well as causing damage to cognitive performance, damage to hearing, and hyperactivity and behavioural problems. Schneider and Freeman (2000: 76) cited estimates that intelligence measured on IQ tests may fall by 2.5 points as a result of an increase of 10 micrograms per decilitre of lead levels in the blood. Around 1.7 million children aged one to five years in the USA may be exposed to lead absorption at or above this level. The level of exposure will depend upon traffic pollution, soil pollution or lead in older buildings. Children living close to major roads, in old industrial areas and in older housing may have relatively high risks of exposure.

The physical threat of violence, terrorist action or war may also affect mental health. Around a third of people exposed to terrorist attacks may develop post-traumatic stress disorder (PTSD) and some may need health care as a result (Lee et al., 2002). Following the attack on the World Trade Center in New York City on 11 September 2001, accounts in the *Journal of Urban Health* (Bulletin of the New York Academy of Medicine) suggested that symptoms of PTSD, such as insomnia and intrusive memories, increased immediately after the attack.

Thus broad scale environmental conditions can have an impact on mental health and some of these aspects of environment are increasingly typical of urban settings. Noise and air pollutants, for example, have become widespread problems associated with urban industrial environments, so that it has been necessary to introduce measures to tackle them, for example, through reduction of the lead content of petrol. The effects of these environmental factors for different populations within the city will depend partly on the efficacy of such policies and also on varying levels of exposure, related to area of residence and movement in the city.

LANDSCAPES OF MATERIAL WEALTH AND POVERTY

Much attention has been focused on how geographical indicators of population mental illness are correlated with socio-economic indicators of poverty and social disadvantage. Many of these studies take the form of ecological analyses testing the association between mental health and socio-economic conditions for aggregated populations of geographically defined areas. Various indicators of mental health have been used in these ecological analyses, including: diagnosed illness recorded in health service registers (information on cases of severe mental illness, such as schizophrenia); mental health service activity, such as psychiatric hospital admissions; outcomes such as suicide; and population survey measures of psychological state or psychosocial health status, which may relate to mild or moderate, as well as severe, forms of mental illness. This diversity of measures and methods of analysis makes it difficult to compare the results of different studies or to discern universal patterns. Nevertheless, some consistency emerges concerning a positive link between mental illness and poor socio-economic circumstances in deprived urban areas (e.g. Johnson et al., 1998b).

Studies of schizophrenia have identified ecological associations with socio-economic deprivation at the scale of small geographical areas. Examples of British studies include early research by Giggs (1973) on the geographical distribution of hospital patients from the city of Nottingham suffering from schizophrenia. Giggs examined the relationship between 'standardized schizophrenia attack rates' for small areas of the city and a range of socio-economic indicators derived from small area census data. This demonstrated a positive association between the geography of urban social deprivation and concentrations of cases of schizophrenia. Dean and James (1981) showed that indicators of poor quality rented accommodation lacking amenities and low social class showed a positive association with the rates of 'first time' hospital admissions of males due to schizophrenia in Plymouth. Harvey et al. (1996) found that in the London borough of Camden, the geographical distribution of cases of schizophrenia showed a significant association with socio-demographic measures of urban deprivation. Research on social conditions for individuals through their life course has also suggested that experience of deprivation early in life may be associated with increased risk for schizophrenia (Dauncey et al., 1993; Castle et al., 1993).

Prevalence of depression also shows links with poverty in residential neighbourhoods. For example, Wilson et al. (1999) demonstrated how depression in a survey sample of elderly people in Liverpool was associated with deprivation, measured using indicators suggested by Townsend et al. (1988) (unemployment, lack of a car, crowded housing, and housing which is not owner occupied). This reinforces findings from earlier research in London concerning the role of social deprivation in depression (e.g. Brown and Harris, 1978).

Congdon (1996a, 1996b) examined suicide rates in London as an outcome indicator of serious mental illness, using data for boroughs and at the more local scale of census wards. He applied Bayesian smoothing methods for the small area suicide data because of the small numbers of suicides occurring at this scale. Suicide was associated with a set of variables which he interpreted as reflecting aspects of 'anomie' (discussed below) and also with social class and a composite measure of material deprivation. Particularly for males, deprivation factors showed a positive association with suicide levels at borough level, as well as for wards. There was also some evidence that in areas of less extreme deprivation, outside the most deprived inner city, local differences in deprivation might have particularly strong effects on variation in suicide. This suggested that socio-economic conditions in the wider area of the city were important for mental health, as well the micro level associations at the scale of very small areas. The finding parallels work on contextual or 'place' effects in general health inequalities discussed in Chapters 1 and 4.

In addition to the studies of particular aspects of mental illness, several other studies in different countries have examined the ecological relationship between psychiatric health service use for mental disorders and socio-economic attributes of urban populations. Dekker et al. (1997) reported associations between deprivation and psychiatric admission rates in Amsterdam. Driessen et al. (1998) reported a multi-level study of individual records and residential area statistics for patients with non-psychotic, non-organic mental disorder in the city of Maastricht. They found that people from deprived areas had greater levels of service use, after adjusting for individual deprivation. This study illustrates again the point that contextual, as well as individual, aspects of poverty may be important for variation in mental health.

Jarman and Hirsch (1992) reported a multivariate regression analysis of acute psychiatric hospital admissions in 185 English health districts in the mid 1980s, which they tested for correlation with a large number of other variables. Among the strongest associations were those with composite and individual measures of urban social deprivation and poverty, notification of drug misuse, health disadvantage reflected in general levels of mortality and morbidity (measured as percentage permanently sick) and high population density. At a finer geographical scale, within parts of London and the outer metropolitan region, Glover et al. (1998) examined variation in small area data on admission rates to hospital for acute psychiatric care. These rates were found to be associated with characteristics of the ward populations, including measures of deprivation such as unemployment and poverty. Some studies in London have also examined variation in reasons for admission between relatively deprived inner urban areas and the more affluent outer city areas. Bristow et al. (2001) examined social and behaviour factors in acute psychiatric admissions in an inner city borough and an outer city borough in London. They found that, in particular, aggression, self-neglect and accommodation problems were more common in the inner city sample.

In New York City, similar relationships are found between socio-economic conditions and rates of hospital admission for mental health problems. Siegel et al. (2000) used a conceptual model to design an index intended to assess mental health and wellbeing in New York as a whole. Socio-economic conditions were included in the index and were measured in terms of relative poverty and unemployment. Almog et al. (2003) examined admission rates to acute psychiatric inpatient care, standardized for age and sex, in zipcode areas of New York City, for men and women aged 15–64. Low median income in zipcodes was associated with higher local admission rates in 1990 and 2000, especially for males aged 15–64. Figure 7.3 presents data compiled by the New York City Turning Point Initiative (1998–2000) (described in Box 9.4). These data are for 59 Health Service Agency areas, showing the relationship between median income (as an indicator of average wealth in the community) and rates of

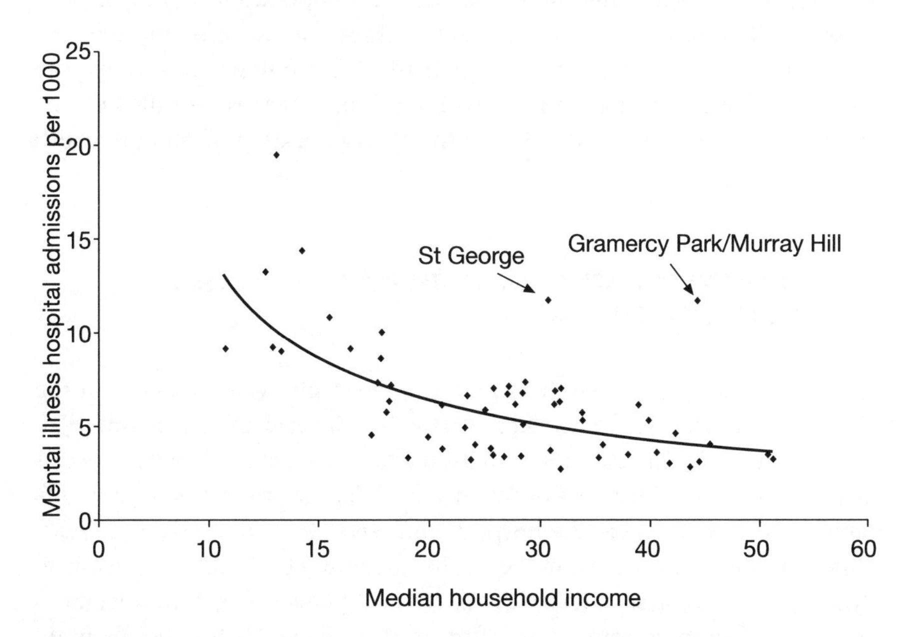

FIGURE 7.3 Health Service Agency neighbourhoods in New York City: hospitalization rates compared with average income of the population, 1990 (New York City Department of Health, 1998)

hospitalization for mental diseases. The diagram illustrates what appears to be an inverse relationship, with poorer areas (with lower income) having generally higher levels of hospital admission. While this simple illustration does not provide enough information to produce any conclusions about causality, it is strongly suggestive of a relationship between poverty and hospital use and suggests that differences in mental illness requiring hospitalization are associated with local poverty levels. However, some areas such as St. George and Gromercy Park/Murray Hill with higher average income also showed relatively high rates of admission. Thus even if poverty is associated with rates of hospitalization in a meaningful way, it is only likely to be a part of the explanation for local differences in this indicator of mental health.

These ecological studies all suggest that for populations in parts of the city where material deprivation is more severe, there are greater risks of mental illness. Whether this is due to 'breeder' effects or 'drift' effects cannot be established from this type of study. However, the evidence reviewed in Chapter 4 suggested that disadvantages in employment and housing are among the 'determinants' contributing to the causal pathways that produce illness, including mental illness. Thus there are strong arguments to support the view that material deprivation found in poor parts of cities is either causing mental illness or exacerbating existing mental illness. Nevertheless, as explained in the following sections, the excess morbidity experienced in deprived urban areas is also likely to be due to other processes associated with the organization of mental health care.

LANDSCAPES OF CONSUMPTION: MENTAL HEALTH AND ACCESS TO MENTAL HEALTH SERVICES

Some of the earliest research in medical geography was by Jarvis in the 19th century, who examined use of psychiatric institutions in America. He postulated 'Jarvis' Law': that there was a universal tendency for use of services to show a 'distance decay effect'. A larger proportion of patients came from areas close to the hospital and fewer from more distant areas. This relationship may to some extent be attributed to 'friction of distance' – the disincentive associated with the 'costs' of covering greater distances to a health facility (discussed in Chapter 5). In fact, Hunter and Shannon (1985) showed that, even in Jarvis' original data, the relationship between use and distance was far from consistent and that for some institutions the distance decay effect was much stronger than for others. Therefore, although proximity to a hospital appeared to be one of the factors affecting hospital use, a number of other factors also were apparently influential, including variable quality of transport routes, differences in admission policies of institutions and variations in the cost of care.

An aspect of the 'drift' hypothesis concerning variation in population mental health is that people with mental illness will tend to be drawn towards areas where services are provided for them. This has implications for ecological studies in the geography of mental health discussed above. It suggests that use of services depends not only on risks to mental health in the local environment, but also on variations in supply of, and access to, health care. Carr-Hill et al. (1994) drew attention to the fact that when patients are cared for in long stay psychiatric institutions they are likely, on discharge, to remain living in the area close to the hospital. This tendency for mental health facilities to 'generate' local populations of mentally ill people in their catchment areas may also result from 'revolving door' cases when patients experience repeated admissions and also may need ongoing support and surveillance while living in the community. Such cases may gravitate towards the places from which they are receiving care. Maylath et al. (1999) showed that in Hamburg, Germany, the risk of hospitalization of psychiatric cases was highly variable between areas. Admission rates for schizophrenics was associated with close proximity to psychiatric units. Risks of admission for neuroses, personality disorders and abuse of drugs and alcohol were concentrated in areas of low social status. Maylath et al. (2000) also showed, using geographical analyses of hospitalization rates, that utilization of addiction-psychiatric facilities was highest by patients from nearby districts. Lamont et al. (2000) showed a high degree of geographical mobility of mental health patients admitted to hospital in London (28% had changed address in the previous year), which may partly reflect this tendency to move towards institutional sources of care and treatment. Patients seen by community psychiatric teams were less likely to have moved. Johnson et al. (1998: 27) also suggested that some of the variability in schizophrenia in London depends on whether or not an area is close to a major rail terminus, which might be the point of arrival for mentally ill people 'drifting' to London from other parts of the country.

It is interesting to speculate whether in future this 'drift' of seriously mentally ill people towards major psychiatric facilities may become less evident as provision of community based mental health care is increasingly developed in patients' original residential areas. At present, community care is not always sufficient to support such mentally ill people in their places of origin, and this may be especially true of isolated rural areas. Techniques in telemedicine, discussed in Chapter 5, have been introduced in remote rural areas in some countries to enhance local mental health care. For example, several authors (Hawker et al., 1998; Dossetor et al., 1999; D'Souza, 2000a; 2000b; Mitchell et al., 2001) have described the use of telepsychiatry in Australia. They report that this approach has the potential to enable people needing specialist psychiatric care to be treated in their local hospitals and to support mental health care staff working in remote settings. If this approach were used more widely in future, it might help to offset some of the impetus toward the service dependent ghetto (discussed above) which is created by concentrations of specialist staff in major urban psychiatric facilities.

Some populations with high levels of geographical mobility may be especially vulnerable to mental health problems and also have poor access to care, and support services, so that they are more susceptible to mental illness crises requiring hospitalization (Sinnerbrink et al., 1996). Work by Fullilove (1996) is discussed below which theorizes the psychological impact of migration due to urban change and renewal. Similar impacts can be important for those who undertake international migration. There are particular psychological impacts of certain types of forced migration, for example, for refugees escaping war and oppression in their countries of origin (e.g. reviewed by Aldous et al., 1999; Watters, 2001). These include problems of adjustment to the refuge setting, uncertainty over their future and persisting reactions to trauma. Rates of reported mental illness such as anxiety and depression are relatively high in refugee populations and they often have poor access to the mental health care that they need.

Homelessness is another important factor in the link between mental illness and geographical mobility. Many people with mental illness living in the community are unable to find and retain adequate and affordable housing and they are at high risk of becoming homeless. The stresses of homelessness may exacerbate mental as well as physical illness and, as discussed in Chapter 5, access to health services is more difficult for homeless people (Rosenheck et al., 2001). Ducq et al. (1997) reviewed studies of mental health among the homeless. Although the diverse nature of these studies made macro level assessment difficult, they concluded that a third of homeless adults have a prior history of psychiatric hospitalization. Estimates of rates of mental illness varied but some studies reported high prevalence of psychoses, affective disorders, alcoholism and substance abuse. Kuhn and Culhane (1998) have documented the association between homelessness and mental illness in New York City and Philadelphia, especially for those who are repeatedly or chronically homeless.

Some British studies suggest that the resource allocation mechanisms currently used to determine the funds available for local mental health services may not be fully sensitive to variations in need for mental health care. As a result, provision may be insufficient to meet all of the demand in some areas. Services may become focused on a relatively small number of patients in the most severe need categories, leaving little resource to respond to mental health care needs of the larger populations with less severe illnesses. Glover et al. (1999) demonstrated in a study of English health authority areas that the rate of use of services for mentally disordered offenders varied by a factor of 20:1 between areas. In areas where there is a very high demand for 'forensic' mental health care, this may consume a large proportion of local mental health resources and services may be insufficient to cope with this and still provide adequately for other patients with mental illness.

Goldberg (2000) commented on the King's Fund report on London's mental health (Johnson et al., 1998a), which described a crisis situation with services struggling to meet high demand due to social deprivation, including high unemployment, and large numbers of people living alone. Emergency care was provided through accident and emergencyservices, and community services were not sufficient in all areas. Proposed solutions include more resources, capital investment in services and better coordination between health and social services. Goldberg (1999) also argued that while community psychiatric care is preferable and beneficial for many patients, some short term inpatient and residential services are required as part of the spectrum of services needed to treat and support people with mental illness effectively. In urban areas like London, the pressure on hospital beds is so great that this inpatient provision is inadequate, making it less likely that the overall care strategy for mental health will succeed. Kisely (1998) argued that other major cities in Britain have similar problems to London.

In the USA, various strategies have been applied in attempts to make mental health care more efficient, and to control costs. One method involves capitation approaches (behavioural managed care organizations compete for contracts to provide mental health care, at a fixed rate per capita, for the populations enrolled with a particular health insurance plan). Another strategy is utilization management which aims to induce medical professionals to provide more cost-efficient care, for example, by reducing the length of time patients spend in hospital (Mechanic, 1995; 1997). However, the introduction of managed care strategies has not proved very successful for people with severe and enduring conditions such as schizophrenia. Boyer and Mechanic (1994) reported an attempt to reduce long hospital stays for psychiatric patients in New York State, which did not produce the expected changes in service provision. One problem was the fragmented nature of hospital and community mental health care. Recently there has been adverse media coverage of serious incidents of failure of care for seriously mentally ill patients, which suggests that mental health services are under significant strain in New York City and the surrounding area (e.g. Levy, 2002a; 2002b).

Thus mental health services in large cities like London and New York City are often depicted as under-resourced, and inadequate to meet the level of need for care. Where there is insufficient provision of community health care and human services support to meet the needs of people with mental illness in the local population, this may operate as a contextual factor, increasing the risk of worsening mental health and leading to higher rates of use of inpatient hospital care. Variation in health care provision and use is therefore a contributor to 'breeder' as well as 'drift' effects.

LANDSCAPES OF POWER IN MENTAL HEALTH CARE

Several studies of the geographies of mental health care illustrate the interplay of processes of power and resistance discussed in Chapter 3. Dear and Wolch's (1987) analysis included discussion of the roles of different groups of stakeholders involved in decisions about mental health services and their location. Typically the position of patients with mental disorders is presented as one of relative weakness in negotiation, as compared with mental health care professionals, planners and wider society. As a wider range of settings, outside institutions, are used to provide care to people with more serious mental illness, the question of who makes decisions about the placement of mentally ill people becomes more complex.

Pescosolido et al. (1998) identified different processes affecting use of mental health services, which they typify as patient's 'choice', 'coercion' of the patient by others and 'muddling through' (where agency is less clear). Patients with larger, closer social networks were more likely to have experienced coercion. Goeres and Gesler (1999) considered the roles of different actors and institutions in determining where people with mental illness are placed and suggested that the mentally ill patient often has limited influence over the decision making process. Parr (1997) examined negotiation between user groups, medical professions and the wider community over mental health service provision and location. While mental health care users often lack influence in these decisions, her account suggested a growing assertiveness on the part of user groups, associated with the introduction of consumerist objectives in the British National Health Service.

The social position, in terms of economic standing, gender and ethnicity, of different patient groups is likely to be associated with their varying patterns of use of mental health care. For example, there is ongoing debate over the reasons for variations among different ethnic groups in use of psychiatric care for serious mental illness (Nazroo, 1997; Preddie and Awai-Boyce, 1999; Bhui and Bhugra, 2001). Explanations which are often put forward are related to the discussion in Chapter 3 concerning socio-economic position, empowerment and health. They include: the effects of various forms of discrimination experienced by minority ethnic groups; cultural variability in the experience and expression of mental states (to which standard psychiatric methods may be insensitive); and varying access to the best mental health care.

Dear and Wolch's model drew upon earlier work concerning the tensions surrounding location of facilities to treat mental illness. Communities may seek to exclude people with mental illness because of the social stigma associated with psychiatric disorders (e.g. Dear and Taylor, 1982). This is interesting for what it reveals about public perception of mental illness and the social relations operating in communities, as well as for the outcomes for mental health facility location. Early work included studies by Smith and Hanham (1981) on proximity to existing mental health facilities as a factor influencing public attitudes towards mental illness. Their study in Oklahoma City, USA, compared two neighbourhoods, one of which was close to a psychiatric facility and the other more distant. They showed that those living close to the psychiatric hospital were in general more accepting of mental illness. However, attitudes to mental illness were strongly associated with overall attitudes to social and welfare issues, and the authors pointed out that those choosing to live close to the hospital might have been more tolerant of mental illness. Also, homeowners who were residentially more permanent in the area were more likely to reject the mentally ill, which the authors interpreted as being due to a 'desire for greater discretion about who is to live in one's neighbourhood' (1981: 161). Furthermore, informants who had direct experience of serious mental illness were less likely to be accepting of the condition.

This sort of research highlights the strong social stigma attached to mental illness, which is a major obstacle to the integration of mentally ill people into communities. More recent work by Takahashi and Gaber (1998) reports data from a national opinion survey of the population in the USA showing the extent to which the respondents would tend to reject or welcome different types of facility in their neighbourhood.

Mental health facilities typically elicited a 'negative neutral' response, tending towards rejection of such facilities (though some other facilities, such as landfill sites, prisons and factories, were likely to be more strongly rejected).

Other studies of institutional landscapes have used Foucauldian ideas of the Panopticon to interpret the power relations reflected in the design and organization of institutional spaces intended to exercise control and surveillance over people with mental illness (Philo, 1989). The nature and degree of control varied according to patients' social position. Parr and Philo (1996: 25–8) describe how, in Nottingham, mentally ill people of different social classes were historically treated in separate facilities. These institutions offered what were seen to be better therapeutic conditions for middle class inmates than those provided for patients in the 'paupers' asylum.

Contemporary models of care in the community offer a range of less institutional spaces for provision of care and support to mentally ill people. Some geographical work has focused on how these combine supportive and therapeutic functions with some of the control and surveillance roles which were associated with institutional settings in the past. For example, Parr (2000) used covert ethnographic observation in a drop-in centre in Nottingham to explore the processes of inclusion and exclusion operating there. She describes 'psychosocial boundary formation' operating in these spaces, whereby 'unusual norms' of behaviour were accepted uncritically by others in the centre, although in 'mainstream' settings they would be considered abnormal and perhaps unacceptable. Examples included making repeated rude gestures to others, standing in the middle of the room staring at nothing, or listening at the door. On the other hand, some behaviour was not accepted so readily, for example, if it disrupted group activities, or involved very aggressive behaviour towards others. Users of the drop-in centre would be included or excluded, depending on whether their behaviour transgressed the accepted psychosocial boundaries in this setting. Parr, like other geographers working in this vein, discussed her findings in the light of Wilton's (1998) interpretation of the Freudian concept of *unheimlich* as the notion of something 'uncanny' (in this case unacceptable behaviour) which threatens our sense of a controlled and predictable environment.

THERAPEUTIC LANDSCAPES: ASYLUM AND POST-ASYLUM GEOGRAPHIES

A further useful perspective on geographies of mental health comes from research on therapeutic landscapes as they relate to psychosocial health.

Fullilove (1996), for example, developed a theoretical perspective on the psychology of place. Individuals require a 'good enough' environment in which to live, determined by degree of attachment, familiarity and identity offered by a place. Attachment involves a mutual bond between a person and a beloved place. Familiarity involves the processes by which a detailed knowledge of a place is built up. Place identity involves derivation of a sense of self from places in which one passes one's life. If these processes are threatened by displacement, problems associated with nostalgia, disorientation and alienation may ensue, with associated risks to mental health. Fullilove used this model to explain the negative psychological impact of migration and displacement due to urban change and renewal. However, the model can also be used to understand the positive or 'therapeutic' aspects of the individual's experience of place.

Institutional spaces to promote wellbeing

The design of asylum spaces and other psychiatric service facilities is often associated with ideas of what comprises a therapeutic space, beneficial for mental health. Gittins (1998) and Parr and Philo (1996) have considered how the changing locational history and design of psychiatric facilities have reflected development in treatment of mental illness over time. Edgington (1997) describes how the design of the York Retreat was influenced by the movement for 'moral treatment' of mental illness in late 18th century England. Features of this asylum included the division of the space into areas for treatment of patients with different types of condition and behaviour (e.g. separating violently disturbed patients from those whose behaviour was calmer) with the aim of encouraging rational behaviour. The retreat aimed to reproduce an idealized form of tranquil family life and social civility. Attention was paid to making the place cheerful, with good lighting, bright and homely furnishings, flowers and domestic animals. There were efforts to reduce the impression of a carceral setting and to create opportunities for recreation and constructive activities both inside and outside the asylum. Prior (1993: 10–11) discusses the layout of the King's County Hospital in New York in 1938, commenting on how space in the hospital was identified for various medical personnel such as psychiatrists, psychologists and nurses and social services personnel, but not for occupational therapists or psychiatric social workers, for example. This probably reflected the medicalized and institutionalized nature and organization of mental health care at the time.

Psychiatric facilities constructed more recently reflect current ideas of what represents a therapeutic space for mental health care. Although

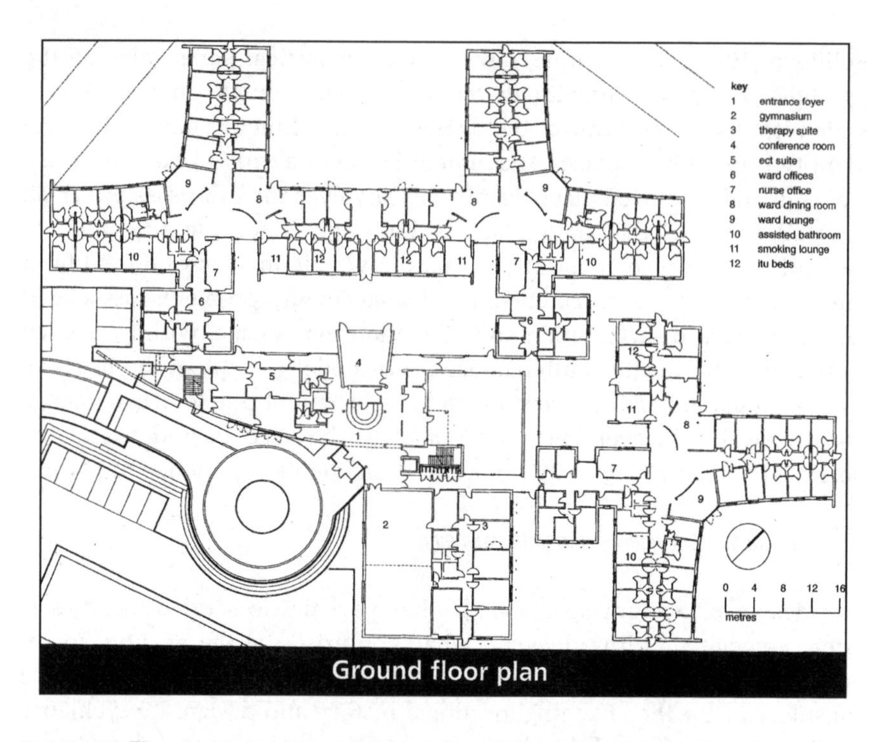

FIGURE 7.4 Floor plan of a new mental health unit in southern England (from Shaw, 2002: 22. Architects, Nightingale Associates; reproduced with permission from *Journal for Healthcare Design & Development*.)

mental health care strategies today are based on the principle of care in the community, some inpatient care is still needed for limited periods and in some parts of Britain there has been investment in new institutional settings to provide this. Shaw (2002) discussed a newly built mental health unit near Worthing, England. The building was designed by a team that included user groups reflecting the emerging view of people with mental illness as consumers, rather than just passive recipients of care. This corresponds to trends toward commodification of health care, discussed in Chapter 5. The designers were said to be trying to create a sense of a 'hotel' space, and they have apparently used various strategies to avoid impressions of regimented and rigid institutional space. The 'wards' comprised clusters of individual bedrooms with some provision for secure intensive observation and treatment (see Figure 7.4). The rooms were carpeted and had en suite facilities. The design recalled the modernist interiors fashionable for other types of public and private spaces in the late 1990s. The structure incorporated curving corridors, glass brick screens and internal courtyards bringing light and space into the heart of the building. There were several small dining and lounge areas to be shared by groups of patients.

In a rather contrasting view of what comprises a therapeutic setting, Bridgman (1999) used ethnographic methods to explore how some of the qualities of life on the street were experienced in a positive way by homeless people with mental illness. She discussed incorporation of these qualities into the design of innovative accommodation to help such people redevelop a sense of 'home'. Features such as a wide central corridor, reminiscent of a 'main street', and retention of elements of old industrial structures from which facilities were converted, such as partly demolished, unpainted brick walls, rough timbering and industrial artifacts, were thought to recall elements of streets outside in run-down urban areas.

All these designs for therapeutic spaces illustrate interesting tensions. On the one hand, they reflect requirements for spaces offering a degree of control and restraint, to create safe and 'healthy' settings encouraging socially acceptable, 'sane' behaviour. On the other hand, they show a desire to recognize human needs and the individuality of patients and to avoid a sense of oppressive control. This is often achieved by incorporating in these institutional settings elements which are seen as homely or familiar, cheerful, reassuring and natural. Recognition of individual identity of patients is expressed through the creation of therapeutic spaces for mental health care which suggest the exercise of 'individual choices' in public and private living environments.

The asylum as refuge

An aspect of institutional care which might be interpreted as 'therapeutic', or at least 'protective', is that of *refuge*. Tomlinson and colleagues described this role of the psychiatric institution as one which 'provided a place where they [the patients] were able to be unaccounted for' (1996: 125–6). The main reasons reported for admission to Friern Psychiatric Hospital in London were 'terror, trauma and trouble making', and long stay patients at the hospital tended to be seen as vulnerable to abuse or public censure. This vulnerability might indicate a need for a refuge away from ordinary life in the community, although these authors also questioned whether the vulnerability itself may have been a result of institutionalization.

Furlong (1996) pointed out that psychiatric hospitals allowed patients to feel part of a community, offering some sense of a shared, unifying identity, belonging and acceptability, as well as continuity of relationships. The key roles of the asylum in providing a refuge included the opportunity to retreat from the general public or from difficult relationships within the home, an escape from responsibilities of adult life, an affirming framework allowing emotional development. Continuing efforts at treatment in a psychiatric institution might also shield patients from the potential reality that their condition was incurable. Furlong argued that these dimensions are sometimes better met in hospital rather than community settings. Tomlinson similarly suggested that: 'the government in Britain has formally defined asylum broadly as a place in which one has privacy and a sense of belonging, and where some form of structured psychiatric regime is provided. Indirectly it has at the same time offered a definition of asylum in narrower terms as a place of residential health care' (1996: 180).

Safe havens outside the asylum

Spaces of refuge may also exist outside the institutional setting. Some geographical studies reflect strategies of resistance exercised by people with mental illness living in the community. Pinfold (2000) described how people with mental illness need 'safe havens' in an often hostile environment, in order to operate survival strategies and protect their fragile mental health. Knowles (2000) explained that these may not always be in 'care facilities'. She discussed how commercial public spaces, such as cheap, fast food outlets, provide relatively safe and comfortable environments which have longer opening hours and are less stigmatized than mental health care facilities. Parr (1999) has also discussed the 'unboundedness' of delusional experience and the unpredictable therapeutic properties of non-medical material spaces, from the perspective of people who experience delusion. Places that are perceived by people with mental illness as safe havens and therapeutic spaces may not be seen by others in the same way. Parr (1997) suggested that the tendency for some people with mental illness to use urban spaces in idiosyncratic, often disruptive, ways may be associated with their need to express resistance to the identity imposed on them by medicine or by society.

Social support

Therapeutic landscapes may also depend on the degree of social support offered by the community in a place (or, conversely, the lack of social support and integration). Early work by Durkheim (1951) suggested that *anomie* (social fragmentation and isolation) was associated with risk of suicide. This would suggest that the level of social support or cohesion is an important dimension of a therapeutic landscape. Berkman et al. (2000)

have reinterpreted Durkheim's ideas and have argued that social networks can affect health behaviour, and mental and physical health. They argue that supportive social networks can have beneficial effects on several dimensions of psychological health. Several geographical studies have demonstrated ecological associations between mental illness and lack of social cohesion or social support in the local community. These often are geographically correlated with measures of urban or rural deprivation, but they do nevertheless appear to have independent associations with mental health of populations.

Thus, for example, Giggs (1973) and Dean and James (1981) reported higher rates of schizophrenia in areas where larger proportions of the population are unmarried. Congdon's (1996b) study of suicide in London showed that suicide rates at the small area level were associated, independently of deprivation, with a measure of 'anomie' comprising information on one-person households, unmarried adults, high population mobility and privately rented accommodation (reflecting a tendency towards low residential stability in the population). Jarman and Hirsch (1992) showed district level psychiatric hospital admissions in English districts were associated with measures of social isolation and illegitimacy. At a finer geographical scale, and for a later period, Carr-Hill et al. (1994: 95–7) analysed variation among English wards in inpatient hospital admissions to psychiatric specialities from 1990 to 1992 (see Figure 7.5). Their approach was designed to control for the effects of supply of beds (see below). Population variables which predicted small area rates of

Lone parents
Proportion in households headed by a lone parent

No carer
Proportion of dependent people with no carer in household

Elderly alone
Proportion of people of pensionable age living alone

New Commonwealth migrants
Proportion of the population born in New Commonwealth countries e.g. Indian subcontinent, Caribbean, African countries

Mortality
Standardized mortality ratio for population under 75 years

Permanent sickness
Proportion of the population permanently sick and unable to work

FIGURE 7.5 Indicators selected for an index of resource needs for psychiatric care (Smith et al., 1996: 313)

admission included: proportion of lone parent households, proportion of dependents with no carer in the household, proportion of people originating from New Commonwealth countries, proportion of elderly people living alone, based on small area statistics drawn from the population census for 1991. These variables are all suggestive of associations between social isolation, lack of family support or mobility and demand for mental health care. Their multi-level modelling suggested that these relationships were variable between districts, suggesting effects operating at various geographical scales.

The MINI (Mental Illness Needs Index) proposed by Glover et al. (1998) includes elements which reflect aspects of social fragmentation or isolation, such as unmarried adults, and accommodation in boarding houses and hostels, which were found to be positively associated with variation in hospital use in London. Also, Siegel et al. (2000) include a dimension of support in an index to help to monitor trends in mental health and wellbeing in New York. The possibility of an 'ethnic density' effect (see Chapter 3) may also be relevant to the mental health and wellbeing of people from minority ethnic groups in urban areas.

These associations between social support and cohesion and mental health are probably not limited to urban areas. Research on social capital and health in areas across Australia provides further evidence of ecological associations between social capital and health indicators including suicide (Siahpush and Singh, 1999). In rural areas, it is often argued that an important aspect of social deprivation is social isolation. Malmberg et al. (1997), Higgs (1999), Charlton (1995), Kelly et al. (1995), and Sauderson and Langford (1996) have discussed the positive association between rural isolation and suicide.

Ecological studies based on aggregated data do not provide clear evidence that anomie is significant for health at the individual level. However, some studies have assessed individual mental health in relation to area social capital and have found evidence of associations (e.g. Kawachi et al., 1999). Individual 'stock' of social capital has also been found to be linked to mental health status (Burdine et al., 1999; Berkman and Syme, 1979). Thus there does appear to be evidence that the ecological associations are not spurious in the sense implied by the ecological fallacy. Greater levels of social support are positively associated with better mental health at the individual level.

It is also difficult to establish with certainty whether lack of social support and social integration factors *cause* mental illness in individuals, or whether the associations observed are due to a process of social disengagement on the part of people who are already mentally ill, particularly associated with geographical 'drift' into socially disadvantaged areas where social capital is relatively low. Again, it is likely that both processes are operating, but there is empirical and theoretical evidence that social support contributes to individual mental (and also physical) health.

The evidence has been comprehensively reviewed, for example, by Stansfeld (1999). One of the key ways that social support may influence mental health is as a protective buffer against the impacts of disruptive and negative life events, while lack of social support in the face of such events may worsen their damaging effects on mental health. Stansfeld (1999) also reviewed evidence that problematic aspects of social support and interaction in the social network may exacerbate existing mental illness. A recent study by Ennis et al. (2000) illustrated the varying ways that social support may buffer the risks associated with material disadvantage. Their survey of low income, single women in America suggested that social support plays a role in offsetting effects of poverty, especially for certain ethnic groups. Social support may protect disadvantaged individuals from the negative impacts of unequal power relations and social exclusion discussed above. Chronic poverty was found to be less distressing than short term material loss. Mastery and social support were more effective in buffering against material loss which created short term needs which could be addressed. Mastery was more important for European American women and social support more significant for African American women. Social support in the workplace, as well as in wider social networks, is also protective of mental health (e.g. Stansfeld et al., 1999).

CONCLUSION: A MODEL OF LANDSCAPES OF MENTAL HEALTH AND ILLNESS

We can therefore understand the local geographical variation in mental health of urban populations in terms of a number of dimensions of the urban landscape, summarized in Figure 7.6. Attributes of landscape associated with higher levels of mental illness in local populations include: physical and biological risks; material deprivation; social fragmentation or anomie. Mental health can be affected by the presence or absence of therapeutic attributes of the landscape that confer positive senses of attachment, familiarity and identity. Problems of access to services can exacerbate mental illness and local rates of illness may be influenced by 'drift' into areas of geographical concentration of mental health services. All of these factors can be understood as operating at the level of places and communities (or the broader scale of whole societies)

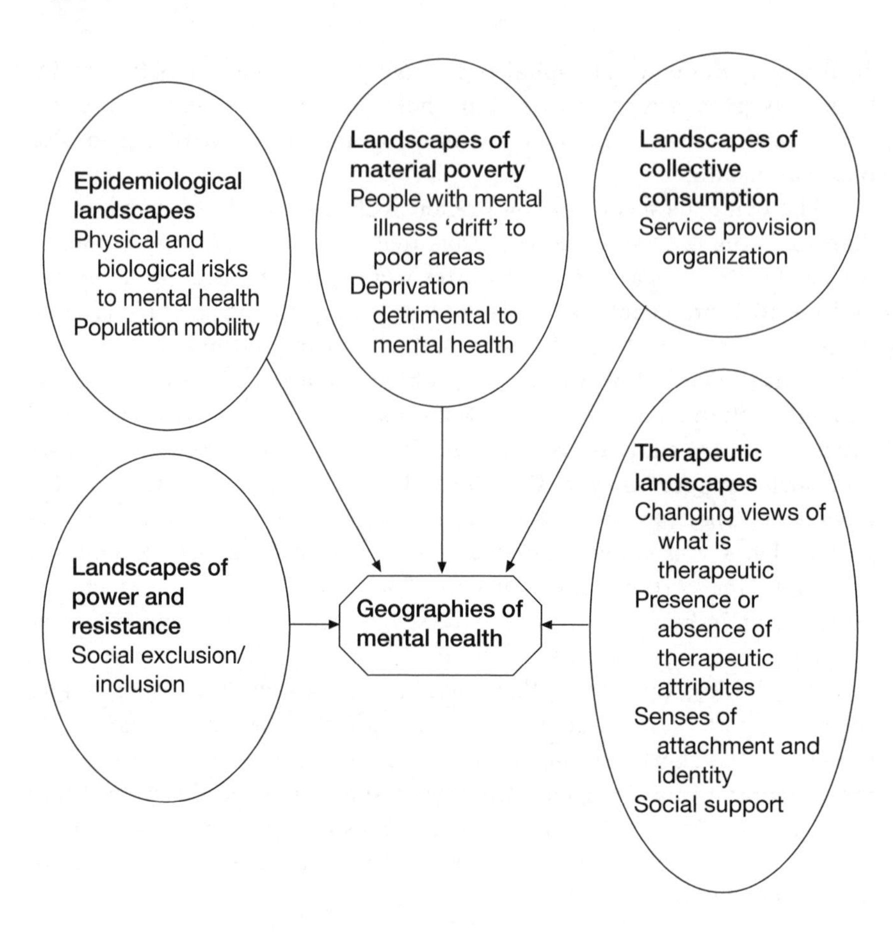

FIGURE 7.6 Geographical landscapes important for mental health in urban areas

and interacting with personal characteristics of individual people. In this sense they are fundamentally geographical dimensions of risk for mental illness.

The research reviewed has commented on various aspects of inequality in mental health. People with mental illness experience disadvantage in society, which includes geographical aspects of exclusion. Also, disadvantaged populations are at greatest risk of mental illness and they may receive worse mental health care. The cumulative effects of these processes is often to generate particularly intense concentrations of demand for mental health care in disadvantaged urban areas. Health systems frequently find it difficult to respond adequately to these needs. This has led to discussion of the 'special' conditions and needs for mental health care in urban settings and the idea of the 'service dependent ghetto'.

While the 'ghetto' model does not appear to be universally or straightforwardly applicable in cities around the world, it has provided a very useful emphasis on the significance for mental health of geographical variation in social, economic, political and physical environmental processes.

FURTHER READING

For those interested to read into the background of the geography of mental health, key references are: Dear and Wolch (1987); Philo (1989) in Wolch and Dear (eds); Parr (1997); Philo (1997).

Bringing the discussion up to date, these authors and several more are represented in an excellent collection of papers on 'Post-Asylum Geographies', published recently in a special issue of the journal *Health and Place* (2000, vol. 3, no. 6), several of which are referred to in this chapter. These include a good review by Wolch and Philo (2000).

For general overviews of mental health and illness and issues of mental health care, good sources are: US Department of Health and Human Services (1999); WHO (2001).

For a discussion of mental health issues as they affect minority ethnic groups I would recommend: Bhui, K. and Olajide, D. (eds) (1999) *Mental Health Service Provision for a Multi-Cultural Society* (London: Saunders).

OBJECTIVES AND QUESTIONS

This chapter is intended to help students to learn about:

- the nature of mental health and mental illness;
- the evolution of ideas about *how* and *where* it is best to care for mental illness;
- concepts used to explain mental health variation such as the idea of 'breeder' and 'drift' effects and the 'service dependent ghetto';
- the importance of places for mental health and mental illness, and the relevance of different 'landscape' perspectives for understanding inequalities in mental health.

Students may wish to think about the following questions to test their knowledge of these issues:

1 Using specific examples, discuss the significance of geographical landscapes for inequalities in mental health.
2 What have 'geographies of the asylum' revealed about the social relations involved in treatment of mental illness?
3 In what ways is geography important for 'post-asylum' mental health care?

8 Geography of an Infectious Disease: the example of tuberculosis

---

This chapter considers geographical dimensions of tuberculosis in urban populations and explains its global significance as a re-emergent infectious disease that has major impacts on public health.

The following discussion explores the relevance of landscapes of health geography, introduced earlier in this book, for our understanding of tuberculosis and its effects on human health. From a geographical perspective, the processes contributing to trends in tuberculosis infection can be interpreted in terms of the five theoretical dimensions of the geography of health reviewed in the first section of this book. This chapter considers epidemiological landscapes relating to the evolution and distribution of the pathogenic agent causing the disease, the distribution and movements of populations at risk of infection and the relationship between HIV and tuberculosis infection. Landscapes of socio-economic inequality, resulting in geographical concentrations of material deprivation, are of great significance for the geography of tuberculosis. Different landscapes of collective consumption are shown to result in variability in use and effectiveness of health services in different settings. Landscapes of power involving social relations between different groups in society are evident in the geography of tuberculosis and treatment of the disease. The responses of medicine and of wider society to tuberculosis in different times and places are also revealed in varying interpretations of therapeutic landscapes and settings for treatment. The chapter concludes by highlighting the health inequalities associated with these geographies of tuberculosis.

As this book is especially concerned with health inequalities in urban areas of high income countries, the examples considered here focus especially on these settings. These countries have well developed health surveillance systems, providing clear data on patterns of tuberculosis, and they demonstrate striking examples of re-emergence of tuberculosis in wealthy societies, where the fundamental difficulties in controlling the disease cannot be due to lack of national resources. The issues raised in the following discussion are, in many respects, even more important in poorer countries and are significant in rural as well as urban areas. WHO (2002) estimates that 80% of tuberculosis cases occur in 22 countries, most of which are poorer nations in Africa and Asia.

THE NATURE OF THE DISEASE

Tuberculosis is caused when a human host is infected by the *Mycobacterium tuberculosis* bacterium (it can also be caused by the bovine form, *Mycobacterium bovis*). Once infected, a person may carry the disease for many years, and, while protected by their immune system, may show no symptoms. The bacterium can lodge in various parts of the body and often affects the respiratory system. Between 5% and 10% of people who are infected with tuberculosis become sick or infectious during their lives. A person who is infectious can be identified as 'sputum smear positive' from microscopic examination of a smear sample of their sputum. Those who become sick experience symptoms such as fever and night sweats, cough, weight loss, blood in the sputum. The disease can result in severe illness and death in some cases, if it is not properly treated. Transmission of the disease is airborne, through the spread of tiny infectious droplets produced when a person who is sick with the bacterium in the lung, or larynx, coughs, sneezes, talks or spits (Snider, 1994; WHO, 2000b).

Tuberculosis (TB) is important because it is very damaging to public health in terms of the burden of illness and death (Blower et al., 1996; Lauzardo and Ashkin, 2000). In 1999 there were 8.4 million cases of tuberculosis worldwide (an increase of 400,000 since 1997) (WHO, 2001b: 8), and the disease is estimated to kill 2 million people a year globally (WHO, 2000b). Worldwide, this one disease causes about 3% of deaths from all causes and about 10% of deaths due to infectious diseases. In areas of Eastern Europe with relatively high adult mortality, tuberculosis accounts for around a third of deaths due to infectious disease (Figure 8.1). Almost half of the loss of disability adjusted life years (DALYs) due to infections are caused by tuberculosis in these countries (Figure 8.1). In 1993, the World Health Organization declared tuberculosis a global emergency because of the grave and worsening nature of the epidemic (WHO, 2000b).

While for every disease the detailed epidemiology and geography are specific, tuberculosis illustrates features common to many re-emergent infectious diseases. Compared with non-infectious diseases such as cancers and cardiovascular disease, or mental illnesses, the causes of infectious diseases like tuberculosis are relatively well understood in medical terms. Like many other infectious diseases, tuberculosis is more common in populations that are relatively disadvantaged socially and economically. As is also the case for many other infectious diseases,

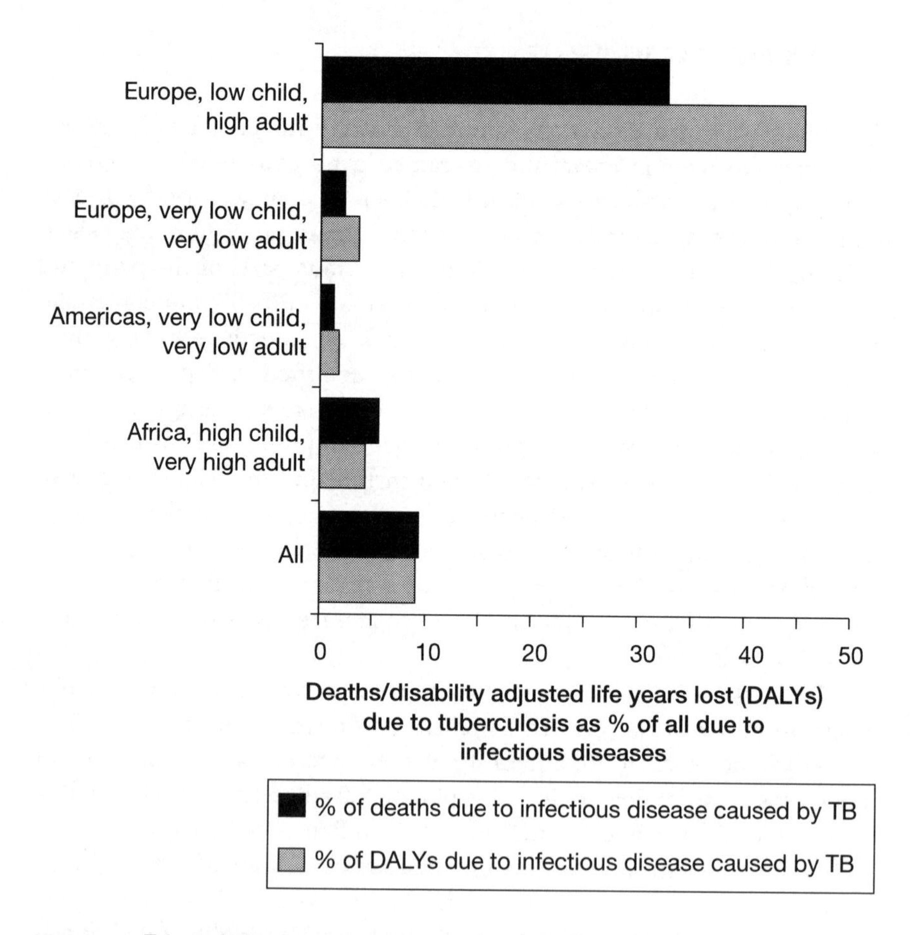

FIGURE 8.1 Tuberculosis as a cause of death and disability from infections in groups of WHO reporting countries, *c.* 1999 (WHO, 2000b: Statistical Annexe, Table 3)

tuberculosis is preventable or curable, so that much of the mortality and morbidity due to these diseases are in principle avoidable.

TUBERCULOSIS AS A RE-EMERGENT DISEASE

The changing pattern of tuberculosis is typical of the phenomenon of re-emergence of infectious diseases where they have previously been controlled, which was described in Chapter 1. For example, until recently, in England and Wales, the rates of tuberculosis in the population had declined steadily for over a century. However in the mid 1980s the improvement levelled off and then began to reverse in the late 1980s, with a 12% increase in rates between 1988 and 1992 (Bhatti et al., 1995).

Figure 8.2, which illustrates data from the WHO *Global Tuberculosis Report* for 2001, shows this trend for the United Kingdom, with the number of cases in each year represented as a percentage of the number in 1980. In the United States a similar and rather more marked trend was evident. In both countries the situation was particularly poor in major cities such as London, where by 1997 the rates of infection were three times the national average (NHSE, 1998), and in New York, where by

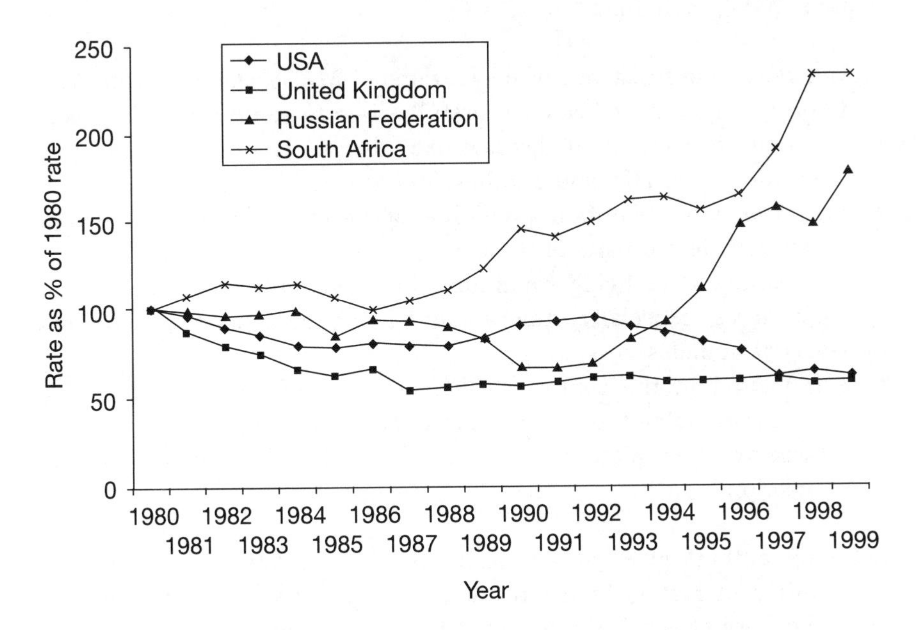

FIGURE 8.2 Trends in numbers of cases of TB notified in selected countries, 1980–99: (WHO, 2001b)

1991 case rates were about five times those for the USA generally (Morse, 1994). Only by the end of the 1990s did the trends once again begin to improve in the USA and the UK, but even then the situation in major cities continued to cause concern. For example, tuberculosis notifications continued to rise in London, in contrast with the situation in Britain generally (NHSE, 1998). Rates in London had been declining until 1987, though they still exceeded the national rates (e.g. around 23 per 100,000 compared with 10 per 100,000). After 1987, the rates in London rose more steeply than in the country as a whole and continued to rise. By 1997 they had reached 35 per 100,000, compared with the national level of 11 per 100,000 in 1996. Years of life lost through tuberculosis in London in 1995 were equivalent to about a fifth of the loss due to traffic accidents. At least 20% of cases were due to transmission rather than reactivation of old cases, indicating that the disease was spreading through the population.

In other countries the worsening situation has been more dramatic and has seen little recent improvement at the national scale. For example, Figure 8.2 shows data for Russia, where national rates in 1999 were 80% above those in 1980, and in South Africa, where rates in 1999 were twice the level in 1980.

The reasons for this re-emergence are complex (e.g. Lauzardo and Ashkin, 2000), including:

- the increasing resistance of new strains of *Mycobacterium tuberculosis* to drugs which have been successfully used to control it in the past;
- the reduced immunity to diseases like tuberculosis among people who have contracted HIV and are therefore affected by AIDS;
- high levels of migration providing opportunity for cross-infection between different parts of the world;
- increasingly poor living conditions for disadvantaged populations in some areas, especially crowded and inadequate housing and high levels of homelessness;
- a reduction in the comprehensiveness and effectiveness of health services providing tuberculosis surveillance and control;
- increasingly complete notification of cases of the disease in the population.

Different authors give varying accounts of the relative importance of these different factors in the resurgence of tuberculosis. This may be partly because of varying approaches to research and different timing of studies. It also seems likely that the processes are variable between different populations in different settings.

CHANGING ECOLOGICAL LANDSCAPES OF TUBERCULOSIS IN CITIES

The biological susceptibility of a population to tuberculosis will depend on the level of exposure to infection (through close proximity to people in whom the disease is active) and the level of resistance to the disease in the population. The physical segregation or proximity of groups of people within an area may be important for disease transmission. (Housing density as a factor in contagion is also discussed in relation to living conditions, later in this chapter.) Factors influencing population resistance include the development of new strains of tuberculosis. In some populations, a significant factor in resistance is the impact of HIV infection, weakening individuals' immune systems. Related factors, discussed later in this chapter, include variation in the physical state of the population (associated with living conditions) and varying effectiveness of health services in terms of local rates of vaccination and treatment of active cases of tuberculosis to minimize the risk of spread of infection.

There is debate over the significance of human migration for the spread of infection and some research has focused on how the presence of immigrant populations may affect the risk of tuberculosis. Cities like San Francisco, New York, and London are subject to significant national and international migration streams, which have often been discussed in relation to the epidemiology of tuberculosis, both today and in the past. Craddock (2000b: 41) suggested that in 19th century America, public health experts in cities like New York tended to present the risks of tuberculosis in quite racialized terms, representing migrants as a 'menace' to public health. In San Francisco, although the problem was seen to be particularly severe among the migrant Chinese population of Chinatown, Craddock argued that the discourse was more focused on poor living conditions in the area. Thus it appears that historically, as well as at present, accounts of risks for tuberculosis, and particularly the risks associated with migrant populations, varied between places according to differences in the social construction of the disease and in social relations between different groups making up the population.

More recent data for urban areas with large immigrant and minority ethnic populations also show high levels of tuberculosis. In London in 1997, the highest rates were in poor areas of inner London with the greatest ethnic diversity (NHSE, 1998); for example, in the borough of Newham the rate was 79 per 100,000 and in the boroughs of Tower Hamlets and Brent it was 77 per 100,000. The highest rate of increase of notifications was in Haringey borough, where it trebled over 10 years. In London, 40% of notifications were in populations ethnically defined as originating from the Indian subcontinent and the city was one of the few parts of the country where notification rates in people of Indian ethnic origin were increasing (from 12 per 100,000 in 1988 to 41 per 100,000 in 1993).

Bhatti et al. (1995) made a detailed study of cases in Hackney (a poor area of London with a large proportion of minority ethnic residents and many migrants and refugees). This showed that locally there was a 77% increase in tuberculosis from 1986 to 1993, to levels which were four times the national average. However, African and refugee populations in the area (originating from areas outside Britain where there are high levels of infection) only accounted for about half of the increase and it was shown that there was little cross-infection between these groups and other ethnic groups of residents. The authors suggested that, although there were higher levels of tuberculosis infection in areas with larger immigrant populations, transmission of the disease by immigrants to Britain was not the crucial factor in the increasing rate of infection in the population as a whole. Similarly, Bakhshi et al. (1997) found rates of notification of tuberculosis in populations of Asian origin in Birmingham to be 17 times higher than those of white residents and, among the Asian population, those born abroad had rates of infection four times those of Asians born in Britain. However, only 4% of Asian patients with tuberculosis had been living in Britain for less than a year, so that for many patients the disease was likely to have become active while they were living in Britain, not abroad. The low rates in the white population showed that there was little evidence of transmission from Asian to other populations.

Elender et al. (1998) made an ecological analysis of mortality due to tuberculosis in 403 local authority district areas in England and Wales between 1982 and 1992. They found that, after controlling for levels of household crowding and poverty, the ethnic composition of the local population was not significantly associated with variation in mortality. This finding also supported the view that the presence of populations with significant proportions born outside the UK is not as significant for the geographical variability of tuberculosis as the relatively disadvantaged conditions in which these populations are living in Britain.

Studies of tuberculosis in urban populations in the USA with large ethnic minority populations also show that tuberculosis incidence rates are high among these groups. Figure 8.3 shows the relative levels of tuberculosis infection among ethnic minority groups in San Francisco, compared with a reference value of 1.0 for the white US born population of the city. These relative rates were calculated for cases that were not part of clusters of infection arising from transmission from a single 'source case' and adjusted for age and sex variation. The rates are

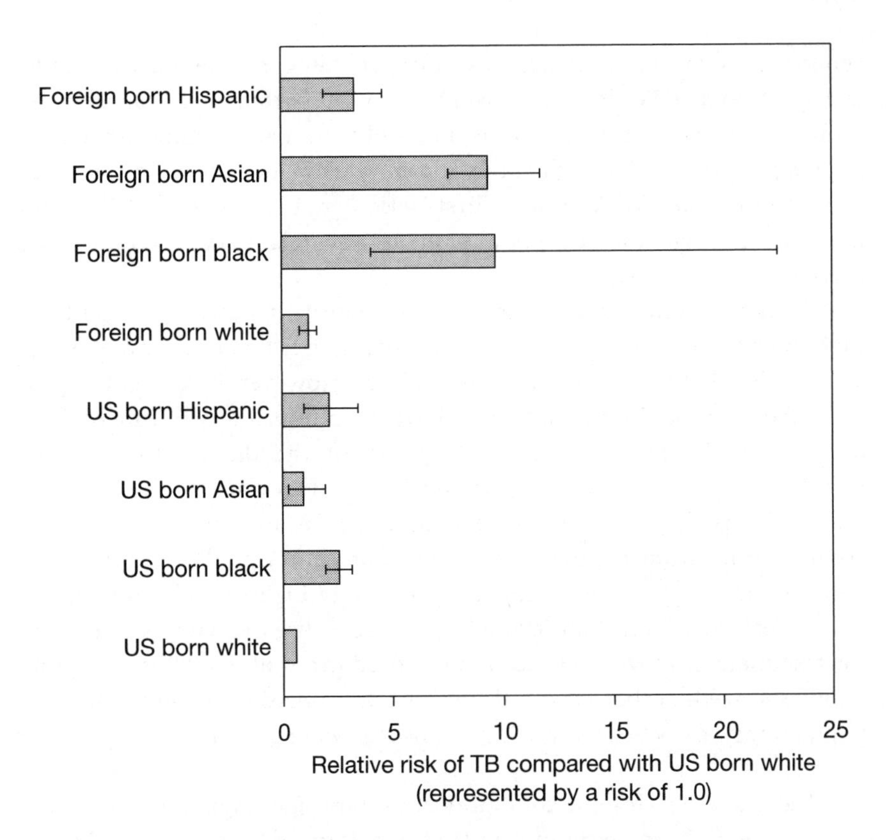

FIGURE 8.3 Relative risk of tuberculosis among different ethnic groups in San Francisco by place of birth (non-clustered cases), 1991–6 (Borgdorff et al., 2000)

particularly high among foreign born black and Asian populations. The rates were also significantly higher among Hispanic populations and US born black populations when compared with US born whites.

As in Britain, the evidence from the USA suggests that higher rates of the disease in some immigrant groups are not particularly important for the spread of the disease to other groups in the population. Borgdorff et al. (2000) reported that people with tuberculosis who were US born generated more secondary cases in the city than immigrants, and that transmission rates were especially high among US born black populations. Jasmer et al. (1997) showed that a fifth of Mexican born patients developing tuberculosis in San Francisco had acquired the infection in that city. Infection of US born persons by the Mexicans studied was rare. De Bruyn et al. (2001) found that in Houston, Texas, factors such as socio-economic conditions, homelessness, overcrowding, frequent use of public transport and HIV infection were more important for variation in tuberculosis than ethnicity. It is not only immigrant minority ethnic populations that show high tuberculosis rates in Britain or in the USA. For example, rates for American First Nation populations are 440% higher than the average for the population as a whole (Sanche Hodge and Fredericks, 1999: 281).

Thus research suggests that ethnic minority groups have relatively high rates of tuberculosis and those immigrating from abroad are particularly at risk, both in the UK and in the USA. However, it does not appear to be the case that immigrants who have been infected with tuberculosis abroad are the primary cause of spread of the disease through the population. The process of migration is less important in the epidemiology of this disease than living conditions for migrants in the 'host' country. Immigrant populations do not represent a public health risk to wider society just because they have relatively high risk of tuberculosis, and tuberculosis is not an 'exotic', 'imported' disease. This is important for treatment strategies. For example, Borgdorff et al. (2000) suggest that emphasis needs to be placed on limiting transmission in some US born populations, as well as on case finding among some foreign born groups.

Factors which inhibit communication between population groups in the city are likely to be significant for the pattern of tuberculosis. The low rate of transmission between ethnic minority groups and other populations in urban areas underlines their socio-geographical separation. Acevedo-Garcia (2000) presents a discussion which emphasizes the importance of residential segregation in the epidemiology of tuberculosis. There is also evidence that geographical barriers can limit communication and transmission. For example, within San Francisco, Bradford et al. (1998) demonstrated that different strains of tuberculosis did not readily spread between areas of the city separated by San Francisco Bay. Casper et al. (1996) showed that the most prevalent strains of tuberculosis in New York were not the same as those found in San Francisco. So, in spite of a significant degree of communication and exchange between these major cities, the level of inter-city diffusion is apparently relatively small.

The epidemiological pattern of tuberculosis in many countries has been importantly affected by the HIV, which weakens the immune system, making an individual more susceptible to infectious diseases. Grange, writing from a global perspective, stated that 'Infection by HIV is now by far the most important of the factors predisposing to the development of overt tuberculosis in those infected by the tubercle bacillus' (1999: 14). The HIV pandemic is having several effects which exacerbate the problems posed by tuberculosis. The chance that a person with HIV, who also has tubercular infection, will develop overt disease is over 50% over their lifespan, compared with only 10% for others infected with tuberculosis. Thus the chance of the infected person developing the active (and contagious) form of the disease in a year is 40 times greater for HIV infected people than others. This leads to a reduction in time interval between infection and manifestation of disease. The presence of HIV infection also modifies the clinical features of tuberculosis and leads to increased transmission because there are more source cases.

The relative importance of HIV infection for risk of tuberculosis appears to be geographically variable. Grange (1999) suggested that about 20% of tuberculosis cases in Africa are AIDS related. Moore et al. (1999) reported that of tuberculosis cases reported in the USA in 1993–4, 14% were also registered as HIV/AIDS patients. The relative numbers of tuberculosis patients who also had HIV infection varied between states from 0% to 31%. In 1995, at least 7% of patients with tuberculosis in London in 1995 also had HIV (NHSE, 1998). However, while the growth in tuberculosis cases in the United States and several African countries was attributed to the effects of the HIV epidemic, in the UK an expert committee concluded that the populations most affected by HIV did not 'overlap' significantly with relatively high levels of tuberculosis (Subcommittee of the Joint Tuberculosis Committee of the British Thoracic Society, 1992). Other factors might be more important in the spread of tuberculosis in Britain, for example, the effects of poor living conditions for some populations. There has also been concern in Britain, as in other countries, about the effectiveness of tuberculosis surveillance and treatment services (considered below).

A further factor in the epidemiological landscape relevant to tuberculosis is the growing prevalence of drug resistant strains of the disease, which are even more difficult to treat. This problem is exacerbated when treatment programmes do not effectively ensure that cases of active tuberculosis are detected and properly treated (Blower et al., 1996; Lauzardo and Ashkin, 2000). Also, patients infected by HIV are more likely to develop drug resistant strains of tuberculosis (Bradford et al., 1996). In New York City in 1991, 23% of new tuberculosis patients had drug resistant strains, compared with 10% seven years earlier, and 7% of cases in 1991 were multi-drug resistant (Morse, 1994). By 1995, 2.5% of cases in London were multi-drug resistant, compared with 1.3% in the UK as a whole (NHSE, 1998).

LANDSCAPES OF MATERIAL POVERTY

Tuberculosis is strongly associated with poverty and marginalized social groups. In 1995, at least 50% of tuberculosis patients in London were not in paid employment, 7.5% abused alcohol and 5% had histories of homelessness. (NHSE, 1998). A survey of people at a Christmas shelter for the homeless in 1992 and 1993 showed that 1.5% had tuberculosis (150 times the national infection rate) (Interdepartmental Working Group on Tuberculosis, 1996: 3). Friedman et al. (1996) reported on a sample of people in New York City abusing alcohol or drugs and claiming welfare benefits, among whom tuberculosis infection was 14.8 times higher than in the general population of the city. Living conditions are often highly differentiated between different parts of major cities, and are strongly associated with the geographical variability of prevalence of the disease.

Poor housing, particularly overcrowding, and homelessness are often reported to be strongly associated with the risk of tuberculosis in the population. This association shows complex interactions with other social and epidemiological factors such as minority ethnic status (as discussed above) and HIV infection. Moss et al. (2000) estimate incidence rates of 270 per 100,000 of the homeless population in San Francisco; the high rates were due to recent transmission in people who were HIV positive and non-white. In England and Wales, Bhatti et al. (1995) and Elender et al. (1998) showed that local levels of household overcrowding were the factor most strongly associated with variation in tuberculosis rates and mortality. In the UK, some of the highest rates of tuberculosis, and the most rapid recent increases in prevalence, occurred in London, and a local study in the borough of Hackney showed that the increase did not seem to be influenced by changing patterns of notification. The numbers of cases among homeless rough sleepers or people with HIV infection did not account for much change. Poor socio-economic conditions in disadvantaged areas were more important for the uneven development of the disease. Mangtani et al. (1995) also showed that, among the boroughs of London, rates of notification of tuberculosis were higher in areas with greater overcrowding and relatively large immigrant populations. However, *change* in the tuberculosis rates between 1982 and 1991 did not show the same association, though there was a positive association with local trends in unemployment. They suggest that this may reflect the importance of socio-economic deprivation factors other than overcrowded housing for trends in this disease.

Poor living conditions and economic deprivation therefore show close associations with the varying pattern of tuberculosis, and it is likely that the relationship is a causal one, since people with active tuberculosis living in overcrowded housing are more likely to pass the disease to others in the household, and poor material living conditions and poverty are likely to reduce a person's resistance to the disease, so it is more likely to become active. This underlines the importance for tuberculosis of the processes discussed in Chapter 4 which produce locally variable landscapes of poverty and material deprivation.

LANDSCAPES OF HEALTH SERVICE ACCESS AND USE

The risk of tuberculosis for local populations is also affected by sociogeographical differences in patterns of access to and use of services (including public health services). The high levels of material deprivation, poverty and social exclusion among those at greatest risk of tuberculosis are often associated with health damaging behaviours and lack of access to welfare services such as health care.

Much of the literature on prevention and care of tuberculosis emphasizes the idea of reducing the impact of the disease by improvement of surveillance and vaccination in the population and treatment of active cases of the disease. Strategies for achieving this often recommend methods for exercising greater medical supervision and control over the behaviours of those who are most at risk, especially ensuring that people with active forms of tuberculosis comply with effective treatment regimes. The significance for the geography of tuberculosis of landscapes of power, surveillance and control, discussed in Chapter 3, is therefore clear.

Since tuberculosis is in most cases treatable and curable, there has been considerable concern internationally over the failure to detect new cases and treat them effectively. The strategy promoted by the World Health Organization includes targets to improve the rate of detection (to 70%) and the extent of successful treatment of detected cases (to 85%). The WHO has placed considerable emphasis on treatment strategies involving directly observed therapy (DOT) whereby a health professional watches the patient take each dose of medication to ensure that the treatment regime is followed completely (Grange, 1999; WHO, 2001b). This requires investment of health service resources to provide the observation as well as the medication, and also demands cooperation from the patient.

By 1999, 82% of the world's population lived in countries which had adopted DOT (WHO, 2001b: 20) and a third lived in areas where DOT programmes were fully developed. For many countries where tuberculosis is highly prevalent, this represented quite a rapid advance in the extent of coverage of DOT programmes during the 1990s. For example, in South Africa, coverage by DOT had progressed from 13% of the population in 1997 to 66% in 1999. It was difficult to judge with certainty what proportion of notified cases of tuberculosis in South Africa were receiving DOT because of inconsistencies in reporting (WHO, 2001b: 34), but it was estimated that before 2005 this country would have achieved the WHO target of 70% case detection and 85% success in treatment.

However, progress in introducing the DOT approach has not been equally rapid in all countries where tuberculosis is most prevalent. The situation in Russia is an illustration. The Russian system historically developed along similar lines to other countries and there was a public health surveillance system which included tuberculosis screening of employees entering new jobs and sanatoria to treat active cases. By the mid 1980s rates had been reduced to a fairly stable level of 30 per 100,000 but then rose again to 67.5 per 100,000 by 1996. The tuberculosis surveillance system was undermined by lack of resources during a period of economic restructuring. Health service reforms did not prioritize the public health system. Sanatoria were becoming overcrowded and no longer able to provide adequate care for growing numbers of patients with active infection (Curtis et al., 1995; 1997; Lygoshina, 1998).

Problems for Russian services stemmed from an expensive, extensive system, which had been developed throughout Russia to manage the disease involving mass screening, diagnosis by chest X-ray, and long periods of inpatient care for those who were infected, with relatively frequent use of surgical treatment. This became untenable in the changed economic and social conditions associated with restructuring. In one instance, a British team from London, working with colleagues in Tomsk, advocated a different approach in line with WHO strategies, concentrating resources on finding and giving short course chemotherapy to most infectious cases (identified using sputum smear tests), using DOT, reducing length of inpatient stay and trying to tackle resource shortages. They found, however, that their Russian colleagues had little knowledge of WHO recommendations (Gleissberg, 1999; Drobniewski et al., 1996). In fact, this lack of knowledge among health professionals is not restricted to countries like Russia: DeRiemer et al. (1999) showed that many physicians in San Francisco were unaware of the best practice for care of tuberculosis.

Given this difficult situation in the Russian Federation, DOT coverage was only extended to 5% of the population in 1999. Two regions, Leningrad (around St Petersburg) and Murmansk, had implemented DOT programmes but otherwise development was limited. Of cases detected in 1998, only 68% were successfully treated, while 8% died and 7% failed to complete the treatment. In 2001 WHO provided aid in the form of a loan to develop DOT with the aim of extending it to 55% of the general population and 45% of the prison population over the following five years (WHO, 2001b: 34).

Thus the national and regional health care systems, varying from country to country, have an impact on the availability and effectiveness of services to detect and treat tuberculosis. This has important implications for the potential to control the disease in countries where it has the greatest impact on human health.

Even relatively rich countries face significant challenges in ensuring effective prevention, detection and treatment of tuberculosis. For example, it has been suggested in various reports (Interdepartmental Working Group on Tuberculosis, 1996; NHSE, 1998; NHSE London Region, 2000) that in London measures to tackle the resurgence of tuberculosis should include better access to treatment for groups at high risk and better use of rapid diagnosis services. For example, 31% of health authorities had no immigrant screening service, and almost 70% lacked an active case finding programme for the homeless. Improved case management was recommended, to make sure all patients complete their treatment, and DOT was recommended for some groups (such as the homeless, alcohol and drug abusers, the seriously mentally ill, patients with multiple drug resistance or with a history of non-compliance). Better recording and monitoring of treatment outcomes were needed (studies showed that up to 19% of cases, and 43% of homeless cases, were lost to follow-up). Better control of hospital infection was advised (38% of hospital trusts had no negative pressure isolation facilities for such cases).

In East London the significance of tuberculosis is reflected in the fact that a recent annual public health report was largely devoted to discussion of the problem and strategies being adopted to address it (ELCHA, 2000). The East London and City Health Authority area includes the London boroughs of Hackney, Tower Hamlets and Newham. As noted above, this area had the highest rates of tuberculosis in London in the late 1990s (Newham had the highest local rates of tuberculosis in the country). The area illustrates how several risk factors for tuberculosis can be concentrated in one area of the city. The report identified deprivation, overcrowding and homelessness as contributory factors. Though most cases of tuberculosis occur among long term residents the report commented on the relatively high rates among some of the many minority ethnic groups living in the area, including large numbers of immigrants and refugees, many of whom had not been vaccinated. The area also had relatively high numbers of people with HIV infection, compared with other parts of the country.

The ELCHA public health report placed a good deal of emphasis on the need to adopt a strategic approach in the local services, well coordinated with national and London-wide initiatives. National and regional strategy included three types of measure: preventing *transmission* by early detection and treatment of infectious cases; prevention of *development* of the disease through chemoprophylaxis; *primary* prevention including vaccination to protect people who may be at risk of infection (e.g. Joint Tuberculosis Committee of the British Thoracic Society, 1994: NHSE London Region, 2000). The objectives for London's tuberculosis service included improvement of coordination and general standards of care through a managed clinical network for the whole north-east sector of London. This part of the strategy aimed to establish clear lines of accountability and leadership and good coordination of different organizations that can play a role in tackling tuberculosis. It was also a way to ensure that the best, most effective practice would be adopted in care and treatment. The regional strategy also included enhancing methods of monitoring tuberculosis in the population, providing adequate numbers of specialist staff to provide care (especially tuberculosis specialist nurses), and setting performance targets for services (e.g. 90% of courses of treatment should be completed and 80% of children should be immunized).

Development of local services in line with national and regional standards for tuberculosis care was clearly important to address the kind of situation prevailing in East London. However, some factors were difficult for the NHS to control at the local level. For example, measures to improve housing conditions and reduce poverty were not conventionally responsibilities of health services, and without addressing these risk factors it would be difficult to limit the spread of the disease. Furthermore, some of the processes important for the local situation in East London operated at the global scale (for example, international migration and movement of refugees). Also, some processes in the NHS were controlled nationally rather than locally. For example, decisions about resource allocation to the local level were largely decided centrally. Also, in East London, development of the vaccination programme was hampered in 1999 by shortages of vaccine, and the public health report questioned whether the national NHS supply strategy for vaccine was effective. This example therefore illustrates the interactions between effects operating at different geographical scales in the context of local services' needs and provision of care for tuberculosis.

Furthermore, some of the difficulties in making tuberculosis control effective in areas like East London stem from the types of socio-geographical conditions prevailing locally where the services are most needed. These often produce circumstances in which conventional approaches to health care delivery are ineffective. More innovative approaches are required that are sensitive to local conditions and the needs and perspectives of users. These conditions can be interpreted in terms of landscapes of power and resistance, and especially in relation to issues of social exclusion considered below.

LANDSCAPES OF SOCIAL AND POLITICAL DIFFERENCE, POWER AND RESISTANCE

The populations most at risk of tuberculosis are those which are frequently disadvantaged and socially excluded. Responses by society to the disease often reflect the social relations between, on the one hand, these groups at risk and, on the other hand, the more affluent and powerful social groups with greater control over the collective resources of society that might be used to combat the disease. There may be reluctance on the part of those with influence to devote resources to services that will most directly benefit disempowered groups. Also, services may find it challenging to deliver effective tuberculosis care to socially excluded populations because their access to health care generally is relatively poor, as shown in Chapter 5. Furthermore, delivery of tuberculosis care requires surveillance and control by health professionals over the behaviour of the populations at risk. This raises issues about the extent to which the 'target populations' are willing and able to comply with these requirements, given that their lifestyles often involve strategies to resist or question the constraints which marginalize and exclude them from mainstream social participation.

Some commentators suggest that tuberculosis has not been generally seen as a problem requiring attention and public resources so long as it was confined to poor and disempowered populations in the city. Morse (1994) cites evidence, similar to that discussed above, that multi-drug resistant strains are relatively common among patients who are homeless, and among infected prison populations. The situation is important for public health as it increases the potential for infection and it should be a general cause of concern. However, Morse observes that the problem of drug resistance only generated a widespread public response after cases were identified in parts of New York State outside New York City, so that the disease began to be seen as a threat to areas inhabited by large numbers of affluent and influential populations. The same attitude had prevailed historically in US cities. Craddock (2000b: 46–7) also discusses the limited response to tuberculosis in 19th century San Francisco, which she refers to as an 'epistemology of neglect'.

Where tuberculosis prevention and control are being given priority, there can be difficulties in making services effective because the groups who need them most often have relatively poor access to mainstream health services, or are mainly in contact with specialized parts of the health service. This implies that tuberculosis care needs to be developed in ways designed to make it more accessible to these users. For example, as noted above, minority ethnic groups including immigrant populations have high rates of tuberculosis, as do poor populations generally. These groups tend not to use health checks or preventive care in the same way as the general population. DeRiemer et al. (1998) showed in a study of tuberculosis care for immigrants and refugees in San Francisco that these groups were less likely to seek medical care and that proactive follow-up care could be a particularly effective strategy in reducing the risks of tuberculosis for this group. Griffiths et al. (1993) showed in a general practice in East London that the most disadvantaged and needy patients are least likely to receive full health checks when they register and this reduces the opportunity for primary care doctors to screen for diseases like tuberculosis among the groups most likely to be affected (Griffiths, 2000).

An illustration of prophylactic tuberculosis care targeted through more specialized health services is discussed by Snyder et al. (1999), who evaluated DOPT (directly observed prevention therapy) delivered in methadone maintenance clinics for drug users. They argued that focusing preventive care on patients already in touch with these specialized services for drug users is cost effective because it reduced the risk of active tuberculosis in patients who were particularly likely to spread infections to others.

Making treatment and prevention services more available to marginalized populations at greatest risk, in specific types of urban settings, may help to reduce the spread of tuberculosis. However, some authors suggest that it will require special strategies to ensure that tuberculosis services are used effectively by patients. For example, Schecter (1997) reported on the DOT approach adopted in San Francisco, commenting that this needed to win patients' cooperation by methods including incentives and enablers, and that there are also questions over how to sustain staff morale while working in this challenging field. This perspective questions whether service users share the objectives of health professionals trying to ensure 'compliance' with tuberculosis care regimes. The situation might also be seen as one in which members of the medical establishment seek to exercise control over the behaviour of socially excluded groups that are seen as 'deviating' from 'correct' health care strategies. This would raise interesting issues about how far individuals have a right to freely determine their own health related behaviour and how far they have a responsibility to take action to protect the public health as well as their own. People who are excluded and disadvantaged by the structures and institutions of society might not consider that they have very strong responsibilities to protect public health and they may not value their own health in the same way as more empowered and advantaged social groups.

However, other authors suggest that patients' failure to use tuberculosis treatment may result not from unwillingness or lack of concern, but from lack of understanding of how to tackle the disease effectively. Tulsky et al. (1999) report on a survey of homeless people in San Francisco which explored their knowledge and understanding of risks of tuberculosis and their attitudes towards control measures. Many of the respondents to the survey were confused about risks associated with HIV transmission and risks for tuberculosis infection, and did not understand the very different pathways by which these two infections are spread. There was, however, a high degree of concern over the risks of tuberculosis and support for measures to ensure infected homeless people received proper therapy (e.g. through DOT). The authors suggest that education campaigns might help to improve knowledge and encourage homeless people needing treatment for tuberculosis to complete courses of treatment. Tulsky et al. (1998) also point out that the incarcerated prisoner population in San Francisco has high levels of tuberculosis infection and significant numbers of cases can be detected by screening of prisoners, but that, in the majority of cases, prisoners are released before their course of treatment is completed and very few (only 3%) then attend clinics to continue their treatment. These authors recommended improvements to continuity of care for these patients to ensure that they completed their treatment. The same group of researchers found that in a later cohort of tuberculosis infected prisoners who received standardized tuberculosis education, a larger proportion (over 23%) attended a clinic after their release. Offering a small financial incentive, in addition to education, did not seem to affect attendance at the clinic, suggesting that lack of motivation was not the main barrier to service use.

Other commentators (e.g. Farmer, 1997) have argued that structural factors associated with material deprivation and poor access to care are paramount in determining behaviour of people with tuberculosis, particularly in very impoverished conditions. He discusses observations in Haiti, and argues that there is too much emphasis on 'cultural factors' influencing use of health care and on educational measures to change behaviour and a relative lack of attention to constraints due to poverty and exclusion.

The models which have conventionally been used to detect and treat new cases of tuberculosis may not be well adapted to the life spaces and lifestyles of the populations they need to reach. Yaganehdoost et al. (1999) reported a study of 48 patients infected with the same strain of tuberculosis in Houston, Texas. They found that most of this group of patients were white, gay men who were HIV positive. Many of them were involved in bar-hopping, frequently visiting several different bars in the same neighbourhood. The pattern of transmission of tuberculosis in this group was complex. The authors suggested that measures to reduce transmission of tuberculosis in this type of population might more effectively focus on these locations rather than a conventional trawl of personal contacts. This is an interesting shift in perspective, from a geographical point of view, since it stresses 'spaces of risk' (see Chapter 1) rather than 'individuals at risk' and the need for sensitivity to the specific action spaces of certain groups in society (as discussed in Chapter 3). There is also an interesting tension involving the aims of health professionals trying to exercise control and surveillance over populations at risk by intervening in settings which can be interpreted as 'spaces of resistance', where most people are using these spaces in order to express lifestyles which challenge conventional behaviours in terms of sexuality and social networking.

THERAPEUTIC LANDSCAPES

The previous section discussed changing views about the appropriate settings for treatment of tuberculosis. Perceptions of therapeutic landscapes, as they relate to tuberculosis, have often reflected prevailing ideas about causes of the disease and the institutional conditions conducive to treatment. These have been associated with wider debates concerning the conditions that are beneficial for public health.

Craddock (2000b) discussed the social and geographical patterning of tuberculosis in the 19th century and how this related to medical theories about tuberculosis and urban salubrity. She focused her discussion especially on San Francisco and showed how the city was presented as being 'salubrious' in the sense of purity of atmosphere, an invigorating climate, with plenty of sunshine and mild winters. These aspects of the local environment were seen to be therapeutic, and therefore attractive, to people with tuberculosis. Reports from the period describe people with tuberculosis moving to the area to seek a healthier climate. As a result there was a tendency to blame relatively high rates of tuberculosis mortality in the city on the migration of infected people into the area. Poor migrant labouring populations were also seen as contributing to the risks of tuberculosis and needing to be segregated from wealthier and healthier groups in the population. The example illustrates similar tensions to those described by Gesler (1998) in Bath, where the healing reputation of the place led to growth which itself threatened the therapeutic nature of the city (see Chapter 2).

A key strategy for care of patients with tuberculosis in the 19th century involved isolation and treatment in sanatoria. Craddock's (2000b) detailed study of 19th century sanatoria, especially in San Francisco, described approaches which are no longer popular today and emphasized how notions of therapeutic spaces have changed over time. She also clearly showed how therapeutic spaces were differently designed for different social groups to protect the prevailing social order. Craddock (2000b: 169–97) describes how the built form and organisation of these sanatoria reflected the treatment favoured by the medical establishment of the time, involving complete supervision and monitoring of the patient by medical staff in ways which might today be seen as infringement by medical authority on the patient's personal space. The regimes emphasized good nutrition, outdoor rest in the fresh air and moderate exercise. Sanatoria were therefore in rural locations, at higher altitudes, and in buildings which included terraces or balconies open to the air and which had very open plan designs to facilitate observation of patients and thorough cleansing of the space.

Craddock (2000b: 172) also comments that the more imposing buildings used for administrative functions reflected the power exercised by the medical establishment in these sanatoria. The high degree of surveillance and control of patients showed similarities with the regimes of mental asylums in the same period (see Chapter 7). Different styles of sanatorium were built for patients from different social classes. Some sanatoria were designed as 'sanctuaries' for middle class patients (2000b: 186) where, while submitting to the medical regime, they could benefit from good diet, educational programmes and access to beautiful rural landscapes and a restful lifestyle. In contrast, an increasing proportion of patients were from the poorer classes, and the models propounded by planners for charitably funded sanatoria caring for these groups stressed modest accommodation and often included a work programme. Apparently sanatoria planners were concerned that conditions in the sanatorium would not be favourably compared with home conditions, lest patients become discontented with their lot in their normal home life. The patients selected for these sanatoria, albeit from the working classes, tended to be those most likely to cooperate with the regime and to recover. Some of the sanatoria included 'improvement' programmes in their regimes to encourage an industrious and hygienic but docile attitude. The most severely ill tubercular patients, and those from the most marginal groups in society at the time, were likely to be provided care in city and county facilities or tubercular wards in public hospitals. In these institutions there was less emphasis on cure and patients were seen to be admitted as a precursor to death from the disease.

Poverty and poor living conditions were also recognized to be contributing to the risk of tuberculosis, and Craddock (2000b: 198–245) described efforts to design a 'healthy city'. Ideas about what would comprise a healthier urban landscape included attention to standards for residential buildings with respect to space and ventilation. Campaigns sought to promote open air porches to homes, larger windows and more streamlined architectural detail that would be easier to clean, as well as open air classrooms in schools. However, there was also emphasis on what were seen to be 'deviant', unhealthy behaviours of poor populations. Craddock questioned whether public health campaigns to combat tuberculosis were motivated by medical objectives of controlling the disease or socio-political objectives of controlling the disadvantaged social classes and maintaining existing class relations.

Historically, therefore, perceptions of therapeutic landscapes, from the point of view of tuberculosis risk, were not only concerned with topographical aspects of environment. These perceptions can be seen as closely bound up with social constructions of what constituted 'healthy' social relations (stable and conservative structures of social position) and healthy behaviour (individual responsibility for public health). The more powerful social groups sought to impose these constructions on the rest of urban society and this reflects the significance of social relations and landscapes of surveillance and control as well as notions of salubrious environments.

Thus some themes apparent in present day notions of therapeutic landscapes were also seen in the past to be important in relation to tuberculosis. There has been continued emphasis on settings with social and physical features which help to ensure surveillance of populations most at risk and compliance with medical regimes by those who are infected. Aspirations for regeneration of urban environments, including improvement of housing conditions, are still presented as producing urban landscapes which promote the healthy city. On the other hand, therapeutic approaches no longer include spatial segregation of populations at risk in institutional sanatoria. Compared with 19th century perceptions of salubrious landscapes, there is less emphasis now on the

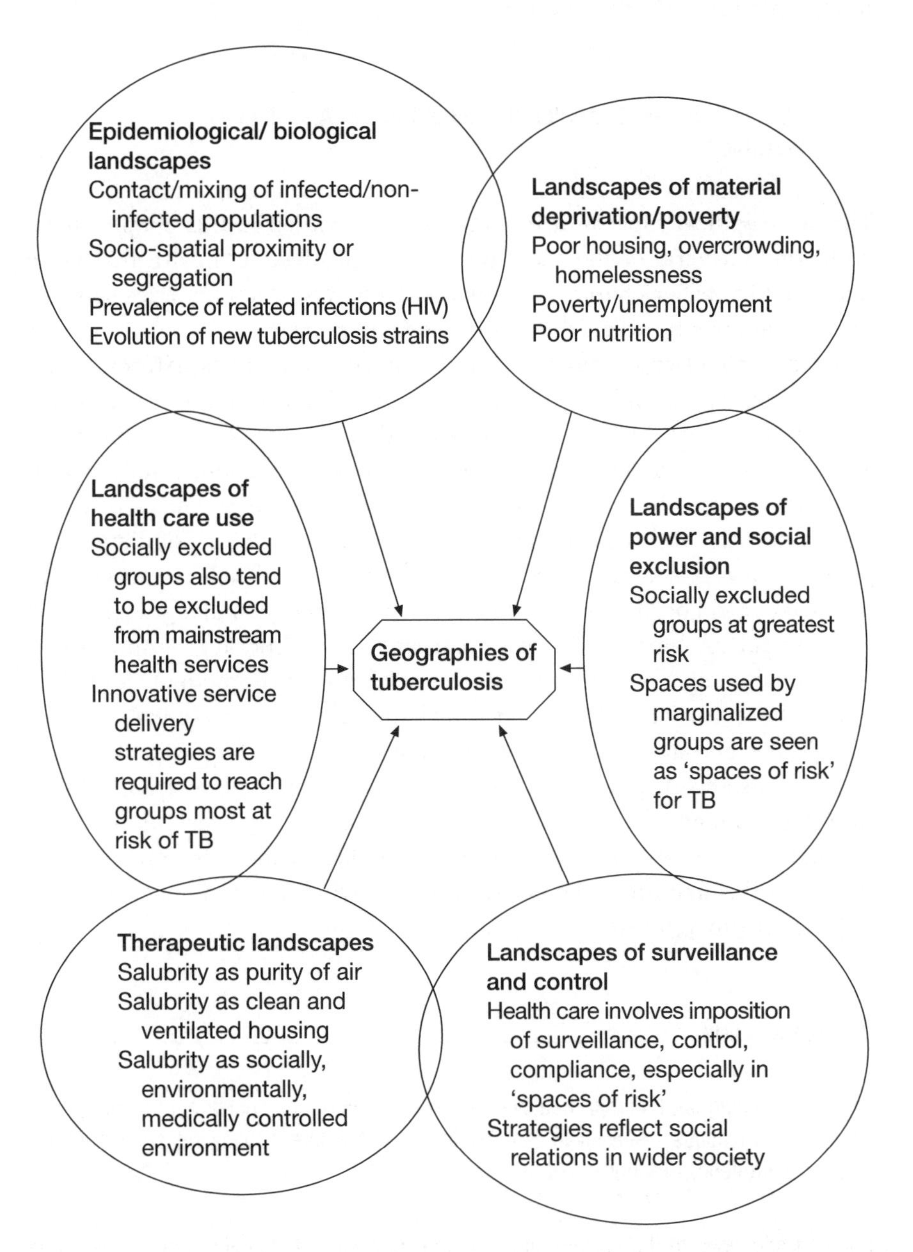

FIGURE 8.4 Geographies of tuberculosis

virtues of a clean, rural atmosphere. Making urban landscapes 'therapeutic' today involves the idea of introducing control and treatment into 'spaces of risk' in the community, rather than in institutional settings. There is more emphasis on good access to biochemical vaccinations and treatments to tackle the disease.

CONCLUSION: GEOGRAPHIES OF TUBERCULOSIS AND HEALTH INEQUALITY

The geographical aspects of tuberculosis considered in this chapter have highlighted several of the geographical perspectives on health that were discussed in earlier chapters in this book. Figure 8.4 summarizes some of the connections between these different perspectives.

Elements of epidemiological and material landscapes determine the risks of contagion and spread of the disease. Varying patterns of health related lifestyle and health care consumption also contribute to the risks, and these are influenced by processes of social exclusion. Public health rhetoric, historically and at present, has constructed accounts of *populations* at risk, and also of *spaces* of risk: places where populations at greatest risk of tuberculosis are concentrated and where social and medical control of the disease is seen to be especially challenging. Models of tuberculosis care, both historical and contemporary, emphasize a requirement for 'compliance' and cooperation by the patient, and they imply strategies for medical and social surveillance and control, both of the disease and also of the populations at risk. This stress on control of socially disadvantaged populations is evident in constructions of the important dimensions of therapeutic landscapes. The ideas of salubrity embodied in representations of therapeutic landscapes also reflect varying views through time, and across space, of what may reduce or increase the risk of tuberculosis.

FURTHER READING

For an engaging and comprehensive historical analysis of social and geographical aspects of infectious diseases in San Francisco in the late 1800s and early 1900s, I would recommend: Craddock (2000).

Good introductions to the current situation with respect to tuberculosis are to be found in Porter and Grange (eds) (1999); Porter and McAdam (eds) (1994); WHO (2001).

OBJECTIVES AND QUESTIONS

This chapter should be useful to students wishing to:

- understand the nature of tuberculosis and national trends in the incidence of tuberculosis;
- understand the factors which have been shown to account for trends in tuberculosis, especially its re-emergence in some countries where it was previously well controlled;
- be able to relate this knowledge to the epidemiological transition model;
- understand how geographical 'landscapes' reflect and influence varying risks for tuberculosis.

Questions to consider which may help students test their knowledge on these points include the following:

1. Why are poor areas of cities most likely to be affected by re-emergence of tuberculosis?
2. With reference to the recent experience of tuberculosis in specific countries, discuss the validity of the epidemiological transition model originally proposed by A. Omran in 1971.
3. 'Sanatoria in the early 20th century provided environments that were controlling as much as they were therapeutic.' Using examples, explain whether you agree with this assertion.

Health Geography in Strategies to Improve Public Health

In this chapter attention is focused on strategies to improve health and address health inequality at various geographical scales. The examples considered are particularly chosen to illustrate the potential for tackling health inequalities in cities by focusing on *places* as well as on people, and in many cases this involves local action, adapted to local conditions. However, as stressed in the earlier chapters of this book, the processes affecting health differences between small areas of cities operate at the national and international scale as well as at the local level, so there is also a need for internationally coordinated action.

The discussion here therefore starts with a focus on the role of *international* organizations in the reduction of health inequalities. Some examples of *national* strategies for reduction of health inequalities are then considered, and illustrations of *local* action in urban areas are reviewed. These demonstrate the importance of action outside the conventional spheres of medicine and health services. While there is considerable scope for action to reduce health inequalities in cities, this is a challenging task and some of the limitations of targeted action for public health are discussed. This chapter concludes with some comments on the key messages from this book, about the relevance of health geography and the ways geographical approaches may develop in future to contribute to the challenges of tackling health inequalities.

HEALTH FOR ALL: A GLOBAL STRATEGY FOR HEALTH IMPROVEMENT

Among the various international organizations and institutions which can play a role in improving the health of the world's populations, the World Health Organization (WHO) is most centrally concerned with this objective. The WHO was constituted in 1946 (WHO, 1946) and established in 1948. It sprang from other international organizations such as the League of Nations, which had originated before World War II (Sze, 1988; Curtis and Taket, 1996: 246–50). The constitution of the WHO (1946) defined health (see Chapter 1) and stated that 'the health of all peoples is fundamental to the attainment of peace and security and is dependent on the fullest cooperation of individuals and states' (WHO, 1998d).

The WHO is a diplomatic organization (part of the UN). It acts through a combination of conventions and agreements (adopted through constitutional processes by the member states), regulations (which countries may opt not to recognize) and, most commonly, recommendations (without legal force) (Curtis and Taket, 1996: 250–3). To pursue its goals, the WHO required a conceptual and strategic framework. This was agreed by the member states of WHO in 1977, when it was decided that the main social target of governments and of WHO should be 'the attainment by all citizens of the world by the year 2000 of a level of health that would permit them to lead socially and economically productive lives' (WHO, 1998d). This became the basis for the programme of work for Health For All (HFA), which is developed through regional HFA programmes.

In 1978, the WHO also adopted the declaration of Alma Ata (named after the venue in the former USSR for an International Conference on Primary Health Care, jointly sponsored and organized by WHO and UNICEF). This declaration stated that primary health care (PHC) was the key to attaining health for all (see also the discussion in Chapter 5). One goal, therefore, is that 'every individual should have access to PHC and through it to all levels of a comprehensive health system, with the objective of continually improving the state of health of the total population' (WHO, 1998d). The formulation of such goals, and the assessment of progress towards their attainment, has evolved over several decades. The approach includes definition of HFA 'targets' for change within different world regions.

Curtis and Taket (1996) have described how HFA targets for states in the European Region of WHO were first agreed in the 1980s. These included a set of measurable indicators or progress in member states toward HFA targets. They have been reviewed periodically to assess progress, and revised in response to changing conditions. In September

1998, as the year 2000 approached, some groups of WHO member states reviewed the HFA strategy and updated it for the new millennium (Herten and Van de Water, 2000). The European member states, for example, agreed a revised Health 21 strategy (WHO, 1998c). Box 9.1 outlines the content of the 21 targets included in this strategy and gives details of those that relate especially to equity.

Groups of member states in other WHO regions are also developing strategies for Health For All. The PanAmerican Health Organization (PAHO), which includes member states in North, Central and South America, has identified a number of initiatives to address poor health, especially in Caribbean and South American countries. For example, Chapter 5 discussed poor access to basic maternity care and high maternal mortality rates in some PAHO countries. The Safe Motherhood Initiative educates and mobilizes communities to work together to make motherhood safe and healthy. The goal is to guarantee that all mothers in Latin America and the Caribbean countries:

- have adequate prenatal care and maintain good hygiene during pregnancy and delivery and after delivery;
- get vaccinated against tetanus and take iron supplements during pregnancy;
- know the danger signs during pregnancy and childbirth and after delivery and will seek help;
- seek the assistance of a skilled birth attendant when it is time to give birth.

The Safe Childhood Campaign (described in Box 9.2) also aims to reduce illness and death among children in Latin American and Caribbean states. The strategy is intended to be sensitive to the broad set of factors that are important for effective child health care in the home, the community and the health services.

The HFA programme illustrates how WHO has promoted international action to reduce health inequality. It acts as a 'health conscience' to prompt continuing international concern for the 'principle of health as a basic human right', collates and exchanges information, and provides evidence based tools for countries to employ in their national strategies. It promotes, monitors and updates HFA policy, and works as a catalyst for action by providing technical support and cooperation through international networks and collaborating centres in the member states, and leadership in efforts to control disease or respond to health emergencies. However, WHO cannot legislate on behalf of European countries, so its actions remain in the diplomatic sphere, acting as innovator, advocate, consultant and facilitator (Kickbush, 1993).

Box 9.1 Examples of Health 21 targets relating to reduction of health inequalities in member states of the European Region

In 1998, the member states of the European Region of WHO agreed a revised set of Health For All (HFA) targets, setting goals to be achieved in the first decades of the 21st century. The Health 21 targets are described by WHO (1998c: Annex 2). The following discussion is partly based on the WHO description.

There are 21 targets in all, relating to the broad scope of the HFA policy. They include goals relating to *equity* in health throughout the life course, from a healthy start in life to health in old age. They cover *various aspects of health* including mental health, communicable diseases, non-communicable diseases, and injury. They also cover the *determinants of health* such as physical environment, health related lifestyles and behaviours, conditions in 'settings for health' such as the home, workplace, schools and communities. The targets also call for improved *coordination* between the different 'sectors' of society (i.e. not just health services) to recognize their responsibilities for, and potential impact on, health. There are targets requiring better integrated health services, to be assessed in terms of quality of care and health outcomes and with resources allocated on the basis of equity of access and cost-effectiveness. Health and other services should be staffed by well trained health professionals, with the right knowledge and skills to protect and promote good health. Also, some targets concern the *research and information needs* of Health For All strategies, the required degree of effective *joint working* between partners in the public, private and voluntary sectors and the sorts of *policies* which individual countries need in order to address Health For All goals.

The first two targets of Health 21 relate to health equity and directly address health inequalities, both at the international and at the intranational level. The *Health 21* document states the targets, and also explains how they are likely to be achieved and how they might be measured.

Target 1 Solidarity for Health in the European Region
By the year 2020, the present gap in health status between member states of the European Region should be reduced by at least one-third.

In particular:

1.1 the gap in life expectancy between the third of countries with the highest and the third of countries with the lowest life expectancy levels should be reduced by at least 30%;
1.2 the range of values for major indicators of morbidity, disability and mortality among groups of countries should be reduced through accelerated improvement of the situation in those that are disadvantaged.

This target can be achieved if:

- all countries contribute to reducing health gaps through international solidarity, mutual support and the sharing of resources, knowledge, information and experience;

continued ⇒

Box 9.1 Continued

- all countries elaborate a comprehensive development plan directly linked to their policy for health for all and a 'common health' vision for Europe;
- international institutions and funding agencies, together with WHO, coordinate their action in health and health related fields;
- all countries ensure that socio-economic, environmental and trade policies are not detrimental to health in other countries and that they contribute as much as possible to the development of disadvantaged countries.

Target 2 Equity in Health
By the year 2020, the health gap between socio-economic groups within countries should be reduced by at least one-fourth in all member states, by substantially improving the level of health of disadvantaged groups.

In particular:

2.1 the gap in life expectancy between economic groups should be reduced by at least 25%;
2.2 major indicators of morbidity, disability and mortality in groups across the socio-economic gradient should be more equitably distributed;
2.3 socio-economic conditions that produce adverse health effects, notably differences in income, educational achievement and access to the labour market, should be substantially improved;
2.4 the proportion of the population living in poverty should be greatly reduced;
2.5 people having special needs as a result of their health, social or economic circumstances should be protected from exclusion and given easy access to appropriate care.

This target can be achieved if:

- public policies are assessed with regard to their impact on equity, are gender sensitive and give higher priority to disadvantaged groups;
- policies ensure that access to educational and other social services does not depend on income;
- policies and legislation are aimed at implementing UN provisions on human rights (including those relating to women, children, people with disabilities, migrants and refugees);
- all sectors in society assume their share of responsibility for reducing social and economic inequalities and alleviating their consequences on health;
- public, private and voluntary resources are available to meet the social and health needs of the most vulnerable groups in society and provide access to appropriate, acceptable and sustainable care for all who need it;
- member states improve and harmonize their health information systems to record important socio-economic variables and analyse their relation to health conditions.

Box 9.2 PAHO Healthy Children Goal 2002

As part of the *Safe Childhood Campaign*, the PanAmerican Health Organization and the World Health Organization, in collaboration with other partners, aim to prevent 100,000 deaths in children under five years of age in the Americas by 2002, using a strategy of Integrated Management of Childhood Illness (IMCI). The Goal 2002 IMCI strategy is described by PAHO (1999).

The problem

In WHO member states in South and Central America, more than 250,000 children die before the age of five years due to illnesses which can be easily prevented or treated. The most common killer diseases for children in these countries are acute respiratory infections, diarrhoeal diseases, malnutrition, intestinal parasites, malaria, and measles. These are easily treatable or preventable by methods such as vaccination. They give rise to a large part of the medical care used by children in these countries. It is estimated that these diseases cause 60% to 80% of the paediatric consultations in health services, and 40% to 50% of hospitalizations of children under five years of age.

Why do these deaths occur?

Some of the reasons cited by PAHO for these high rates of childhood illness and death are outlined below.

Poor access to health care and information Large numbers of families with young children in these countries do not have access to appropriate preventive information or to health services because of geographical, social, economic, and cultural barriers. Parents may lack knowledge of the early danger signs of illness requiring treatment or of appropriate preventive measures.

Limitations of vertical treatment strategies In recent years disease control efforts have mainly used 'vertical treatment approaches' which concentrate on specific diseases and medical treatment for children. This highly focused approach has been seen as a good way to ensure the delivery of certain types of health care. However, vertical treatment approaches have been associated with missed opportunities for other forms of care. Vertical treatment may not detect early signs of diseases unless these are the principal reason for a medical consultation, so health problems may be allowed to become more severe, making it harder to treat them effectively. Vertical methods are also criticized because they tend not to educate parents on providing better general care for the child through breastfeeding, nutrition practices, preventive measures like vaccination or the promotion of child health through hygiene and stimulation.

Health workers may not have the right skills Training for health workers often does not enable them to identify early danger signs of diseases or to diagnose and provide effective treatment.

continued ⇒

Box 9.2 Continued

Some use of medical treatment is inappropriate Some medicines are prescribed unnecessarily (e.g. antibiotics used in excess, cough syrups, anti-diarrhoeals). Medical laboratory services are not always used appropriately and hospitalization for some cases of pneumonia and diarrhoea is unnecessary.

The solution
The Integrated Management of Childhood Illness (IMCI) strategy promotes improvement of three components of child health care:
- the skills of health professionals;
- the quality of the health system;
- family and community practices.

Following the IMCI, the health professional will take advantage of a child's visit to a health facility to assess his or her overall health status and identify any problems or disorders by:
- identifying critically sick children who require hospital treatment and using improved methods of referral to hospital care;
- focusing attention on the child's integrated care, not just on the original reason for the consultation;
- evaluating the child's nutritional status, vaccination record, and growth and development, as well as preventive practices used in the home;
- improving the quality and utilization of child health care in the health services and in the community.

IMCI is also said to optimize the use of community resources because:
- It identifies and explains healthy practices in the home, focuses on improving family and community practices important to the child, and provides preventive information to the parents and community.
- It actively involves the participation of the community so that preventive actions to support the wellbeing of the child are taken.
- It adapts to the reality of local health conditions and the operational capacity and cultural realities of each country.

The ICMI strategy is based on a holistic approach to child health, taking into account the child's family and home environment and all aspects of the child's health, rather than focusing on a specific disease. It aims to reduce the main causes of illness and death in the child population as a whole, rather than the more unusual diseases. It aims to tackle aspects of health systems that are seen to need improvement in the countries concerned. Also it tries to adapt the strategy to local conditions, being sensitive to the local context. Because the perspective used is concerned with the health of children in their wider communities and is sensitive to local variations in context, the approach reflects several concepts that are also important in health geography.

INTERNATIONAL MULTI-SECTORAL ACTION TO IMPROVE HEALTH

The Health 21 policy for the WHO European Region advocates 'multi-sectoral strategies to tackle the determinants of health, taking into account physical, economic, social, cultural and gender perspectives' (WHO, 1998c: Chapter 1, Section 1.2). International action and policy on these determinants require involvement of a range of other international institutions apart from WHO. For example, in the European Region, the European Union, the World Bank, the European Bank for Reconstruction and Development, the Council for Europe, and a number of international charitable organizations need to work together with WHO to achieve Health For All goals (WHO, 1998c: Section 1.3). The roles of the European Union and the World Bank provide illustrations of the potential and limitations for action by such organizations.

The European Union (EU) can intervene more directly than WHO to pass legislation determining action in EU member states. However, health gain was not a primary aim for the EU when it was originally established to promote economic cooperation. Objectives for greater integration in matters of social welfare and health have more recently been incorporated within the scope of action of the EU (Moon and Curtis, 1998). The involvement of the EU in health policy stems from the relationships between health and economies in Europe and the need to support and regulate health related industries (Ashton, 1992a; Verwers, 1992). Issues of most direct concern for the EU include, for example: environmental pollution by industry; consumer protection; the facilitation of more open labour markets for medical workers across the EU, which requires mutual recognition of qualifications; and regulation of the pharmaceutical industry. The Maastricht Treaty, agreed in December 1991, paved the way for greater economic integration in the European Union, and placed more emphasis on health protection as a focus for EU policy (particularly in Article 129). It recognized the need to consider the health impacts of different social and political sectors.

The Social Chapter of the Maastricht Treaty has gone some way towards helping to reduce variability across Europe in the factors that affect health. For example, European directives on working conditions now have some influence on the situation across the whole region. However, the level of integration of EU policy making is variable (Williams, 1998). Only some areas of EU policy can be agreed by a majority vote of countries; others require unanimous agreement. The principle of subsidiarity ensures that some issues are matters for national decision making. Thus, for example, health service organization and funding remain largely outside the scope of integrated EU policy. Moon and Curtis (1998: 304) argued that this may have some advantages, allowing countries to respond flexibly to their particular national conditions and needs, but that there are also disadvantages. Differences in access to and quality of health care across the EU, for example, are unlikely to be beneficial in terms of achieving greater equity in health across the region. Countries choosing to invest more in health and social measures to improve health may appear, at least in the short term, less competitive on narrow economic criteria than countries spending less of their national income on health care. A further complication arises from the fact that the EU region is not currently as extensive as the WHO European Region (the latter includes countries from the ex-Soviet block in Eastern Europe, Scandinavian and Mediterranean areas which do not belong to the EU). Ashton (1992a) drew attention to the need for more coordinated partnership between the WHO and the EU. On many measures, Eastern European countries have a worse health experience than Western countries (e.g. Moon and Curtis, 1998; and see Chapter 1 of this book). Effective action on health within the current boundaries of the EU may result in a widening of this regional east–west health divide in Europe unless similar action can be brought to bear in the east. Stein (1993) suggested that broader European coalitions such as the Council for Europe have a role to play and that policy needs to be sensitive to professional and public views, as well as international politicians and administrators.

De Beyer et al. (2000) described the role of the World Bank, suggesting that the rationale for its involvement in international health issues springs from the importance of population health for a healthy workforce and for socio-economic development. The World Bank seeks to improve health, nutrition and population outcomes of the poor, enhance the performance of health care systems and secure sustainable health care funding. These objectives are being pursued through several different types of activity, including extending loans and credits and making development grants to assist the development of health services. The World Bank also offers policy advice and research expertise, especially with respect to how to use resources in a cost-effective way (e.g. World Bank, 1993). In 1999, the Bank was supporting projects in 84 countries.

De Beyer et al. (2000) also commented on factors which limit the effectiveness of World Bank activity in the health sphere. Interventions to improve public health need to be long term to be effective because a person's health is influenced by conditions experienced throughout their life course. Such action is diffuse, and often intensely political, involving a range of partners and activities in different settings. It is difficult for an organization like the World Bank to engage with such processes and to monitor and evaluate them. Health objectives may not always be compatible with other objectives of World Bank policy. It is acknowledged that the kinds of macroeconomic structural adjustments advocated by the Bank lead to reduced health care spending by governments and that this is damaging for health (Van der Gaag and Barham, 1998). There are major questions over whether free market processes, fundamental to much of World Bank policy, are effective in tackling health variation. In fact, they are often cited as a cause of health inequality, as discussed in Chapters 4 and 5. There is a need to pool risk to insure against the high cost of illness, and, although the private sector is important for health care development, it needs to be subject to regulation by governments, which often lack the necessary institutional capacity to play this important role.

Action on health by international agencies such as the EU and the World Bank illustrates the contribution to health change of economic globalization, international economic regulation and supranational social policy making. These organizations may seem potentially well placed to influence economic development that is important for health (as discussed in Chapters 1 and 4). However, their mandates and responsibilities concerning economic wellbeing of countries do not always result in a focus on public health as a prime objective. Some of their other actions in pursuit of economic development have seemed in conflict with health goals. Their powers are often restricted in some crucial areas, and this also limits their effectiveness as agents for health improvement and reduction of health difference.

NATIONAL TARGET SETTING TO IMPROVE PUBLIC HEALTH

Individual countries have also set national policies and targets that reflect WHO principles, adapted to national conditions and priorities. In England, for example, policies on public health produced by successive governments have identified targets for health improvement (Fulop and Hunter, 2000). The *Health of the Nation* strategy (Department of Health, 1992) concentrated on average health conditions in the country as a whole and set targets for reductions of some diseases which contributed most significantly to the burden of illness and death. These included cardiovascular disease, a major killer in late middle aged populations in England, and suicide, which is an important cause of death in younger adults. One target called for a 40% reduction in coronary heart disease (CHD) and stroke in those under 65 by the year 2000, and another specified a 15% reduction in the suicide rate by 2000. This approach was criticized for failing to address *inequalities* in health as well as average conditions. Bryce et al. (1994) showed that while overall reductions in death due to CHD were achieved in the 1980s, the geographical disparities widened, because health improvements started earlier and progressed most rapidly in areas with initially better mortality rates.

The public health strategy of the following government (Department of Health, 1999a: *Saving Lives: Our Healthier Nation*) included aims to improve life expectancy and the number of healthy years of life, free from illness. It also aimed to improve the health of the poorest groups more quickly, to reduce the health gap between disadvantaged populations and the rest, though an explicitly quantified target for reduction in inequality was not originally specified (Fulop and Hunter, 2000). This strategy continued to focus on major causes of illness, disability and death in England: CHD and stroke; cancer, mental health and accidents. There was emphasis on action to improve health by a range of partners, at different levels: government and organizations at national level; local health agencies, local government and other actors in the community; and individual members of the public.

One national initiative in England designed to implement these government public health policies is the Health Action Zone initiative (HAZnet, 2002). In 1998–9 the government established Health Action Zones (HAZs) in areas of England which were particularly disadvantaged in terms of health, and where it was thought that innovative strategies might help to address poor health and health inequalities more effectively. Most of the HAZs were located in urban areas where there were particular concentrations of deprivation including parts of large cities like London and Manchester, a diverse range of smaller cities such as Luton, Wolverhampton, Sheffield, Bradford, and Plymouth, and some more rural areas such as Cornwall and the Scilly Isles. They were variable in geographical area and population size, including populations ranging from 180,000 to 1.4 million. In total 13 million people were living in HAZs (about a quarter of the total population in England).

The HAZs were intended to use innovative methods to improve health and tackle health inequalities. They were also expected to promote government objectives for the modernization of services, making them more effective, efficient and responsive. They were seen as representing a new approach to public health, which would link activity in different sectors such as health, education, regeneration, employment and housing. They were therefore expected to work collaboratively across different statutory agencies and involve local communities. A HAZ would often extend over the boundaries of administrative authorities for health and local government. There were 34 health authority districts and 73 local government authorities represented in HAZs. The HAZ programme was meant to try new schemes, which, if successful, might later be introduced into mainstream provision. The HAZ programme was allocated a relatively modest budget, equivalent to about 1% of the total NHS budget (Baud et al., 2000). The HAZ scheme has given rise to many different types of initiative. The National Evaluation Team, commissioned by government to assess the HAZ strategy, identified 582 different key 'purposes' of HAZ programmes (Baud et al., 2000). Some illustrations are considered in the next section.

Similarly, in the USA, Health For All objectives are the basis for the Healthy People 2010 initiative which is promoted by the US government, and this has also used a target based approach. Box 9.3 illustrates selected examples of the targets which have been identified as relevant for developing healthy communities. The targets listed reflect an emphasis on aspects of *places* (communities, workplaces, schools, local environments). Several of these targets concern action to address aspects of the geography of health which have been discussed in earlier chapters of this book. Examples include: reducing environmental pollution and exposure to chemicals such as lead; increasing access to health services for groups that are most likely to be excluded from mainstream services; improving workplace conditions; and improving levels of food security. Examples of action to achieve these sorts of targets at the local level in the USA are discussed below.

LOCAL APPLICATION OF POLICIES FOR HEALTHY CITIES: THE 'SETTINGS' APPROACH

The HFA strategy has a 'top-down' approach, involving international agreement by WHO and the member states to pursue jointly agreed targets. By contrast, the Healthy Cities initiative of WHO has developed as a more flexible and varied 'bottom-up' approach with many different local actions identifying themselves under the Healthy Cities banner (Ashton, 1992b). Healthy Cities initiatives put broad objectives of Health For All into action at a local level, recognizing that a significant part of the illness and death suffered by populations of cities throughout the world could be reduced by improving physical environment and social and economic conditions of their daily lives. The aim is to put health on the agenda of decision makers, build a strong lobby for public health in cities and develop holistic, popular and participative approaches to improvement of health for urban populations (WHO, 1999a). WHO (1999b) defines a healthy city as 'one that is continually creating and improving those physical and social environments and expanding those community

Box 9.3 Examples of Healthy People 2010 targets relevant for community planning

A recently published *Community Planning Guide* drew attention to 50 targets, selected from a longer list identified for the Healthy People 2010 initiative. A few of these targets are reproduced below. These relate to action to change the places and settings where people live and work. They aim to address some of the contextual effects that places can have on health, as described in geographical health research. The list is sourced from US Department of Health and Human Services (2001: Appendix B).

Communities
Target 7–11 Increase the proportion of local health departments that have established culturally appropriate and linguistically competent community health promotion and disease prevention programmes.
Target 19–18 Increase food security among US households and in so doing reduce hunger.
Target 21–9 Increase the proportion of the US population served by community water systems with optimally fluoridated water.
Target 27–13 Establish laws on smoke-free indoor air that prohibit smoking or limit it to separately ventilated areas in public places and worksites.

Access to health care
Target 1–6 Reduce the proportion of families that experience difficulties or delays in obtaining health care or do not receive needed care for one or more family members.

continued ⇒

resources which enable people to mutually support each other in performing all the functions of life and in developing to their maximum potential'.

From a geographical perspective, an interesting aspect of the Healthy Cities approach is the way that it aims to promote initiatives in local *settings*. Settings are defined as 'major social structures that provide channels and mechanisms of influence for reaching defined populations' (WHO, 1999c). A relevant setting might be a whole city, or a particular type of place within a city, such as home, school, workplace, or marketplace. Settings are spaces where regular interaction takes place between people, where information can be exchanged, gatekeepers can be accessed, specific populations can be contacted. Certain professional identities and particular practices and training traditions are linked to some settings (e.g. teachers delivering education in schools, vendors working in marketplaces). It may be possible to enlist the support of these workers in action for health promotion. These settings are good points of

Box 9.3 Continued

Target 14–19 Increase the proportion of adults who are vaccinated annually against influenza and ever vaccinated against pneumococcal disease.

Worksites
Target 7–5 Increase the proportion of worksites that offer a comprehensive employee health promotion programme to their employees.
Target 20–2 Reduce work related injuries resulting in medical treatment, lost time from work, or restricted work activity.
Target 20–9: Increase the proportion of worksites employing 50 or more persons that provide programmes to prevent or reduce employee stress.

Schools
Target 14–23 Maintain vaccination coverage levels for children in licensed day care facilities and children in kindergarten through the first grade.
Target 22–8 Increase the proportion of the nation's public and private schools that require daily physical education for all students.
Environmental Health
Target 8–5 Increase the proportion of persons served by community water systems who receive a supply of drinking water that meets the regulations of the Safe Drinking Water Act.
Target 8–12 Minimize the risks to human health and the environment posed by hazardous sites.
Target 8–22 Increase the proportion of persons living in pre–1950s housing that has been tested for the presence of lead based paint.

access to the populations being targeted and they offer good potential for wide social influence. Also they provide good conditions for participation and cooperation among different types of stakeholder. The idea of a 'settings' approach is consistent with theoretical perspectives discussed earlier in this book including Giddens' (1984) formulation of structuration theory, examined in Chapter 3 (especially his notions of the relevance of time-space paths for human interaction), and Bourdieu's (2000) concept of the habitus, influencing patterns of consumption, considered in Chapter 5. The concept is inherently geographical because of the emphasis on structures associated with particular types of *place*.

Schools, for example, are seen to be good settings for health promotion targeted at young people and their families. Schools bring together a range of different stakeholders in local community health, apart from children themselves: parents, teachers, education authorities, for example, all have a normal role to play in the school setting and they will be concerned about the school and willing to work together to improve conditions there. A range of other relevant partners, such as community groups, voluntary organizations, health care providers, other government agencies, can be engaged in healthy schools projects. Healthy schools projects aim to tackle a range of factors important for health of both pupils and staff of schools as well as their wider community. Strategies go beyond health education in schools and school health services to include efforts to create healthy environments, to improve nutrition and food safety, or to make opportunities for physical education and recreation, counselling, social support.

Similarly, *marketplaces* have been identified as important settings for health promotion through Healthy Cities initiatives. We saw in Chapter 5 that food retailing was an important consumption process for health of the population and this is one reason why marketplaces are identified by WHO as 'essential settings' for health promotion. Food security can be improved for the population using the market by making sure that the food sold is affordable, safe and nutritious. Furthermore markets are commercial and social centres; they are strategic sites at which to influence commercial practices and to communicate knowledge about health and health care. The WHO recommends that action taken in a 'healthy marketplaces' programme should relate to: the administration of markets; inspection of products and the conditions under which they are sold; water, sanitation and drainage, including toilets and washing facilities; environmental management such as waste disposal, pest and noise control; training and education for food vendors and consumers.

Earlier chapters of this book cited examples of strategies to tackle particular diseases that reflect a 'settings approach'. For example, in the fight against tuberculosis, schools in the most affected communities or social venues used by certain high risk groups can be efficient sites for screening. Arguments in favour of a 'settings approach' have also been put forward in national policy debates. Macintyre (1999) presented evidence to the Independent Inquiry into Inequalities in Health in Britain that made the case for such approaches in terms of geographical variations in health and of contextual effects in health variation. She gave illustrations of measures to affect the social environment in ways that build social capital and reduce the impacts of unhealthy social conditions in poor areas. They include: zero tolerance strategies for crime; community policing to reduce fear of crime in public spaces; increased local opportunities for healthy recreation and exercise; greater use of local schools outside normal school hours for social activities involving both children and adults. These would enhance social interaction, and provide safe, stimulating places for children whose parents work, but find it hard to afford the best child care services. This type of approach was reflected in British government proposals for changing conditions affecting health in the poorest areas (e.g. in Health Action Zones) (Department of Health, 1999a). Various settings for initiatives were proposed. For example, child health could be tackled through 'healthy schools', adult health through 'healthy workplaces' and health of older people, in particular, though 'healthy communities'. Examples of specific local interventions to improve health in the most disadvantaged places in Britain are discussed later in this chapter.

ILLUSTRATIONS OF LOCAL, AREA BASED STRATEGIES TO IMPROVE HEALTH

There are many examples of policies and interventions focused on geographical areas. This section considers some of these 'area based' strategies. Some of them seem consistent with 'whole system' perspectives discussed in Chapter 1, giving consideration to aspects of the local political economy and social life, which influence the determinants of health, as well as the conventional medical health service sector.

In the USA, an example of a city-wide strategy to improve public health in a large urban area was the New York City Public Health agenda, published in 2000 by the Turning Point (New York City Public Health) partnership. Some key aspects of this initiative are summarized in Box 9.4. The Turning Point strategy illustrated the WHO Healthy Cities principles of working in partnership.

In England, the Health Action Zone programme, described above, has included many different local initiatives. Specific illustrations of the types of activities undertaken in East London and City HAZ included action to encourage people to eat more fruit in their daily diet, such as an advertising campaign on local buses, and fruit tuckshops in local schools. Another initiative in East London was a breakfast club providing a venue for children who had to leave home early because their parents were working. It provided a safe place to be before school and an opportunity to eat a healthy breakfast to start the school day. Other initiatives were also started to help black people with mental health problems including one that aimed to provide new services to help young black men with mental illness. A third type of strategy aimed to increase levels of employment of local people in primary care, in order to tackle high local rates of unemployment, and also to make the health service more responsive to the needs of the patients it serves. Action has included a

Box 9.4 The Turning Point public health initiative in New York City

The Turning Point initiative aimed to address various aspects of public health in the city. Issues of concern included: large disparities in health between rich and poor people in the city; lack of access to medical care and community health promotion and illness prevention; lack of attention to determinants of health such as education, income, social position; inadequate mechanisms for community participation; and lack of coordination in approaches to health improvement. It was recognized that health needed to be considered in terms of the WHO definition of complete physical, social and mental wellbeing (see Chapter 1). Collaboration of both the health services and other, 'non-health' sectors such as housing, education, economic development and faith based organizations was seen to be necessary to improve health. New methods of working together were sought, to increase community control over health determinants. Turning Point aimed to create public health partnerships at various levels: city-wide, within each of the five boroughs of New York City, and in local communities. These partnerships would identify health concerns in the city and develop strategies to address them. The objective was to find new ways of joint working to improve public health.

The first newsletter published by the Turning Point project cited the WHO Healthy Cities programme and invoked the WHO definition of a 'healthy community', which offers:

- a clean, safe physical environment;
- a stable and sustainable ecosystem;
- a strong, mutually supportive and non-exploitative community;
- high levels of public participation and control over decisions affecting their lives;
- an environment which meets people's basic needs and gives access to a wide variety of contact, interaction and communication;
- a diverse, vital, innovative urban economy;
- the best health services accessible to all;
- high levels of wellbeing and low levels of disease.

Community profiles of each of the five boroughs of New York were compiled, showing aspects of population health and health care. Community fora were set up, which identified what were seen in each borough as 'top priorities' for health. Issues identified in all the boroughs included access to health care, mental health, environmental health, nutrition and obesity. Some other issues were only identified in some boroughs. For example, the Bronx was the only area to stress diabetes as a priority public health issue, and domestic violence was identified as important in Queens.

Partners in the Turning Point project were involved in a number of initiatives that were publicized through Turning Point. For example, a forum for leaders of the faith communities in New York City considered issues of wellness and ways to promote healthier lifestyles for minority ethnic groups. A city-wide initiative aimed to enrol uninsured New Yorkers in public health insurance programmes such as Medicaid.

handbook describing good practice for primary care employers, designed to improve practices of recruitment and working conditions so that local people in East London are more likely to be employed effectively in primary care. All of these projects sought to make changes to the settings in which people carried out their daily lives in particular places, and thereby to produce changes at the individual level in conditions and behaviours which affect health (East London and City HAZ, 2000; 2001).

THE HEALTH IMPACT OF URBAN CHANGE AND REGENERATION

The initiatives discussed so far have health improvement as their main goal. However, many other policies and interventions, that are not primarily intended to influence health, produce change in the social and physical environment of cities that may be equally important for health. Policies for regeneration and renewal of declining urban areas illustrate the types of action that can produce such change. Furthermore, when major reconstruction takes place in declining urban areas, the resources invested and the scale of change, are generally greater than those achieved through the local public health initiatives discussed above.

Internationally, there is a trend towards consideration of the health outcomes resulting from processes of urban change and urban renewal. This has led to a growth of interest in health impact assessment (HIA) of policies and projects for urban development and renewal (Frankish et al., 1996; Scott-Samuel et al., 1998; British Medical Association, 1998; WHO European Centre for Health Policy, 1999). The WHO European Centre for Health Policy (1999) defines HIA broadly as 'a combination of procedures, methods and tools by which a policy, programme or project may be judged as to its potential effects on the health of a population, and the distribution of those effects within the population'.

Recent British policy on public health has reflected this trend by emphasizing the need to make health impact assessments of non-health sector policies and actions, and there has been rapid growth since 1998 in efforts to develop the capacity for HIA of urban regeneration initiatives. The Merseyside group (Scott-Samuel et al., 1998; Hendley et al., 1999) laid important foundations for this work and reported some of the earliest explicit attempts at health impact assessment of development and regeneration schemes in Britain. HIA of urban regeneration has included a number of exercises carried out in London and reviewed by Ison (2000), Cave and Curtis (2001), Curtis et al. (2002) and Joffe and Mindell (2002), as well as work in other parts of the country, such as Scottish studies on transport and housing strategies for Edinburgh and Glasgow reported by Mcintyre and Pettigrew (1999) and Douglas et al. (2001).

Regeneration programmes and schemes in Britain have a long history, reviewed in detail by authors such as Lawless (1989), Atkinson and Moon (1994), Hill (2000) and Foley and Martin (2000). The development of these programmes has reflected changing views in government about the processes causing urban decline and deprivation. Although the strategies employed have varied, there has been a fairly consistent emphasis on action in particular geographical areas. During the 1990s, persistent inequality and poverty in parts of British cities led to consideration (e.g. by Murray, 1990; Mead, 1997) of the concept of an 'underclass', put forward in the USA (e.g. Wilson, 1987; Osterman, 1991). This idea of a 'culture of poverty' in some communities is quite contentious. However, it has usefully served to draw attention to the very deep seated disadvantages experienced by some populations, which exclude them from key structures and institutions, such as labour market opportunities, housing markets, mainstream health and social services. In the 1990s, a programme of urban regeneration schemes in England funded through the Single Regeneration Budget has sought to address these aspects of disadvantage, with emphasis especially on action relating to employment, housing and environmental renewal. Local agencies and communities were expected to bid competitively for these funds, to support regeneration initiatives in their areas. This type of strategy has been discussed, for example, by Atkinson and Moon (1994).

A change of government from Conservative to Labour in 1997 was associated with a further policy shift, and a renewed emphasis on social and community factors important for urban renewal. This reflected the Labour government rhetoric of a 'third way' in social policy, which represented work as a route to moral and cultural citizenship, and emphasized enhanced community involvement and responsibility in urban renewal (Atkinson, 1999; Foley and Martin, 2000). This 'third way' philosophy has been criticized for failing to recognize or accord value to unpaid work (Haylett, 2001), and for underestimating the problems of achieving effective community involvement and intersectoral partnership (discussed below). In addition to 'action zone' initiatives (including Health Action Zones, considered above), which have focused on specific aspects of deprivation including health or education, there has been major investment in targeted New Deal for Communities initiatives which aim to tackle a range of conditions in the poorest neighbourhoods (Wallace, 2001).

Urban regeneration initiatives have been led centrally by government ministries with responsibility for environment and local government, and locally either by local government or by quasi-autonomous agencies responsible for the local administration of government programmes for urban renewal. These administrative bodies are separate from the health service in England and in the past they have not considered health to be central to their responsibilities. Knowledge about health services and public health has not been very accessible to professionals and communities working in urban regeneration. Regeneration programmes have therefore tended in the past to pay relatively little attention to the potential health effects of their activities. Increasing emphasis on health impact assessment has now started to clarify these effects and may help to maximize the potential health gain from urban regeneration.

Cave and Curtis (2001) reported examples of applied HIA in East London which have focused on specific types of regeneration schemes and which have been used to build local capacity to carry out HIA of regeneration schemes. This developmental work was partly supported by the HAZ in East London, which illustrated the links which can be made between area based action in the health sector and in other sectors of public policy. In one demonstration, involving a housing renewal and refurbishment project in East London, a community consultation group was invited to participate in an HIA workshop. Details are given in Box 9.5. This exercise helped participants to appreciate the potential health impacts of the housing initiative. As a result some changes were made to the implementation of the plan in order to maximize potential health gains and mitigate or reduce possible negative impacts on health.

LIMITATIONS OF STRATEGIES TO IMPROVE HEALTH AND REDUCE INEQUALITIES

The examples considered above illustrate the types of policies and interventions that are being used at the global, national and local scale to improve health and reduce health inequalities in cities. However, these strategies have a number of limitations. Some of these relate, for example, to the problems of using target setting, challenges for the achievement of effective public participation, limitations of geographically targeted initiatives in local areas, and issues concerning the evidence base for planning multi-sectoral action to improve health. The following sections review some of these difficulties.

Box 9.5 Example of a participative strategy for health impact assessment: a workshop on potential health outcomes of a housing improvement and refurbishment project in East London

An HIA facilitator made contact with key informants involved in this project and agreed a process for prospective HIA. He made an initial 'scoping' analysis of documented plans and a series of interviews with key informants involved in the scheme. Following an approach adapted from the strategy of *realistic evaluation* (Pawson and Tilley, 1997), he made an initial summary of the *context* in which the project was operating, the *mechanisms* by which the project would work and the *outcomes* for local housing expected from the project. A structured review of the findings of research on the links between housing and health was also prepared to inform the HIA (Cave et al., 2001), including some of the findings summarized in Chapter 4.

In the areas targeted by this project, significant numbers of dwellings are in low quality, privately rented houses. These houses offer poor living conditions with very high levels of overcrowding. The tenants are typically very poor, marginalized populations with large numbers from minority ethnic and refugee groups. Most of the tenants receive state subsidy: this money is paid directly to the landlord. The private landlords have no incentive to improve the conditions of their properties. In neighbourhoods nearby there are also concentrations of deprived populations in social housing.

This project was financed by the Single Regeneration Budget programme operated by central government as part of its urban renewal programme. The project aimed to work with private landlords to achieve improvements in the standard of local properties and to reduce the number of dwelling units in multiple occupation. It was setting up a non-commercial housing company to acquire and refurbish properties: these will be rented out according to a community lettings policy. This type of policy encourages 'key workers', such as low paid public sector employees, to remain in the area. The company planned to work with major providers of public housing in the programme area and to use any surplus it generated to establish a community investment fund.

The facilitator drew on the existing consultative networks, established through the programme, to convene a half-day workshop to consider the possible health impacts of the housing project. This was intended to achieve wider participation, including representation of local communities affected by the housing project. Those attending the workshop included local residents in the intervention area, community development workers, voluntary group representatives, environmental health officers, housing department representatives, project managers and local councillors, reflecting some of the range of different stakeholders in this project.

Participants expected that various positive changes in health might result from the housing project. They anticipated enhanced community empowerment, solidarity and social support, resulting from a more settled population. A generally less deprived, as well as better housed, population would be associated with improvement in general health status and reduction in infectious diseases and in accidents at home. With reduced residential mobility, there would be potential to improve access to NHS care

continued ⇒

Box 9.5 Continued

and deliver more effective prevention and treatment of illness to a less transient population. Also, a generally wealthier population would be better able to afford to purchase alternative medicines. Participants also recognized some potential for negative changes in health due to this housing scheme, resulting from stress due to the upheaval of rehousing and hazards of building operations for both residents and building workers.

Context

Young, mobile, ethnically diverse population, high levels of deprivation
Poor living conditions in privately rented houses
Private landlords have little incentive to maintain their properties
Large proportion of the tenants are state benefit claimants

Mechanisms
Housing company

Acquire and refurbish properties, community letting, key workers important
Work with the council and RSLs in directing capital spending programmes
Use surpluses to build a community investment fund

Outcomes

Residents feel more positive about area
Area will become more attractive to tenants who want permanent homes and are able to pay higher rents in the private market

Building works
Renovation process
Some people might have to move to new accommodation outside programme area

Health outcomes

Strengthen social capital
Positive health outcomes from improved housing
Improve average measures of health but may not reduce health inequalities

Mental stresses, increased pollution and accident hazards from building works
Negative health outcomes associated with continuing problems with housing and possibly homelessness for some groups

FIGURE 9.1 Summary of findings from a participative health impact assessment in East London (from Cave and Curtis, 2001)

The participants did question whether everyone in the area would benefit equally from the scheme. However, the workshop did not include members of the most

continued ⇒

Box 9.5 Continued

excluded and deprived communities, currently occupying the worst private sector housing in the area. They might feel the most direct impacts of the scheme, and might in some cases be displaced by the housing project to other parts of London. It was noted that those taking part in the workshop were not very inclined to consider the situation of these groups. This may have limited the capacity of the meeting to consider the potential impacts on health inequalities. It seemed to illustrate some of the problems of achieving comprehensive community participation.

The project managers tended at first to react defensively to the idea that there might be negative health impacts of this regeneration scheme. This may have reflected the institutional constraints within which they were working. Nevertheless, the findings from this workshop (summarized in Figure 9.1) contributed to decisions to modify the project design in ways intended to reduce risks associated with building works, provide support for families during refurbishment or rehousing, and take additional measures to foster social capital. Better integration of the housing project with plans for health care development was also proposed. These changes might be expected to help to mitigate potential negative impact on health and increase the balance of health gain from the housing scheme.

PROBLEMS OF TARGETS FOR PUBLIC HEALTH

The development of specific goals and measurable targets at various geographical scales is widely recognized as an effective way to channel effort into achieving certain health outcomes. For example, internationally and nationally agreed targets for clarifying HFA objectives and monitoring change were discussed above. Target setting has often employed a fundamentally geographical approach, involving international comparison of indicators for different countries, showing how they are progressing towards shared health goals. Ritsatakis (2000) explained the rationale for the target based approach employed by WHO. It helps to focus attention on the scientific basis for HFA objectives, helps to communicate important aspects of the health situation in a country, and can provide a rallying point for groups organizing to improve health. The monitoring of progress towards targets may be useful for evaluating action relating to health and health care and it may improve accountability of agencies whose activities affect health.

The WHO European Region is a good illustration of the complexity of implementing international strategies across large and varied regions. By the year 2000, the European Region included 51 different states and there was a good deal of variation between them in the extent to which HFA targets had been adopted, and in the strategies used to address those targets (Ritsatakis, 2000). Ritsatakis (2000) reported that, at the end of the 1990s, of the 51 member states in the European Region, only 27 had approved or were formulating targets for health at the national, regional or city level.

Difficulties involved in target setting by national governments are illustrated by British health policy. Fulop and Hunter (2000), for example, argued that the targets used in British policy emphasized performance management by central government, using a 'command and control' approach. They have been designed as much to monitor progress as to inspire new action in the required areas. Used in this way, targets may have drawbacks. Perverse incentives and manipulation of target indicators occur where managers concentrate on conforming to the targets, rather than focusing on the intentions behind the targets (this was illustrated in Chapter 3, where the inappropriate manipulation of waiting list data was considered). Focusing on specific diseases rather than overall health can also be too narrow to optimize health gain, as argued in the critique of vertical child health programmes in Box 9.2.

There is also a risk that target based performance management will only pay attention to health outcomes which are measurable. This could mean, for example, that reduction of numbers of people with particular diseases is prioritized over improvement in wellbeing and quality of life. More attention may be paid to delivery of services for which there is good routine information (e.g. hospital inpatient services) rather than those where data gathering is more complex and less well developed (e.g. community services).

Target setting also needs to be evidence based, otherwise there is a risk that resources will be invested in health care and treatments that are not necessarily effective in improving health. Targets are unlikely to be useful as forms of performance management if they are unrealistically ambitious, or if those responsible for achieving them do not command the necessary power or resources. On the other hand, if government policy concentrates on very achievable targets, it may be doing no more than acknowledging a level of progress that would have been reached in any case.

From a geographical point of view, it is also interesting to consider the potential for tensions between central target setting, based on national priorities, and local target setting, which may be more sensitive to local concerns and local health status. Thus, for example, East London and City Health Authority (ELCHA, 1999) identified some local priorities for health improvement, such as reduction in cardiovascular disease and mental illness, which corresponded to national priorities. However, the

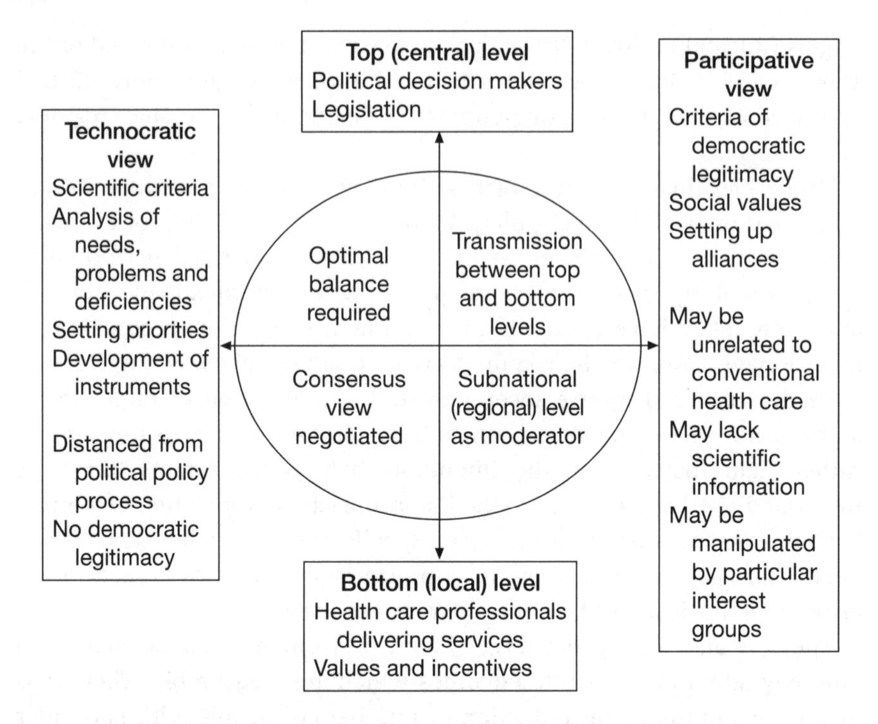

FIGURE 9.2 The political coordinates of health target programmes (after Wismar and Busse, 2000: Figure 3, with permission of Oxford University Press and the European Health Association)

health authority also prioritized some illnesses of particular concern locally, such as tuberculosis.

Similar observations are made in other countries. For example, Wismar and Busse (2000) consider the German experience of targeting for Health For All objectives and comment on the politically delicate balance which needs to be struck between the priorities of the federal government, the *Länder* (regions) and the local institutions. They suggested that feasible programmes require compromises between these different administrative and geographical levels.

Figure 9.2 is based on their model of the political coordinates of health target programmes, which emphasized the need to build political alliances and consensus between the top level (central) and bottom level (local) perspectives and between technocratic and participative strategies. The top level is dominated by central political considerations and dependent on legislation to enable change. The bottom level is influenced, at least in part, by the experiences of local health care practitioners, patients and service users. The *technocratic view* is driven by 'scientific enlightenment' – an evidence based approach aiming to identify which methods can be scientifically shown to be effective, and to develop instruments to assess effectiveness and measurable health variation as a basis for setting priorities. However, scientific criteria may not have democratic legitimacy if they are inconsistent with considerations, such as social values, moral criteria or ethics, which are also important for society in general. The *participative view* involves setting up alliances between different stakeholders in the health planning process. It will be influenced by democratic criteria of legitimacy, which are not necessarily scientific, and it may be dominated or manipulated according to the interests and views of certain particularly politically powerful groups.

Finding the right balance between these different 'poles' will depend on negotiation at various geographical scales, and the preferred solution may vary locally from one area to another. This poses some challenging questions concerning how social justice in framing health policy and allocating resources can be resolved 'universally', for the country as a whole, and 'locally', respecting the need for equity and sensitivity to differences between different areas. Wismar and Busse (2000) suggested that this requires negotiation to reach a consensus and that a relatively independent regional level of government has a useful role in moderating between the central and the local level.

In countries such as England, the regional tier of health service and government administration has seen its independent powers and democratic standing progressively eroded over the last 20 years. Mohan, for example, commented on the reduction in powers of regional health authorities in England, stating that 'the overwhelming impression is of centralisation and weakened representation' (1995: 214). This may make it more difficult to achieve the optimal balance between central and local perspectives.

LIMITATIONS OF LOCAL AREA BASED STRATEGIES

Area based strategies have been used to tackle poverty for several decades and they have often been criticized for various reasons. Areas with the largest *concentrations* of people in poverty do not contain the majority of poor people, so initiatives focused on certain *areas* will miss a large proportion of poor *individuals* who need to be targeted (e.g. Holterman, 1975; Maclaran, 1981). Also, policies focused on areas may be blunt instruments for directing support to the most needy people, even if they are living within the targeted area.

Realistic expectations of the likely outcomes of area based schemes also need to take into account their other limitations. Local schemes will have little impact on regional, national and global structures important for health inequalities. They tend to focus on selected *symptoms* such as local problems of social exclusion, poverty and poor health, rather than tackling structural factors in the wider political economy which influence these effects. They often target specific populations in restricted areas, so that other disadvantaged areas do not benefit and inequalities may worsen.

The changes produced in area initiatives within a limited time may not be sufficiently large and long term to make a perceptible impact in terms of health improvement (Curtis et al., 2002). As illustrated in this book, health inequalities are often entrenched and established over the life course (Benzeval and Judge, 2000). To reverse them requires significant, sustained action and the effects of change will not always be immediately evident.

Maclaran (1981) discussed the view that area based policies have been popular with governments because, although they are not actually very likely to make a big impact on poverty, they produce the appearance of action in a limited number of places where the effects of poverty are most obvious, and at lower cost than would be necessary to address the full extent of poverty. However, Powell et al. (2001) have argued that poverty of *people* (measured in terms of individual indicators such as income) is not identical with poverty of *places*. The latter is better measured, not just in terms of geographical concentrations of poor people, but also in terms of access to welfare services and other benefits in kind which accrue from living in a particular place. Powell and Moon (2001) therefore suggested that the English HAZ programme was an improvement on earlier area based social policies addressing poverty and exclusion because it primarily addressed *place* poverty, rather than *people* in poverty. It might be more effective than earlier initiatives in coming to grips with aspects of local context in deprived areas (the landscapes of health reviewed in this book), which add to the health disadvantage of poor individuals.

One interesting finding from the national evaluation of the English HAZ programme is that HAZ initiatives not only influence the local context in which they are operating, but are importantly influenced *by* this context. Thus Barnes et al. (2001) report that the response of key organizations to HAZ initiatives is influenced by 'locality' factors. These include the tension between balancing varying needs of urban and rural areas or inner and outer city areas. The attitudes of partners involved in HAZ projects were influenced by allegiances and sense of belonging to particular places such as villages and towns, as well as to communities which were not necessarily defined geographically. HAZ projects were also affected by high rates of population turnover due to migration in and out of HAZ areas, and by the ethnic diversity of the population. Barnes et al. (2001) reported on the difficulties of operating HAZ initiatives successfully where administrative boundaries of health and local authorities do not match. When boundaries were not coterminous this increased the number of statutory agencies required to collaborate for action to be successful. Furthermore the relationships between central and local government were variable and were important for the processes of joint working required by HAZs. Different area initiatives (such as HAZ and urban renewal schemes) have not always been very well coordinated centrally, or at the local level (Painter and Clarence, 2001; Powell and Moon, 2001).

CHALLENGES FOR PARTICIPATIVE STATEGIES

Several of the policies and initiatives described above have aimed for increased participation and involvement of a range of actors, including members of local communities, as well as professionals working in different sectors. The approach draws on concepts which have received attention in the geography of health, such as the idea of a geographically defined 'community', and the potential for health gain which may flow from enhancing the social environment and social capital of an area.

Many commentators have emphasized the difficulty of achieving this. Painter and Clarence (2001), for example, considered the 'action zone' strategy employed in Britain, including HAZs. They showed that efforts at intersectoral collaboration in 'action zones' have been hampered by a number of obstacles. These have included a paradoxical lack of coordination of the central government departments trying to promote 'joined-up' working at the local level, and a proliferation of different types of action zone, in overlapping but differently defined geographical areas.

Furthermore, it is a long and difficult process to realize effective partnership working among professionals employed by different organizations with varying cultures and objectives. For example, workers in the National Health Service in Britain operate according to models drawn from medicine, professions allied to medicine and public health, with a primary focus on health and health care in their working objectives. They are employed by organizations managed by executives who are appointed, rather than democratically elected. Workers in social services, urban regeneration, housing, on the other hand, are employed either by local governments, which are democratically elected, multi-purpose public agencies, or in private, voluntary or quasi-governmental organizations. Health is not normally the main objective of these organizations and their professional models are not based in medicine or related disciplines, but in social sciences, management and social care. Joint working means that conventional structures of line management are lost in the more complex arrangements needed for different organizations to work in parallel. It may therefore be more difficult for those with authority to control the work of all the more junior staff in these collaborative ventures.

Curtis et al. (2002) also discuss the obstacles arising from defensive reasoning (Argyris, 1993) which can result from the institutional frameworks within which key actors are operating. Such institutional and individual modes of thinking may tend to impose solutions already espoused by the most influential stakeholders and limit the potential to address possible drawbacks of these strategies, or consider alternative approaches. Defensive reasoning could place structural constraints on innovative action to improve health, as anticipated by regulation theory and regime theory (Stone, 1993; Stoker and Mossberger, 1994). Box 9.5, for example, illustrates the initial reluctance on the part of managers to consider potentially negative health impacts of a regeneration scheme.

Added to these difficulties are the challenges faced in achieving community involvement. These have been reviewed, for example, by Atkinson (1999), Geddes (2000) and Foley and Martin (2000). They have observed that it is difficult to achieve participation and empowerment, for members of local communities (especially those from the most marginalized groups), on an equal footing with stakeholders from the statutory agencies and private industry. Groups such as women, ethnic minorities and those who are socially excluded and precariously situated in communities, are often not able to participate in processes which depend on the existing power relationships, and stress agendas and perspectives of the dominant groups in society. Although different types of stakeholder may be able to participate, there is a risk that only the elite groups will be represented. Furthermore the most powerful stakeholders (e.g. major actors in the local economy or in mainstream public sector institutions) may not share the same agenda with the other stakeholders who are less influential. There may be more emphasis placed on strategies favoured by the majority of people, such as improving 'mainstream' services. Less attention may be paid to issues that may also be important for disadvantaged minority communities, such as alternative health and social services, designed to meet their specific needs. Solutions which are beneficial for the majority of the population, in the 'mainstream' of society, may also cause 'displacement' of excluded groups, moving them out of the community to their further disadvantage (see Box 9.5). Furthermore, the amount of resource devoted to social participation has been typically quite small, compared with funds allocated to initiatives to develop economic infrastructure. Considerable effort and time is needed to build effective community participation, so especially in the short term, there may seem to be small added value resulting from such limited investment of resources.

There are characteristics of community and voluntary groups that can also limit their power to influence the development agenda. They may have internal divisions and conflicts, and they may prioritize concerns which are too parochial to affect larger scale processes, or are contrary to national efforts to maintain universal standards. Identifying 'the community' is also problematic (Moon, 1990; Atkinson, 1999). Communities in poor areas of major cities today are complex. They are not only traditional, 'place based' coalitions. They include, for example, minority ethnic group networks which are often based on ties stretching across continents. Other groups may be orientated around marginalized lifestyles such as use of illicit drugs, or travelling. In order to participate in dominant discourses concerning regeneration, excluded groups may need to adopt the values and modes of thinking typical of dominant groups, but this may mean that they lose touch with perspectives important for their own community (Atkinson, 1999). It may be important for excluded groups to have realistic goals, but managing their expectations may restrict community participation to what the dominant stakeholders see as legitimate (Atkinson, 1999).

However, while these considerations may all limit the real potential of the participative model as a means to change the structural factors important for health, the process of participation in itself may have psychosocial health benefits. Cattell (2001) described a qualitative study which showed how social networks contributed to the health of socially excluded individuals on a deprived housing estate in Newham, East London. People felt that they benefited from emotional support, information and practical mutual aid, through a number of different networks, including extended families, work based networks, neighbours and voluntary organizations representing shared interests. If these types of network and the social supports they provide are enhanced through efforts to promote community involvement and participation in health improvement, there may be positive health effects and reductions in health inequality.

PROBLEMS OF EVIDENCE AND EVALUATION

Effective action to address health disadvantage needs complex multi-faceted action over long time frames. This type of initiative is difficult to establish and hard to sustain in the long term. Governments and other funding agencies often find it hard to justify expenditure of sufficient resources to establish these initiatives and maintain them, especially as it can be difficult to assess what the effects have been. This is one reason to promote a wider engagement of society with the public health sciences, including research in geography of health reviewed here. If information about 'what works' could be made available to a wider group of stake-holders, this would empower them to make more informed judgements about the health effects of urban regeneration and public health improvement projects. However, there are major challenges for evaluation in this sphere and the evidence base to inform policy is incomplete.

Efforts to evaluate programmes such as the HAZ programme in England (Baud et al., 2000) and improve the evidence base to support local HIA (McIntyre and Pettigrew, 1999; Curtis et al., 2002; Joffe and Mindel, 2002) have illustrated these problems. Evaluation is challenging because such a wide range of factors influence health that it is difficult to attribute health change to a particular scheme. The most powerful evidence often comes from longitudinal studies of individuals and communities, showing whether health changes follow from improvements in determinants of health such as housing, employment, transportation infrastructure, environmental quality or other aspects of living conditions in the city. Even when such studies are undertaken, it is often very difficult to provide 'control' areas, outside the intervention zone, which would help to verify whether change is likely to be due to the intervention of interest (Rogers et al., 2001). Also some of the outcomes (such as psychosocial benefits) may be hard to measure and variable in different situations and for different populations.

There is a large amount of evidence on the association between socio-economic disadvantage and poor living conditions and health, and also a number of studies illustrating changes in health as such conditions worsen. Many studies have been reviewed in earlier chapters of this book and have documented the health effects of redundancy or homelessness (for example see Boxes 4.1 and 4.2). However, there is less firm evidence on how health changes when living conditions *improve*. The available evidence suggests that although some health benefits can be demonstrated, the health changes are not always universally positive, especially in the short term. This is partly due to the limitations of local initiatives considered above.

This evidence is therefore quite complex and somewhat ambivalent in its findings. It has been carried out using a number of different methods and the research does not lend itself to formal systematic review methods. This makes it particularly challenging to present the findings of the research to the audiences who need to use the evidence to take local policy decisions. They are often not specialists in public health, and some of them have little or no experience in interpreting scientific evidence. Curtis et al. (2002) argue that it is important to pay attention to the views of these stakeholders in regeneration schemes, including regeneration project managers and members of local communities, and to consider the extent to which their knowledge of regeneration corresponds with the evidence from public health research. If health impact assessment is to be based in scientific evidence, it is also necessary to present research findings in formats that are accessible and relevant for these users.

THE CONTRIBUTION OF HEALTH GEOGRAPHY TO ACTION FOR PUBLIC HEALTH

This chapter has reviewed some of the links between geography of health and strategies to address health inequalities in cities. The findings of research in health geography, as well as other types of public health research, contribute to the evidence base for policy making and practice to reduce health inequality.

This book as a whole has emphasized the importance for health inequality of processes operating at different geographical scales. The relationships between action at the global, national and local level demonstrate the importance of scale for public health strategies. Some of the processes influencing health inequalities are operating at the macro scale (globally or nationally) so it is not feasible to tackle health inequalities by means of local action alone. At the same time, action that is insensitive to the particular context of different places is unlikely to be effective, so that international or national strategies must be adapted to local conditions. The failure, at all geographical scales, to coordinate initiatives of different types of organization, and the difficulties of integrating 'top level' and 'bottom level' action, present obstacles to real progress in reducing health inequalities. Research in health geography is continuing to clarify how macro scale processes influence, and are mediated by, local conditions, and how health policy might respond better to the relationships between processes operating at these different levels.

The processes affecting health are dynamic. Geographers have shown a growing concern with *change* in the landscapes influencing health inequality, focusing on health variation in both time and space. This book has reviewed examples of research taking a life course perspective on factors which produce health differences, and considering the life experiences of individuals in different urban settings. While a good deal of evidence has now built up demonstrating that these associations are important for health inequality, more needs to be done to *explain* these relationships and the causal pathways which produce health difference. We also need more evidence from *evaluation* of strategies to influence these causal pathways and reduce health inequality. The theoretical frameworks discussed in this book illustrate that progress is being made in this direction by many researchers in health geography as well as in other disciplines.

Geography is increasingly concerned with concepts of health which extend beyond medical definitions. Health geography has been using a growing range of different theoretical perspectives and methods to develop knowledge in this field. It is drawing on theories that relate to human perception, social construction of health, political ecology and relationships of power and resistance. It is simultaneously employing concepts relating to the physical and biological processes that are important for health in different settings. The range of methods is also broad, including intensive and qualitative as well as extensive and statistical strategies. There needs to be a continuing effort to consider the complementarity of these different theoretical positions and methodological strategies, rather than considering them as competing paradigms.

A major contribution of recent research in health geography has been to highlight the varying impact of *places* and *context* on health and health inequality. The thinking behind area based strategies to reduce health inequality shows strong parallels with arguments in health geography about 'spaces of risk' and why places are important for health experience. Many initiatives to improve the health of disadvantaged populations are 'area based', focused on particular geographical areas such as deprived inner districts. They also often have a focus on certain types of setting. Existing research in health geography suggests that, while area based strategies may have certain limitations, they do have potential to influence 'place effects' producing health difference, especially if they are well coordinated with national initiatives and sustained over time. We need more evaluative studies to improve the evidence on this point. Continuing work in this field should also improve our knowledge of how the influence of place interacts with individual experience. This would provide a more integrated evidence base to inform future action to reduce health inequalities in different settings.

FURTHER READING

Readers may wish to follow up themes in this chapter relating to international organizations and their potential to influence health inequality. Relevant publications include: Moon and Curtis (1998).

There are also several useful articles on the European experience of health policy in a special issue of the *European Journal of Public Health* (2000, vol. 10, no. 4 supplement).

Also relevant here, of course, are publications by the World Health Organization. They include: WHO (1998), and various documents available on the WHO website, especially: WHO (1999). A good introduction to the Healthy Cities programme is also provided by: Ashton, J. (1992) *Healthy Cities* (Milton Keynes, Open University Press).

Interesting discussion of local area based initiatives in Britain from a health geography perspective is provided by: Powell and Moon (2001), Powell, Boyne and Ashworth (2001).

Readers should also look out for the following forthcoming publication, which will provide a wealth of valuable insights into the Health Action Zone programme in Britain: Barnes, Baud, Benzeval, Judge, Mackenzie and Sullivan (eds) (2004).

OBJECTIVES AND QUESTIONS

This chapter has aimed to bring together some key themes which students may wish to focus on, especially:

- the differing roles of international agencies and local agencies in tackling health inequalities;
- the potential and limitations of target setting as a strategy to reduce geographical variation in health;
- the relevance of a 'settings' approach for tackling health inequality at the local level;
- the critique of area based policy and why this is relevant for policy to improve public health;
- the future agenda for health geography and its potential to contribute to key policy questions as well as theoretical and factual knowledge about health variation.

Key questions to consider after reading this book are:

1 Geographers have argued that *places*, as well as *people*, are important for health differences. Discuss the significance of this argument for policies designed to tackle health inequality.

2 Discuss the relationships between the 'landscapes' perspective in health geography and the 'settings' approach put forward by the World Health Organization as public health strategy.
3 'Geography is all about maps.' In the light of the discussion in this book, do you think this is an accurate view of geography as it relates to health and health care?

References

Aakaster, C. (1986) Concepts in alternative medicine. *Social Science and Medicine*, 22, 265–274.

Aday, L. and Andersen, R. (1974) A framework for the study of access to medical care. *Health Services Research*, 9, 208–220.

Acevedo-Garcia, D. (2000) Residential segregation and the epidemiology of infectious diseases. *Social Science and Medicine*, 51, 8, 1143–1161.

Ahmad, W. (ed.) (1993) *'Race' and Health in Contemporary Britain*. Buckingham, Open University Press.

Ahmad, W. (1996) The trouble with culture. In Kelleher, D. and Hillier, S. (eds) *Researching Cultural Differences in Health*. London, Routledge. Chapter 9, 190–219.

Aldous, J., Bardsley, M., Daniell, R., Gair, R., Jacobsen, B., Lowdell, C., Morgan, D., Storkey, M. and Taylor, G. (1999) *Refugee Health in London: Key Issues for Public Health*. London, Health of Londoners Project.

Almog, M., Curtis, S., Copeland, A. and Congdon, P. (2003) Geographical variation in acute psychiatric admissions within New York City 1990–2000: growing equalities in service use? *Social Science and Medicine*.

Altenstetter, C. and Björkman, J. (1997) *Health Policy Reform: National Variations and Globalization*. Basingstoke, Macmillan.

Annandale, E. and Hunt, K. (eds) (2000) *Gender Inequalities in Health*. Milton Keynes, Open University Press.

Arber, S. (1999) Gender. In Gordon, D., Shaw, M., Dorling, D. and Davey-Smith, G. (eds) *Inequalities in Health: the Evidence Presented to the Independent Inquiry into Inequalities in Health, Chaired by Sir Donald Acheson*. Bristol, Policy. Chapter 14, 197–206.

Arber, S. and Cooper, H. (1999) Gender differences in health in later life: a new paradox? *Social Science and Medicine*, 48, 1, 63–78.

Argenti, O. (2000) Feeding the cities: food supply and distribution. In *IFPRI 2020 Focus 3*, Brief 5 of 10. Washington, DC, International Food Policy Research Institute.

Argyris, C. (1993) *Knowledge for Action: a Guide to Overcoming Barriers to Organizational Change*. San Francisco, Jossey-Bass.

Armstrong, D., Michie, S. and Marteau, T. (1998) Revealed identity: a study of the process of genetic counselling. *Social Science and Medicine*, 47, 11, 1653–1658.

Ashcroft, D. (1999) Herbal remedies: issues in licensing and economic evaluation. *Pharmacoeconomics*, 16, 4, 321–328.

Ashton, J. (1992a) Setting the agenda for health in Europe: what the United Kingdom could do with its presidency of the European Community. *British Medical Journal*, 304, 1643–1644.

Ashton, J. (1992b) *Healthy Cities*. Milton Keynes, Open University Press.

Asthana, S. (1998) The relevance of place in HIV transmission and prevention: the commercial sex industry in Madras. In Kearns, R. and Gesler, W. (eds) *Putting Health into Place: Landscape, Identity and Well-Being*. Syracuse, NY, Syracuse University Press. Chapter 9, 168–187.

Asthana, S., Gibson, A. and Parsons, E. (1999) The geography of fundholding in southwest England: implications for the evolution of primary care groups. *Health and Place*, 5, 4, 271–278.

Astin, J. (1998) Why patients use alternative medicine: results of a national study. *Journal of the American Medical Association*, 279, 1533–1548.

Atkinson, R. (1999) Discourses of partnership and empowerment in contemporary British urban regeneration. *Urban Studies*, 36, 1, 59–72.

Atkinson, R. and Moon, G. (1994) *Urban Policy in Britain: the City, the State and the Market.* London, Macmillan.

Atkinson, T. (1991) *Principles of Political Ecology.* London, Belhaven.

Ayanian, J., Weissman, J., Schneider, E., Ginsburg, J. and Zaslavsky, A. (2000) Unmet health needs of uninsured adults in the United States. *Journal of the American Medical Association*, 284, 16, 2061–2069.

Bakhshi, S., Hawker, J. and Ali, S. (1997) The epidemiology of tuberculosis by ethnic group in Birmingham and its implications for future trends in tuberculosis in the UK. *Ethnicity and Health*, 2, 3, 147–153.

Balzi, D., Geddes, M., Brancker, A. and Parkin, D. (1995) Cancer Mortality in Italian Migrants and their offspring in Canada. *Cancer Causes and Control*, 6, 1, 68–74.

Bardsley, M., Jones, I.R., Kemp, V., Dodhia, H., Aspinall, P. and Bevan, P. (1998) *Housing and Health in London: a Review by the Health of Londoners Project.* London, East London and The City Health Authority.

Barnes, M., Sullivan, H. and Matka, E. (2001) *National Evaluation of Health Action Zones: Context, Strategy and Capacity. Initial Findings from the Strategic Level Analysis.* Birmingham, University of Birmingham.

Barnes, M., Baud, L., Benzeval, M., Judge, K., Mackenzie, M. and Sullivan, H. (eds) (2004) *Health Action Zones: Partnerships for Health Equity.* London, Routledge, forthcoming.

Barnett, P. (1995) Unemployment, work and health: opportunities for healthy public policy. *New Zealand Medical Journal*, 108, 998.

Barrett, F. (2000) *Disease and Geography: the History of an Idea.* Geographical Monographs vol. 23. Toronto, Atkinson College, York University.

Bartley, M. (1994) Unemployment and ill health: understanding the relationship. *Journal of Epidemiology and Community Health*, 48, 333–337.

Bartley, M. and Plewis, I. (1997) Does health selective mobility account for socioeconomic differences in health? *Journal of Health and Social Behaviour*, 38, 376–386.

Bartley, M., Blane, D. and Davey-Smith, G. (eds) (1998) *The Sociology of Health Inequalities.* Oxford, Blackwell.

Bartley, M., Ferrie, J. and Montgomery, S. (1999) Living in a high-unemployment economy: understanding the health consequences. In Marmot, M. and Wilkinson, R. (1999) *Social Determinants of Health.* Oxford, Oxford University Press. Chapter 5, 81–104.

Bastos, F., Barcellos, C., Lowndes, C. and Friedman, S. (1999) Co-infection with malaria and HIV in injecting drug users in Brazil: a new challenge to public health? *Addiction*, 94, 8, 1165–1174.

Battram, A. (1998) *Navigating Complexity.* London, Industrial Society.

Baud, L., Judge, K., Lawson, L., Mackenzie, M., Mackinnon, J. and Truman, J. (2000) *Health Action Zones in Transition: Progress in 2000. Executive Summary.* Glasgow, Department of Social Policy and Social Work.

Baudrillard, J. (1970) *La Société de consommation: ses mythes, ses structures.* Paris, Gallimard.

Baxter, J. and Eyles, J. (1997) Evaluating qualitative research in social geography: establishing 'rigour' in interview analysis. *Transactions of the Institute of British Geographers*, 22, 4, 515–525.

Beale, N. and Nethercott, S. (1987) The health of industrial employees four years after compulsory redundancy. *Journal of the Royal College of General Practitioners*, 37, 390–394.

Beck, L., Lobitz, B. and Wood, B. (2000) Remote sensing and human health: new sensors and new opportunities. *Emerging Infectious Diseases*, 6, 3, 217–227.

Beck, U. (1992) *Risk Society: Towards a New Modernity.* London, Sage.

Becker, K., Glass, G., Brathwaite, W. and Zenilman, J. (1998) Geographic epidemiology of gonorrhea in Baltimore, Maryland, using a geographic information system. *American Journal of Epidemiology*, 147, 7, 709–716.

Bell, D. and Valentine, G. (1997) *Consuming Geographies.* London, Routledge.

Bell, M. (1999) Rehabilitating Middle England: integrating ecology, aesthetics and ethics. In Williams, A. (ed.) *Therapeutic Landscapes: the Dynamic between Place and Wellness.* New York, University Press of America. Chapter 2, 15–27.

Bell, M. and Evans, D. (1997) Greening 'the heart of England': redemptive science, citizenship and 'symbol of hope for the nation'. *Environment and Planning D: Society and Space*, 15, 257–279.

Bellander, T., Berglind, N., Gustavsso, P., Jonson, T., Nyberg, F., Pershagen, G. and Jarup, L. (2001) Using geographic information systems to assess individual historical exposure to air pollution from traffic and house heating in Stockholm. *Environmental Health Perspectives*, 109, 633–639.

Bentham, C.G. and Haynes, R. (1992) Evaluation of a mobile branch surgery in a rural area. *Social Science and Medicine*, 34, 1, 97–102.

Bentham, C.G., Hinton, J., Haynes, R., Lovett, A. and Bestwick, C. (1995) Factors affecting non-response to cervical cytology screening in Norfolk, England. *Social Science and Medicine*, 40, 1, 131–135.

Benzeval, M. and Judge, K. (2000) Income and health: the time dimension. *Social Science and Medicine*, 52, 1371–1390.

Benzeval, M., Judge, K. and Whitehead, M. (1995) *Tackling Health Inequalities: an Agenda for Action.* London, King's Fund.

Bergquist, N. (2001) Vector borne parasitic diseases: new trends in data collection and risk assessment. *Acta Tropica*, 79, 1, 13–20.

Berkman, L. and Syme, S. (1979) Social networks, host resistance and mortality: a nine year follow up study of Almeda County residents. *American Journal of Epidemiology*, 109, 2, 186–204.

Berkman, L., Glass, T., Brissette, I. and Seeman, T. (2000) From social integration to health: Durkheim in the new millennium. *Social Science and Medicine*, 51, 843–857.

Best, R. (1995) The housing dimension. In Benzeval, M., Judge, K. and Whitehead, M. (eds) *Tackling Inequalities in Health: an Agenda for Action.* London, King's Fund. 53–68.

Best, R. (1999) Health inequalities: the place of housing. In Gordon, D., Shaw, M., Dorling, D. and Davey-Smith, G. (eds) *Inequalities in Health: the Evidence Presented to the Independent Inquiry into Inequalities in Health, Chaired by Sir Donald Acheson.* Bristol, Policy. 45–67.

Bettcher, D. and Wipfli, H. (2001) Towards a more sustainable globalisation: the role of the public health community. *Journal of Epidemiology and Community Health*, 55, 9, 617–618.

Beyea, J. and Hatch, M. (1999) Geographic exposure modelling: a valuable extension of geographic information systems for use in environmental epidemiology. *Environmental Health Perspectives*, 107, 181–190.

Bhatti, N., Morris, J., Halliday, R. and Moore-Gilon, J. (1995) Increasing incidence of tuberculosis in England and Wales: a study of the likely causes. *British Medical Journal*, 310, 967–969.

Bhopal, R. (1997) Is research into ethnicity and health racist, unsound or important science? *British Medical Journal*, 314, 7096, 1751–1756.

Bhui, K. and Bhugra, D. (2001) Transcultural psychiatry: some social and epidemiological research issues. *International Journal of Social Psychiatry*, 47, 3, 1–9.

Bhui, K. and Olajide, D. (eds) (1999) *Mental Health Service Provision for a Multi-Cultural Society*. London, Saunders.

Bines, W. (1994) *The Health of Single Homeless People*. York, Centre for Housing Policy, University of York.

Blakely, T., Lochner, K. and Kawachi, I. (2002) Metropolitan area income inequality and self rated health: a multi-level study. *Social Science and Medicine*, 54, 1, 65–77.

Blane, D., Harding, S. and Rosato, M. (1999) Does social mobility affect the size of the socioeconomic mortality differential? Evidence from the Office for National Statistics Longitudinal Study. *Journal of the Royal Statistical Society Series A*, 162, 1, 59–70.

Blatchford, O., Capewell, S., Murray, S. and Blatchford, M. (1999) Emergency medical admissions in Glasgow: general practices vary despite adjustment for age, sex, and deprivation. *British Journal of General Practice*, 49, 551–554.

Blaxter, M. (1990) *Health and Lifestyles*. London, Tavistock/Routledge.

Bloor, K. and Maynard, A. (1995) *Equity in Primary Care*. Discussion Paper 141. York, Centre for Health Economics Consortium, University of York.

Blower, S., Small, P. and Hopewell, P. (1996) Control strategies for tuberculosis epidemics: new models for old problems. *Science*, 273, 497–500.

Bolton, W. and Oatley, A. (1987) A longitudinal study of social support and depression in unemployed men. *Psychological Medicine*, 17, 453–460.

Booman, M., Durrheim, D., La Grange, K., Martin, C., Mabuza, A., Zitha, A., Mbokazi, F., Fraser, C. and Sharp, B. (2000) Using a geographical information system to plan a malaria control programme in South Africa. *Bulletin of the World Health Organization*, 78, 12, 1438–1444.

Borgdorff, M., Behr, M., Nagelkerke, N., Hopewell, P. and Small, P. (2000) Transmission of tuberculosis in San Francisco and its association with immigration and ethnicity. *International Journal of Tuberculosis and Lung Disease*, 4, 4, 287–294.

Bosma, H., Stansfeld, S. and Marmot, M. (1998) Job control, personal characteristics, and heart disease. *Journal of Occupational Health Psychology*, 3, 4, 402–409.

Bosma, H., van de Mheen, H., Borsboom, G. and Mackenbach, J. (2001) Neighborhood socioeconomic status and all-cause mortality. *American Journal of Epidemiology*, 153, 4, 363–371.

Botti, C., Comba, P., Forastiere, F. and Settimi, L. (1996) Causal inference in environmental epidemiology: the role of implicit values. *Science of the Total Environment*, 184, 1/2, 97–101.

Bourdieu, P. (2000) *Distinction: a Social Critique of the Judgement of Taste*. London, Routledge.

Bowater, M. (2001) The experience of a rural general practitioner using videoconferencing for telemedicine. *Journal of Telemedicine and Telecare*, 7, 2, 24–25.

Bowen, W. (2002) An analytical review of environmental justice research: what do we really know? *Environmental Management*, 29, 1, 3–15.

Bowling, A. (1997) *Measuring Health: a Review of Quality of Life Measurement Scales*, 2nd edn. Buckingham, Open University Press.

Boyce, J., Klemer, A., Templet, P. and Willis, C. (1999) Power distribution, the environment, and public health: a state-level analysis. *Ecological Economics*, 29, 1, 127–140.

Boyer, C. and Mechanic, D. (1994) Psychiatric reimbursement reform in New York State: lessons in implementing change. *Milbank Quarterly*, 72, 4, 621–651.

Boyle, P., Gattrel, A. and Duke-Williams, O. (1999) The effect on morbidity of variability in deprivation and population stability in England and Wales: an investigation at small area level. *Social Science and Medicine*, 49, 791–799.

Bradburn, N. (1969) *The Structure of Psychological Well-Being*. Chicago, Aldine.

Bradford, W., Martin, J., Reingold, A., Schecter, G., Hopewell, P. and Small, P. (1996) The changing epidemiology of acquired drug-resistant tuberculosis in San Francisco, USA. *Lancet*, 348, 928–931.

Bradford, W., Koehler, J., El-Hajj, H., Hopewell, P., Reingold, A., Agasina, C., Cave, M., Rane, S., Yan, Z., Crane, C. and Small, P. (1998) Dissemination of mycobacterium tuberculosis across the San Francisco Bay area. *Journal of Infectious Diseases*, 177, 4, 1104–1107.

Bridgman, R. (1999) The street gives and the street takes: designing housing for the chronically homeless. In Williams, A (ed.) *Therapeutic Landscapes: the Dynamic between Place and Wellness*. New York, University Press of America. 153–166.

Brimblecombe, N., Dorling, D. and Shaw, M. (1999) Where the poor die in a rich city: the case of Oxford. *Health and Place*, 5, 4, 287–300.

Bristow, M., Kohen, D. and O'Mahony (2001) Effects of social and behavioural factors in acute psychiatric admissions: a comparison between inner and outer London. *Journal of Mental Health*, 10, 1, 109–113.

British Medical Association (1987) *Deprivation and Ill-Health*. Discussion Paper. London, British Medical Association Board of Science and Education.

British Medical Association (1990) *Living with Risk*. London, Penguin.

British Medical Association (1993) *Complementary Medicine: the BMP Guide to Good Practice*. Oxford, Oxford University Press.

British Medical Association (1998) *Health and Environmental Impact Assessment: an Integrated Approach*. London, Earthscan.

Brown, D. (1999) *Care in the Country: Inspection of Community Care in Rural Communities*. London, Department of Health.

Brown, G. and Harris, T. (1978) *The Social Origins of Depression: a Study of Psychiatric Disorders in Women*. London, Tavistock.

Brown, T. and Duncan, C. (2000) London's burning: recovering other geographies of health. *Health and Place*, 6, 4, 363–375.

Bryce, C., Curtis, S. and Mohan, J. (1994) Coronary heart disease: trends in spatial inequalities and implications for health care planning in England. *Social Science and Medicine*, 38, 5, 677–690.

Bunker, J. (1995) Medicine matters for all. *Journal of the Royal College of Physicians of London*, 29, 2, 105–112.

Burchell, B. (1994) The effects of labour market position on job insecurity and unemployment on psychological health. In Gallie, D., Marsh, C. and Vogler, C. (eds) *Social Change and the Experience of Unemployment*. Oxford, Oxford University Press. 188–212.

Burdine, J., Felix, M. et al. (1999) Measurement of social capital. *Annals of the New York Academy of Sciences*, 896, 393–395.

Burridge, R. and Ormandy, D. (1993) *Unhealthy Housing: Research, Remedies, and Reform*. London, Spon.

Butler, R. and Bowlby, S. (1997) Bodies and spaces: an exploration of disabled people's experiences of public space. *Environment and Planning D: Society and Space*, 15, 411–433.

Butler, R. and Parr, H. (eds) (1999) *Mind and Body Spaces: Geographies Illness, Impairment and Disability.* London, Routledge.

Byrne, D. (1997) Social exclusion and capitalism: the reserve army across time and space. *Critical Social Policy*, 17, 27–51.

Calnan, M. (1987) *Health and Illness: the Lay Perspective.* London, Tavistock.

Cameron, E. (1998) Gender and disadvantage in health: men's health for a change. In Bartley, M., Blane, D. and Davey-Smith, G. (eds) *The Sociology of Health Inequalities.* Oxford, Blackwell. Chapter 5, 115–134.

Campbell, J., Ramsay, J. and Green, J. (2001) Age, gender, socioeconomic, and ethnic differences in patients' assessments of primary health care. *Quality in Health Care*, 10, 2, 90–95.

Cant, S. (1999) *A New Medical Pluralism? Alternative Medicine, Doctors, Patients and the State.* London, UCL Press.

Capra, F. (1997) *The Web of Life: a New Synthesis of Mind and Matter.* London, Harper Collins.

Carrasquillo, O., Himmelsteint, D., Woolhandler, S. and Bor, D. (1999) Trends in health insurance coverage, 1989–1997. *International Journal of Health Services*, 29, 467–483.

Carr-Hill, R., Hardman, G., Martin, S., Peacock, S., Sheldon, T. and Smith, P. (1994) *A Formula for Distributing NHS Revenues Based on Small Area Use of Hospital Beds.* York, Centre for Health Economics, University of York.

Carstairs, V. and Morris, R. (1991) *Deprivation and Health in Scotland.* Aberdeen, Aberdeen University Press.

Carter, H. (1981) *The Study of Urban Geography*, 3rd edn. London, Arnold.

Carter, R., Mendis, K. and Roberts, D. (2000) Spatial targeting of interventions against malaria. *Bulletin of the World Health Organization*, 78, 12, 1401–1411.

Carter, Y., Curtis, S., Harding, G., Maguire, A., Meads, G., Riley, A., Ross, P. and Underwood, M. (2000) *National Evaluation of Primary Care Act Pilots: Addressing Inequalities.* London, Department of General Practice and Primary Care, Queen Mary, University of London.

Casper, C., Singh, S., Rane, S., Daley, C., Schecter, G., Riley, L., Kreiswirth, B. and Small, P. (1996) The transcontinental transmission of tuberculosis: a molecular epidemiological assessment. *American Journal of Public Health*, 86, 4, 551–553.

Castells, M. (1977) *The Urban Question: a Marxist Approach.* London, Arnold.

Castells, M. (1989) *The Informational City: Information, Technology, Economic Restructuring and the Urban–Regional Process.* Oxford, Blackwell.

Castells, M. (1996) *The Rise of the Network Society.* Oxford, Blackwell.

Castle, D., Scott, K., Wesseley, S. and Murray, R. (1993) Does social deprivation during gestation and early life predispose to later schizophrenia? *Social Psychiatry and Psychiatric Epidemiology*, 28, 1–4.

Cattell, V. (2001) Poor people, poor places and poor health: the mediating role of social networks and social capital. *Social Science and Medicine*, 52, 10, 1501–1516.

Cave, B. and Curtis, S. (2001) Developing a practical guide to assess the potential health impact of urban regeneration schemes. *Promotion and Education*, 8, 12–16.

Cave, B., Curtis, S., Coutts, A. and Aviles, M. (2001) *Health Impact Assessment for Regeneration Projects. Vol. II: Selected Evidence Base.* London, Queen Mary, University of London and East London and City Health Action Zone.

Charlton, J. (1995) Trends and patterns in suicides in England and Wales. *International Journal of Epidemiology*, 24, 1, s45–s52.

Chenhuei Chi (1994) Integrating traditional medicine into modern health care systems: examining the role of Chinese medicine in Taiwan. *Social Science and Medicine*, 39, 3, 307–321.

Chouinard, V. (1999) Body politics: disabled women's activism in Canada and beyond. In Butler, R. and Parr, H. (eds) *Mind and Body Spaces: Geographies of Illness, Impairment and Disability*. London, Routledge. Chapter 15, 269–294.

Clark, D. (1996) *Urban World/Global City*. London, Routledge.

Cliff, A. and Haggett, P. (1988) *Atlas of Disease Distributions: Analytic Approaches to Epidemiological Data*. Oxford, Blackwell.

Cobb, S. (1974) Physiological changes in men whose jobs were lost. *Journal of Psychosomatic Research* 18, 245–258.

Cockerham, W., Rutten, A. and Abel, T. (1997) Conceptualizing contemporary lifestyles: moving beyond Weber. *Sociological Quarterly*, 22, 321–342.

Coetzee, M., Craig, M. and Le Sueur, D. (2000) Distribution of African malaria mosquitoes belonging to the *Anopheles gambiae* complex. *Parasitology Today*, 16, 2, 74–77.

Cohn, S., Klein, J., Mohr, J., Vanderhorst, C. and Weber, D. (1994) The geography of AIDS: patterns of urban and rural migration. *Southern Medical Journal*, 87, 6, 599–606.

Congdon, P. (1996a) The incidence of suicide and parasuicide: a small area study. *Urban Studies*, 33, 137–138.

Congdon, P. (1996b) The epidemiology of suicide in London. *Journal of the Royal Statistical Society, Series A*, 159, 515–533.

Congdon, P. (1996c) General linear gravity models for the impact of casualty unit closures. *Urban Studies*, 33, 9, 1707–1728.

Congdon, P. (2000) A Bayesian approach to prediction using the gravity model, with an application to patient flow modelling. *Geographical Analysis*, 32, 2, 205–224.

Congdon, P., Shouls, S. and Curtis, S. (1997) A multi-level perspective on small area health and mortality: a case study of England and Wales. *International Journal of Population Geography*, 7, 35–51.

Congdon, P., Campos, R., Curtis, S., Southall, H., Gregory, I. and Jones, I.R. (2001) Quantifying and explaining changes in geographical inequality of infant mortality in England and Wales since the 1890s. *International Journal of Population Geography*, 7, 35–52.

Corner, J. (2001) *Between You and Me: Closing the Gap between People and Health Care*. Queen Elizabeth the Queen Mother Fellowship Lecture, Royal Society of Arts, London, 20 June.

Cornia, G. (2001) Globalization and health: results and options. *Bulletin of the World Health Organization*, 79, 9, 830–846.

Cornwell, J. (1984) *Hard Earned Lives: Accounts of Health and Illness from East London*. London, Tavistock.

Cosgrove, D. and Jackson, P. (1987) New directions in cultural geography. *Area*, 19, 2, 95–101.

Coughlin, S. (1996) Environmental justice: the role of epidemiology in protecting unempowered communities from environmental hazards. *Science of the Total Environment*, 184, 1/2, 67–76.

Courtenay, W. (2000) Constructions of masculinity and their influence on men's well-being: a theory of gender and health. *Social Science and Medicine*, 50, 10, 1385–1401.

Coutts, A. and Curtis, S. (2002) *Stakeholder Perceptions of the Health Impacts of Active Labour Market Interventions*. Research Paper. London, Queen Mary, University of London.

Coward, R. (1993) The myth of alternative health. In Beattie, A., Gott, M., Jones, L. and Sidell, M. (eds) *Health and Wellbeing: a Reader*. Milton Keynes, Open University. Chapter 10, 94–101.

Craddock, S. (2000a) Disease, social identity and risk: rethinking the geography of AIDS. *Transactions of the Institute of British Geographers*, 25, 2, 153–168.

Craddock, S. (2000b) *City of Plagues: Disease, Poverty and Deviance in San Francisco*. Minneapolis, MN, University of Minnesota Press.

Craig, G. and Dowler, E. (1997) Let them eat cake! Food poverty in the UK. In Riches, G. (ed.) *First World Hunger.* Basingstoke, Macmillan.

Crail, M. (1999) Travel sickness. *Health Service Journal*, September, 11.

Crang, P. (1996) Displacement, consumption and identity. *Environment and Planning A*, 28, 47–67.

Cravey, A., Washburn, S., Gesler, W., Arcury, T. and Skelly, A. (2001) Developing socio-spatial knowledge networks: a qualitative methodology for chronic disease prevention. *Social Science and Medicine*, 52, 12, 1763–1775.

Cromley, E. and McLafferty, S. (2002) *GIS and Public Health.* New York, Guilford.

Cummins, S. and Macintyre, S. (2002) A systematic study of the foodscape: the price and availability of food in Greater Glasgow. *Urban Studies*, 39, 11, 2115–2130.

Cunnan, P. (2002) The health of urban poor, black women street traders in Durban, South Africa. Unpublished PhD thesis. London, Queen Mary, University of London.

Curtis, S. (1980) Spatial access, need and equity: an analysis of the accessibility of primary health facilities for the elderly in parts of East Kent. Unpublished PhD thesis, University of Kent.

Curtis, S. (1982) Spatial analysis of surgery locations in general practice. *Social Science and Medicine*, 16, 303–313.

Curtis, S. (1987) Self reported morbidity in London and Manchester: inter-urban and intra-urban variations. *Social Indicators Research*, 19, 255–272.

Curtis, S. (2001) Health in London. *Area*, 33, 1, 84–92.

Curtis, S. and Jones, I.R. (1998) Is there a place for geography in the analysis of health inequality? *Sociology of Health and Illness*, 20, 5, special issue, 645–672.

Curtis, S. and Lawson, K. (2000) Gender, ethnicity and self-reported health: the case of African Caribbean populations in London. *Social Science and Medicine*, 50, 365–385.

Curtis, S. and Taket, A. (1996) *Health and Societies: Changing Perspectives.* London, Arnold.

Curtis, S., Eames, M., Ben-Shlomo, Y., Marmot, M., Mohan, J., Callam, C. and Killoran, A. (1993) Geography of coronary heart disease in England: the implications for local health planning. *Health Education Journal*, 52, 2, 72–78.

Curtis, S., Petukhova, N. and Taket, A. (1995) Health care reforms in Russia: the example of St Petersburg. *Social Science and Medicine*, 40, 6, 755–765.

Curtis, S., Pethukhova, N., Sezonova, G. and Netsenko, N. (1997) Caught in the traps of managed competition? Examples of Russian health care reforms from St Petersburg and the Leningrad Oblast. *International Journal of Health Services*, 27, 4, 661–686.

Curtis, S., Gesler, W., Smith, G. and Washburn, S. (2000) Approaches to sampling and case selection in qualitative research: examples in the geography of health. *Social Science and Medicine*, 50, 1001–1014.

Curtis, S., Cave, B. and Coutts, A. (2002) Is urban regeneration good for health? Perceptions and theories of the health impacts of urban change. *Environment and Planning C*, 20, 517–534.

Curtis, S., Southall, H., Congdon, P. and Dodgeon, B. (2003) Area effects on health variation over the life-course: analysis of the longitudinal study sample in England using new data on area of residence in childhood. *Social Science and Medicine*, in press.

Cutchin, M. (2000) Telemedicine and regionalization: conceptualizing the medical geography of a new frontier. Paper presented to the 9th International Symposium in Medical Geography, 3–7 July, Montreal.

Cutter, S. (1995) Race, class and environmental justice. *Progress in Human Geography*, 19, 1, 111–122.

Cutter, S. and Solecki, W. (1996) Setting environmental justice in space and place: acute and chronic airborne toxic releases in the southeastern United States. *Urban Geography*, 17, 5, 380–399.

Cutter, S., Holm, D. and Clark, L. (1996) The role of geographic scale in monitoring environmental justice. *Risk Analysis*, 16, 4, 517–526.

Cutter, S., Hodgson, M. and Dow, K. (2001) Subsidized inequities: the spatial patterning of environmental risks and federally assisted housing. *Urban Geography*, 22, 1, 29–53.

Dandeker, C. (1990) *Surveillance, Power and Modernity*. Cambridge, Polity.

Dartonhill, J., Mandryk, J., Mock, P., Lewis, J. and Kerr, C. (1990) Sociodemographic and health factors in the well-being of homeless men in Sydney, Australia. *Social Science and Medicine*, 31, 5, 537–544.

Dauncey, K., Giggs, J., Baker, K. and Harrison, G. (1993) Schizophrenia in Nottingham: lifelong residential mobility of a cohort. *British Journal of Psychiatry*, 163, 613–619.

Davey-Smith, G., Neaton, J., Wentworth, D. and Stamler, J. (1996) Socioeconomic differentials in mortality risk among men screened for the multiple risk factor intervention trial. I: White men. *American Journal of Public Health*, 86, 4, 486–496.

Davey-Smith, G., Dorling, D. and Shaw, M. (Eds) (2001) *Poverty, inequality and health in Britain 1800–2000: A reader*. Bristol, Policy.

Davies, B. (1968) *Social Needs and Resources in Local Services*. London, Joseph.

Davis, M. (1990) *City of Quartz: Excavating the Future in Los Angeles*. London, Verso.

De Beyer, J., Preker, A. and Feacham, R. (2000) The role of the World Bank in international health: renewed commitment and partnership. *Social Science and Medicine*, 50, 2, 169–176.

De Bruyn, G., Adams, G., Teeter, L., Soini, H., Musser, J. and Graviss, E. (2001) The contribution of ethnicity to mycobacterium tuberculosis. *International Journal of Tubercular Lung Disease*, 5, 7, 633–641.

Dean, K. and James, H. (1981) Social factors and admission to psychiatric hospital: schizophrenia in Plymouth. *Transactions of the Institute of British Geographers*, NS, 6, 39–52.

Dear, M. and Taylor, S. (1982) *Not on Our Street: Community Attitudes to Mental Health Care*. London, Pion.

Dear, M. and Wolch, J. (1987) *Landscapes of Despair: from Institutionalization to Homelessness*. Oxford, Polity.

Dear, M., Gaber, L., Takahashi, L. and Wilton, R. (1997) Seeing people differently: the socio-spatial construction of disability. *Environment and Planning D: Society and Space*, 455–480.

Dekker, J., Peen, J., Goris, A., Heijnen, H. and Kwakman, H. (1997) Social deprivation and psychiatric admission rates in Amsterdam. *Social Psychiatry and Psychiatric Epidemiology*, 32, 8, 485–492.

Department of Health (1992) *The Health of the Nation: a Strategy for England*. Cm 1986. London, HMSO.

Department of Health (1996) *The Spectrum of Care: Local Services for People with Mental Health Problems*. London, Stationery Office.

Department of Health (1998) *Independent Inquiry into Inequalities in Health*. London, Stationery Office.

Department of Health (1999a) *Saving Lives: Our Healthier Nation*. Cm 4386. London, Stationery Office.

Department of Health (1999b) *Modernising Mental Health Services: Safe, Sound and Supportive*. London, Stationery Office.

Department of Health (1999c) *A National Service Framework for Mental Health*. London, Department of Health.

Department of Health (2001) *Department of Health Waiting List Data: Hospital Inpatient Statistics* (Yellow Book). http://www/doh/gov.uk/waiting times.

DeRiemer, K., Chin, D., Schecter, G. and Reingold, A. (1998) Tuberculosis among immigrants and refugees. *Archives of Internal Medicine*, 158, 7, 753–760.

DeRiemer, K., Daley, C. and Reingold, A. (1999) Preventing tuberculosis among HIV-infected persons: a survey of physicians' knowledge and practices. *Preventive Medicine*, 28, 4, 437–444.

Diez-Roux, A., Nieto, F., Caulfield, L., Tyroler, H., Watson, R. and Szklo, M. (1999) Neighbourhood differences in diet: the artherosclerosis risking communities, (ARIC) study. *Journal of Epidemiology and Community Health*, 53, 55–63.

Diggle, P. and Rowlingson, B. (1994) A conditional approach to point process modeling of elevated risk. *Journal of the Royal Statistical Society Series A*, 157, 433–440.

Dollar, D. (2001) Is globalization good for your health? *Bulletin of the World Health Organization*, 79, 9, 827–833.

Dolk, H., Mertens, B., Kleinschmidt, I., Walls, P., Shaddick, G. and Elliott, P. (1995) A standardization approach to the control of socio-economic confounding in small area studies of environment and health. *Journal of Epidemiology and Community Health*, 49, 2 supplement, s9–s14.

Donovan, J. (1984) Ethnicity and health. *Social Science and Medicine*, 19, 663–670.

Donovan, J. (1986a) *We Don't Buy Sickness, It Just Comes: Health, Illness and Health Care in the Lives of Black People in London*. Aldershot, Gower.

Donovan, J. (1986b) Black people's health: a different approach. In Rathwell, T. and Phillips, D. (eds) *Health, Race and Ethnicity*. Beckenham, Croom Helm. Chapter 5, 117–136.

Dorn, M. and Laws, G. (1994) Social theory, body politics and medical geography: extending Kearns's invitation. *Professional Geographer*, 46, 1, 106–110.

Dossetor, D., Nunn, K., Fairley, M. and Eggleton, D. (1999) A child and adolescent psychiatric outreach service for rural New South Wales: a telemedicine pilot study. *Journal of Paediatrics and Child Psychiatry*, 35, 6, 525–529.

Douglas, M., Conway, L., Gorman, D., Gavin, S. and Hanton, P. (2001) Developing principles for health impact assessment. *Journal of Public Health Medicine*, 23, 2, 148–154.

Dowler, E. (1997) Budgeting for food on a low income in the UK: the case of lone-parent families. *Food Policy*, 22, 5, 405–417.

Dowler, E. and Calvert, C. (1995) Looking for 'fresh' food: diet and lone parents. *Proceedings of the Nutrition Society*, 54, 3, 759–769.

Dowler, E. and McConnell, C. (1993) Nutrition and migration: people on the move. *Proceedings of the Nutrition Society*, 52, 1, 265–266.

Dowler, E. and Raats, M. (1998) *Report on Literature Regarding Access to Food Shops*. London, HEA.

Downey, D. and McGuigan, J. (eds) (1999) *Technocities*. London, Sage.

Downey, L. (1998) Environmental injustice: is race or income a better predictor? *Social Science Quarterly*, 79, 4, 766–778.

Drgona, V. and Turnock, D. (2002) Slovakia. In Carter, F. and Turnock, D. (eds) *Environmental Problems of East Central Europe*, 2nd edn. London, Routledge. Chapter 10, 207–227.

Driessen, G., Gunther, N. and Van Os, J. (1998) Shared social environment and psychiatric disorder: a multilevel analysis of individual and ecological effects. *Social Psychiatry and Psychiatric Epidemiology*, 33, 12, 606–612.

Drobniewski, F., Tayler, E., Ignatenko, N., Paul, J., Connolly, M., Nye, P., Lyagoshina, T. and Besse, C. (1996) Tuberculosis in Siberia. 2: Diagnosis, chemoprophylaxis and treatment. *Tubercle Lung Disease*, 77, 297–301.

D'Souza, R. (2000a) Telemedicine for intensive support of psychiatric inpatients admitted to local hospitals. *Journal of Telemedicine and Telecare*, 6, 1, 26–28.

D'Souza, R. (2000b) A pilot study of an educational service for rural mental health practitioners in South Australia using telemedicine. *Journal of Telemedicine and Telecare*, 6, 1, 187–189.

Dubos, R. (1960) *The Mirage of Health*. London, Allen and Unwin.

Ducq, H., Guesdon, I. and Roelandt, J. (1997) Psychiatric morbidity of homeless persons: a critical review of Anglo-Saxon literature. *Encephale*, 23, 6, 420–430.

Duhme, H., Wiland, S. and Keil, U. (1998) Epidemiological analyses of the relationship between environmental pollution and asthma. *Toxicology Letters*, 103, 307–316.

Duncan, C. and Jones, K. (1995) Individuals and their ecologies: analysing the geography of chronic illness within a multi-level modeling framework. *Health and Place*, 1, 27–40.

Duncan, C., Jones, K. and Moon, G. (1996) Health-related behaviour in context: a multilevel modelling approach. *Social Science and Medicine*, 42, 6, 817–830.

Duncan, C., Jones, K. and Moon, G. (1999) Smoking and deprivation: are there neighbourhood effects? *Social Science and Medicine*, 497–505.

Dunn, C. and Kingham, S. (1996) Establishing links between air quality and health: searching for the impossible? *Social Science and Medicine*, 42, 6, 831–841.

Dunn, C., Kingham, S., Rowlingson, B., Bhopal, R., Cockings, S., Foy, C., Acquilla, S., Halpin, J., Diggle, P. and Walker, D. (2001) Analysing spatially referenced public health data: a comparison of three methodological approaches. *Health and Place*, 7, 1, 1–12.

Dunn, J. and Hayes, M. (2000) Social inequality, population health and housing: a study of two Vancouver neighborhoods. *Social Science and Medicine*, 51, 563–587.

Durkheim, E. (1951) *Suicide*. New York, Free.

Dyck, I. (1998) Women with disabilities and everyday geography: home space and the contested body. In Kearns, R. and Gesler, W. (eds) *Putting Health into Place: Landscape, Identity and Well-Being*. Syracuse, NY, Syracuse University Press.

Dyck, I. (1999) Body troubles: women, the workplace and negotiations of a disabled identity. In Butler, R. and Parr, H. (eds) *Mind and Body Spaces: Geographies Illness, Impairment and Disability*. London, Routledge. Chapter 6, 119–137.

Easterlin, R. (2001) Income and happiness: towards a unified theory. *Economic Journal*, 111, 465–484.

East London and City HAZ (2000) *East London and City HAZ Update*. Summer.

East London and City HAZ (2001) *East London and City HAZ Update*. Summer.

Eastwood, H. (2000) Why are Australian GPs using alternative medicine? Postmodernisation, consumerism and the shift towards holistic health. *Journal of Sociology*, 36, 2, 133–156.

Ecob, R. and Macintyre, S. (2000) Small area variations in health related behaviours: do these depend on the behaviour itself, its measurement, or personal characteristics? *Health and Place*, 6, 4, 261–274.

Edgington, B. (1997) Moral architecture: the influence of the York Retreat on asylum design. *Health and Place*, 3, 2, 91–100.

Edwards, D. and Watt, R. (1997) Oral health care in the lives of gypsy travellers in East Hertfordshire. *British Dental Journal*, 183, 7, 252–257.

ELCHA (1999) *Health in the East End: Annual Public Health Report 1999/2000*. London, East London and The City Health Authority.

ELCHA (2000) *Health in the East End: Annual Public Health Report 2000/2001*. London, East London and The City Health Authority.

Elender, F., Bentham, G. and Langford, I. (1998) Tuberculosis mortality in England and Wales during 1982–1992: its association with poverty, ethnicity and AIDS. *Social Science and Medicine*, 46, 6, 673–681.

Ellaway, A. and Macintyre, S. (1996) Does where you live predict health related behaviours? A case study in Glasgow. *Health Bulletin*, 54, 6, 443–446.

Ellaway, A., Macintyre, S. and Kearns, A. (2001) Perceptions of place and health in socially contrasting neighbourhoods. *Urban Studies*, 38, 12, 2299–2316.

Elliot, S., Taylor, S., Walter, S., Stieb, D., Frank, J. and Eyles, J. (1993) Modelling psychosocial effects of exposure to solid waste facilities. *Social Science and Medicine*, 37, 791–805.

Ellis, M. and Muschkin, C. (1996) Migration of persons with AIDS: a search for support from elderly parents? *Social Science and Medicine*, 43, 1109–1118.

Elpers, J. and Levin, B. (1996) Mental health services: epidemiology, prevention and service delivery systems. In Levin, B. and Petrila, J. (eds) *Mental Health Services: a Public Perspective*. Oxford, Oxford University Press.

Engebretson, J. (1999) Alternative and complementary healing: implications for nursing. *Journal of Professional Nursing*, 15, 4, 214–223.

Ennis, N.E., Hobfoll, S.E. and Schroder, K. (2000) Money doesn't talk, it swears: how economic stress and resistance resources impact inner-city women's depressive mood. *American Journal of Community Psychology*, 28, 2, 149–173.

Ensign, J. and Gittelsohn, J. (1998) Health and access to care: perspectives of homeless youth in Baltimore City. *Social Science and Medicine*, 47, 12, 2087–2099.

Esmen, N. and Marsh, G. (1996) Applications and limitations of air dispersion modeling in environmental epidemiology. *Journal of Exposure Analysis and Environmental Epidemiology*, 6, 3, 339–353.

Evans, G. and Maxwell, L. (1997) Chronic noise exposure and reading deficits: the mediating effects of language acquisition. *Environment and Behaviour*, 29, 5, 638–656.

Exworthy, M., Powell, M. and Mohan, J. (1999) The NHS: quasi-market, quasi-hierarchy and quasi-network? *Public Money and Management*, 19, 4, 15–22.

Eyles, J. (1985) *Senses of Place*. Warrington, Silverbrook.

Eyles, J. and Donovan, J. (1990) *The Social Effects of Health Policy*. Aldershot, Avebury.

Eyles, J., Taylor, S., Johnson, N. and Baxter, J. (1993) Worrying about waste: living close to solid waste disposal facilities in southern Ontario. *Social Science and Medicine*, 37, 805–812.

Fairley, D. (1999) Daily mortality and air pollution in Santa Clara County, California, 1989–1996. *Environmental Health Perspectives*, 107, 8, 637–641.

Farmer, P. (1997) Social scientists and the new tuberculosis. *Social Science and Medicine*, 44, 3, 347–358.

Fearn, R., Haynes, R. and Bentham, G. (1984) The role of branch surgeries in a rural area. *Journal of the Royal College of General Practitioners*, 34, 488–491.

Feder, G., Salkind, M. and Sweeney, O. (1989) Traveller gypsies and general practitioners in East London: the role of the traveller health visitor. *Health Trends*, 21, 93–94.

Feder, G., Vaclavik, T. and Streetly, A. (1993) Traveller gypsies and childhood immunization: a study in East London. *British Journal of General Practice*, 372, 281–284.

Fernandez-Ballesteros, R., Zamarron, M. and Ruiz, M. (2001) The contribution of socio-demographic and psychosocial factors to life satisfaction. *Ageing and Society*, 21, 1, 25–43.

Fidler, D. (2001) The globalization of public health: the first 100 years of international health diplomacy. *Bulletin of the World Health Organization*, 79, 9, 842–849.

Fielder, J. (1981) A review of the literature on access and utilization of medical care with special emphasis on rural primary care. *Social Science and Medicine*, 15C, 129–142.

Fiscella, K. and Franks, P. (1997) Poverty or income inequality as predictor of mortality: Longitudinal cohort study. *British Medical Journal*, 314, 1724–1727.

Fiscella, K. and Franks, P. (2000) Individual income, income inequality, health and mortality: what are the relationships? *Health Services Research*, 35, 1, 307–318.

Fitzpatrick, J. and Jacobson, B. (2001) *Mapping Health Inequalities across London*. London, London Health Observatory.

Fitzpatrick, R., Hinton, J., Newman, S., Scambler, G. and Thompson, J. (1984) *The Experience of Illness*. London, Tavistock.

Foley, P. and Martin, S. (2000) A new deal for the community? Public participation in regeneration and local delivery. *Policy and Politics*, 28, 4, 479–491.

Forde-Roberts, V. (1999) Working for patients? An analysis of some effects of the National Health Service reforms in South Buckinghamshire. Unpublished PhD thesis. London, Queen Mary, University of London.

Forsyth, A., Macintyre, S. and Anderson, A. (1994) Diets for disease? Intra-urban variation in reported food consumption in Glasgow. *Appetite*, 22, 259–274.

Foster, H. (1992) *Health, Disease and Environment*. London, Belhaven.

Foucault, M. (1979) *Discipline and Punish*. Harmondsworth, Penguin.

Foucault, M. (1993) *Naissance de la Clinique*, 3rd edn (1st edn 1963). Paris, Quadridge/Presses Universitaires de France.

Fox, A. and Goldblatt, P. (1982) *The Longitudinal Study: Socio-Demographic Mortality Differentials*. OPCS Series L5.1. London, HMSO.

Francis, S. and Glanville, R. (2001) *Building a 20/20 Vision: Future Health Care Environments*. London, Nuffield Institute.

Frankel, M. (1996) Guidelines/codes of ethics: merging process and content. *Science of the Total Environment*, 184, 1/2, 13–16.

Frankish, C.J., Green, L.W., Ratner, P.A., Chomik, T. and Larsen, C. (1996) *Health Impact Assessment as a Tool for Population Health Promotion and Public Policy. A Report Submitted to the Health Promotion Development Division of Health Canada*. Institute of Health Promotion Research, University of British Columbia.

Frese, M. and Mohr, G. (1987) Prolonged unemployment and depression in older workers: a longitudinal study of intervening variables. *Social Science and Medicine*, 25, 2, 173–178.

Friedman, L., Williams, M., Singh, T. and Frieden, T. (1996) Tuberculosis, AIDS and death among substance abusers on welfare in New York City. *New England Journal of Medicine*, 334, 828–833.

Fry, P. (2000) Religious involvement, spirituality and personal meaning for life: existential predictors of psychological wellbeing in community-residing and institutional care elders. *Aging and Mental Health*, 4, 4, 375–387.

Fuller, T., Edwards, J., Vorakitophokatorn, S. and Sermsri, S. (1996) Chronic stress and psychological well-being: evidence from Thailand on household crowding. *Social Science and Medicine*, 42, 2, 265–280.

Fullilove, M. (1996) Psychiatric implications of displacement: contributions from the psychology of place. *American Journal of Psychiatry*, 153, 12, 1516–1523.

Fulop, N. and Hunter, D. (2000) The experience of setting health targets in England. *European Journal of Public Health*, 10, 4-supplement, 20–24.

Furlong, R. (1996) Haven within or without the hospital gate: a reappraisal of asylum provision in theory and practice. In Tomlinson, D. and Carrier, J. (eds) *Asylum in the Community*. London, Routledge. Chapter 7, 135–168.

Furnham, A., Cheng, H. and Shirasu, Y. (2001) Lay theories of happiness in the east and west. *Psychologia*, 44, 3, 173–187.

Gabe, J. and Williams, P. (1993) Women, crowding and mental health. In Burridge, R. and Ormandy, D. (eds) *Unhealthy Housing: Research, Remedies and Reform.* London, Spon. 191–208.

Gatrell, A. (2002) *Geographies of Health.* Oxford, Blackwell.

Gatrell, A., Thomas, C., Bennett, S., Bostock, L., Popay, J., Williams, G. and Shatahmasebi, S. (2000) Understanding health inequalities: locating people in geographical social spaces. In Graham, H. (ed.) *Understanding Health Inequalities.* Milton Keynes, Open University Press. 156–169.

Geddes, M. (2000) Tackling social exclusion in the European Union? The limits to the new orthodoxy of local partnership. *International Journal of Urban and Regional Research*, 24, 4, 782–800.

Geores, M. (1998) Surviving on metaphor: how 'Health = Hot Springs' created and sustained a town. In Kearns, R. and Gesler, W. (eds) *Putting Health into Place: Landscape, Identity and Well-Being.* Syracuse, NY, Syracuse University Press. Chapter 3, 17–35.

Gesler, W. (1992) Therapeutic landscapes: medical geographic research in light of the new cultural geography. *Social Science and Medicine*, 34, 7, 735–746.

Gesler, W. (1993) Therapeutic landscapes: theory and a case study of Epidauros, Greece. *Environment and Planning D: Society and Space*, 11, 171–189.

Gesler, W. (1996) Lourdes: healing in a place of pilgrimage. *Health and Place*, 2, 2, 95–105.

Gesler, W. (1998) Bath's reputation as a healing place. In Kearns, R. and Gesler, W. (eds) *Putting Health into Place: Landscape, Identity and Well-Being.* Syracuse, NY, Syracuse University Press. Chapter 2, 17–35.

Gesler, W. and Ricketts, T. (eds) (1992) *Health in Rural North America.* New York, Rutgers University Press.

Gesler, W.M., Jordan, J.M., Dragomir, A., Luta, G. and Fryer, J.G. (1999). A geographic assessment of health care coverage in two 'rural' North Carolina communities. *Southeastern Geographer,* 39, 2, 127–144.

Giddens, A. (1984) *The Constitution of Society: Outline of a Theory of Structuration.* Cambridge, Polity.

Giggs, J. (1973) The distribution of schizophrenics in Nottingham. *Transactions of the Institute of British Geographers,* 59, 55–76.

Giggs, J. (1975) The distribution of schizophrenics in Nottingham: a reply. *Transactions of the Institute of British Geographers*, 64, 150–156.

Gillam, S. (1992) Provision of health promotion clinics in relation to population need: another example of the inverse care law. *British Journal of General Practice*, 42, 54–56.

Gittins, D. (1998) *Madness in its Place: Narratives of Severalls Hospital 1913–1997.* London, Routledge.

Glanville, R. (2001) *50 Years of Ideas in Health Care Buildings.* London, Nuffield Institute.

Glasner, P. and Rothman, H. (2001) New genetics, new ethics: globalisation and its discontents. *Health Risk and Society,* 3, 3, 245–259.

Gleeson, B. (1999) *Geographies of Disability.* London, Routledge.

Gleissberg, V. (1999) A response by nurses to the challenge of tuberculosis in the United Kingdom and Russia. In Porter, J. and Grange, J. (eds) *Tuberculosis: an Interdisciplinary Perspective.* London, Imperial College Press.

Glover, G., Robin, E., Emami, J. and Arabscheibani, G. (1998) A needs index for mental health care. *Social Psychology and Psychiatric Epidemiology,* 33, 2, 89–96.

Glover, G., Leese, M. and McCrone, P. (1999) More severe mental illness is more concentrated in deprived areas. *British Journal of Psychiatry,* 175, 544–548.

Goeres, M. and Gesler, W. (1999) Compromised space: contests over the provision of a therapeutic environment for people with mental illness. In Williams, A (ed.) *Therapeutic Landscapes: the Dynamic between Place and Wellness*. New York, University Press of America. 99–122.

Gold, M. (1998) The concept of access and managed care. Beyond coverage and supply: measuring access to health care in today's market. *Health Services Research*, 33, 3, 625–652.

Goldberg, D. (1999) The future pattern of psychiatric provision in England. *European Archives of Psychiatry and Clinical Neuroscience*, 249, 3, 123–127.

Goldberg, D. (2000) Findings from *London's Mental Health*: a service in crisis. *Acta Psychiatrica Scandinavica*, 101, 57–60.

Goldberg, D. and Williams, P. (1988) *A User's Guide to the General Health Questionnaire*. Windsor, NFER–Nelson.

Goldman, B. (1996) What is the future of environmental justice? *Antipode*, 28, 2, 122.

Golledge, R. (1993) Geography and the disabled: a survey with special reference to vision impaired and blind populations. *Transactions of the Institute of British Geographers*, 18, 63–85.

Gordon, D., Shaw, M., Dorling, D. and Davey-Smith, G. (1999) *Inequalities in Health: the Evidence Presented to the Independent Inquiry into Inequalities in Health, Chaired by Sir Donald Acheson*. Bristol, Policy.

Gould, M. and Jones, K. (1996) Analysing perceived limiting long-term illness using UK census microdata. *Social Science and Medicine*, 42, 6, 857–869.

Gould, P. (1993) *The Slow Plague: a Geography of the AIDS Pandemic*. Oxford, Blackwell.

Graham, E., MacLeod, M., Johnston, M., Dibben, C., Morgan, I. and Briscoe, S. (2000) Individual deprivation, neighbourhood and recovery from illness. In Graham, H. (ed.) *Understanding Health Inequalities*. Milton Keynes, Open University Press. 170–185.

Graham, H. (2000) The challenge of health inequalities. In Graham, H. (ed.) *Understanding Health Inequalities*. Buckingham, Open University Press.

Graham, R., Forrester, M., Wysong, J., Rosenthal, T. and James, R. (1995) HIV/AIDS in the rural United States: epidemiology and health services delivery. *Medical Care Research Review*, 52, 435–452.

Graham, S. (1997) Imagining the real-time city: telecommunications, urban paradigms and the future of cities. In Westwood, S. and William, J. (eds) *Imagining Cities: Scripts, Signs, Memory*. London, Routledge. Chapter 2, 31–49.

Grandjean, P. and Sorsa, M. (1996) Ethical aspects of genetic predisposition to environmentally related diseases. *Science of the Total Environment*, 184, 1/2, 37–43.

Grange, J. (1999) The global burden of tuberculosis. In Porter, J and Grange, J. (eds) *Tuberculosis: an Interdisciplinary Perspective*. London, Imperial College Press. Chapter 1, 3–32.

Great Britain Parliament (1997) *The New NHS: Modern, Dependable*. White Paper, Cm 3807. London, Stationery Office.

Great Britain Parliament (2000) *Our Countryside: the Future. A Fair Deal for Rural England*. Cm 4909. London, Stationery Office.

Green, A. (1994) *The Geography of Poverty and Wealth: Evidence on the Changing Spatial Distribution and Segregation of Poverty and Wealth from the Census of Population 1991 and 1981*. Warwick, Institute for Employment Research.

Greenburg, M. (1991) Urban/rural differences in behavioural risk factors for chronic diseases. *Urban Geography*, 8, 2, 146–151.

Gregory, D. (1981) Human agency and human geography. *Transactions of the Institute of British Geographers*, NS, 5, 1–16.

Gregory, D. (1989) Presence and absences: time–space relations and structuration theory. In Held, D. and Thompson, J. (eds) *Social Theory of the Modern Societies: Anthony Giddens and His Critics*. Cambridge, Cambridge University Press.

Gregory, D. (1994) *Geographical Imaginations*. Oxford, Blackwell.

Grembowski, D., Cook, K., Patrick, D. and Roussel, A. (2002) Managed care and the US health care system: a social exchange perspective. *Social Science and Medicine*, 54, 1167–1180.

Griffiths, C. (2001) TB prevention, screening and treatment: what role for primary care? In ELCHA (ed.) *Health in the East End: Annual Public Health Report 2000/2001*. London, East London and The City Health Authority. Chapter 6, 50–56.

Griffiths, C., Cooke, S. and Toon, P. (1993) Registration health checks: inverse care in the inner city? *British Medical Journal*, 44, 201–204.

Gross, P. (1972) Urban health disorders, spatial analysis and the economics of health facility location. *International Journal of Health Services*, 2, 1, 63–84.

Gross, R., Langfried, B. and Herman, S. (1996) Height and weight as a reflection of the nutritional situation of school-aged children working and living in the streets of Jakarta. *Social Science and Medicine*, 43, 4, 453–458.

Gruffudd, P. (2001) 'Science and the stuff of life': modernist health centres in 1930s London. *Journal of Historical Geography*, 27, 3, 395–416.

Gulube, S. and Wynchank, S. (2001) Telemedicine in South Africa: success or failure? *Journal of Telemedicine and Telecare*, 7, 2, 47–49.

Gurling, H., Kalsi, G., Brynjolfson, J., Sigmundsson, T., Sherrington, R., Mankoo, B., Read, T., Murphy, P., Blaveri, E., McQuillin, A., Petursson, H. and Curtis, D. (2001) Genomewide genetic linkage analysis confirms the presence of susceptibility loci for schizophrenia, on chromosomes 1q32.2, 5q33.2, and 8p21–22, and provides support for linkage to schizophrenia, on chromosomes 11q23.3–24 and 20q12.1–11.23. *American Journal of Human Genetics*, 68, 3, 661–673.

Guy, C. (1991) Urban and rural contrasts in food prices and availability: a case study in Wales. *Journal of Rural Studies*, 7, 3, 311–325.

Hägerstrand, T. (1952) *The Propagation of Innovation Waves*. Lund, Sweden, Gleerup.

Hägerstrand, T. (1975) Space, time and human conditions. In Karlqvist, A., Lunqvist, L. and Snickars, F. (eds) *Dynamic Allocation of Urban Space*. Farnborough, Saxon House.

Hägerstrand, T. (1982) Diorama, path and project. *Tidschrift vor Economische en Sociale Geografie*, 73, 323–339.

Haggett, P. (1976) Hybridizing alternative models of an epidemic diffusion process. *Economic Geography*, 52, 136–146.

Haggett, P. (1994) Prediction and predictability in geographical systems. *Transactions of the Institute of British Geographers*, 19, 1, 6–20.

Haines, M., Stansfeld, S., Job, R., Berglund, B. and Mead, J. (2001) Chronic aircraft exposure, stress responses, mental health and cognitive performance in school children *Psychological Medicine*, 31, 265–277.

Hajioff, S. and McKee, M. (2000) The health of the Roma people: a review of the published literature. *Journal of Epidemiology and Community Health*, 54, 11, 864–869.

Hall, E. (2000) Blood, brain and bone. *Area*, 32, 1, 21–29.

Halpern, D. and Nazroo, J. (2000) The ethnic density effect: results from a national community survey of England and Wales. *International Journal of Social Psychiatry*, 46, 1, 34–46.

Ham, C. (1997) *Health Care Reform: Learning from International Experience*. Buckingham, Open University Press.

Hammer, T. (1993) Unemployment and mental health among young people: a longitudinal study. *Journal of Adolescence*, 16, 407–420.
Hanson, M. (1999) Biotechnology and commodification within healthcare. *Journal of Medicine and Philosophy*, 24, 3, 267–287.
Harpham, T. (1994a) Cities and health in the Third World. In Phillips, D. and Verhasselt, Y. (eds) *Health and Development*. London, Routledge. 111–121.
Harpham, T. (1994b) Urbanization and mental health in developing countries. *Social Science and Medicine*, 39, 2, 233–245.
Hart, G., Ecob, R. and Smith, G. (1997) People, place and coronary heart disease risk factors: a multilevel analysis of the Scottish Heart Health Study. *Social Science and Medicine*, 45, 6, 893–902.
Harvey, C., Pantelis, C., Taylor, J., McCabe, P., Lefevre, K., Campbell, P. and Hirsch, S. (1996) The Camden Schizophrenia Surveys. 2: High prevalence of schizophrenia in an inner London borough and its relationship to socio-demographic factors. *British Journal of Psychiatry*, 169, 4, 418–426.
Harvey, D. (1982) *The Limits to Capital*. Oxford, Blackwell.
Harvey, D. (1989a) *The Condition of Post Modernity*. Oxford, Blackwell.
Harvey, D. (1989b) *The Urbanisation of Capital*. Oxford, Blackwell.
Hawker, F., Kavanagh, S., Yellowlees, P. and Kalucy, R. (1998) Telepsychiatry in South Australia. *Journal of Telemedicine and Telecare*, 4, 4, 187–194.
Haybron, D. (2001) Happiness and pleasure. *Philosophy and Phenomenological Research*, 62, 3, 501–528.
Haylett, C. (2001) Modernization, welfare and 'third way' politics: limits to theorizing in 'thirds'? *Transactions of the Institute of British Geographers*, 26, 1, 43–56.
Haynes, R. (1987) *The Geography of Health Services in Britain*. London, Croom Helm.
Haynes, R. (1991) Inequalities in health and health care use: evidence from the General Household Survey. *Social Science and Medicine*, 33, 4, 361–368.
Haynes, R. and Bentham, G. (1979) *Community Hospitals and Rural Accessibility*. Farnborough, Saxon House.
Haynes, R. and Bentham, G. (1982) The effects of accessibility on general practitioner consultations, out-patient attendances and in-patient admissions in Norfolk, England. *Social Science and Medicine*, 16, 561–569.
Haynes, R. and Gale, S. (1999) Mortality, long-term illness and deprivation in rural and metropolitan wards of England and Wales. *Health and Place*, 5, 4, 301–312.
Haynes, R., Bentham, G., Lovett, A. and Eimmermann, J. (1997) Effect of labour market conditions on reporting of limiting long term illness and permanent sickness in England and Wales. *Journal of Epidemiology and Community Health*, 51, 3, 282–288.
Haynes, R., Bentham, G., Lovett, A. and Gale, S. (1999) Effects of distances to hospital and GP surgery on hospital inpatient episodes, controlling for needs and provision. *Social Science and Medicine*, 49, 3, 425–433.
HAZnet (2002) A fully interactive site for sharing the experiences of Health Action Zones in tackling health inequalities. http://www.haznet.org.uk.
Health Services Research (1989) *A Rural Health Services Agenda*. *Health Services Research*, special issue, 23, 6.
Helfand, G. and Peyton, L. (1999) A conceptual model of environmental justice. *Social Science Quarterly*, 80, 1, 68–83.
Helman, C. (1984) *Culture, Health and Illness*. Bristol, Wright.
Hemon, D. (1995) Research in environmental epidemiology: selected methodological aspects. *Revue d'Epidémiologie et de Santé Publique*, 43, 5, 395–411.

Hendley, J., Barnes, R., Hirschfield, A. et al. (1999) *What is HIA and How Can It Be Applied to Regeneration Programmes?* Working Paper 1. Liverpool, Departments of Civic Design and Public Health, University of Liverpool.

Herten, L. and Van de Water, H. (2000) Health policies on target? Review on health target setting in 18 European countries. *European Journal of Public Health*, 10, 4, 11–16.

Higgs, G. (1999) Investigating trends in rural health outcomes: a research agenda. *Geoforum*, 30, 3, 203–221.

Higgs, G., Senior, M. and Williams, H. (1998) Spatial and temporal variation of mortality and deprivation. 1: Widening health inequalities. *Environment and Planning A*, 30, 9, 1661–1682.

Hill, D. (2000) *Urban Policy and Politics in Britain*. Basingstoke, Macmillan.

Hills, J. (1995) *Joseph Rowntree Foundation Inquiry into Income and Wealth. Vol. 2: A Summary of the Evidence*. York, Joseph Rowntree Foundation.

Hills, J. (1998) *Income and Wealth: the Latest Evidence*. York, Joseph Rowntree Foundation.

Hinchcliffe, S. (2001) Indeterminancy in decisions: science, policy and politics in the BSE (bovine spongiform encephalopathy) crisis. *Transactions of the Institute of British Geographers*, NS, 26, 2, 182–204.

Holterman, S. (1975) Areas of urban deprivation in Great Britain: an analysis of 1971 census data. *Social Trends*, 6, 33–47.

Horton, B. (1996) Going to work in genes catches on. *Nature*, 383, 739–740.

Howe, G.M. (1997) *People, Environment, Disease and Death: a Medical Geography of Britain throughout the Ages*. Cardiff, University of Wales Press.

Humphreys, K. and Carr-Hill, R. (1991) Area variation in health outcomes: artefact or ecology? *International Journal of Epidemiology*, 20, 1, 251–258.

Hunt, C. (1996) Social vs. biological: theories on the transmission of AIDS in Africa. *Social Science and Medicine*, 42, 9, 1283–1296.

Hunt, S., McEwen, J. and McKenna, S. (1986) *Measuring Health Status*. London, Croom Helm.

Hunter, J. and Shannon, G. (1985) Jarvis revisited: distance decay in service areas of mid 19th century asylums. *Professional Geographer*, 37, 3, 296–302.

Huston, J. (1999) Telemedical record documentation: a preliminary survey. *Journal of Telemedicine and Telecare*, 5, 1, 6–8.

Hyndman, S. (1990) Housing dampness and health among British Bengalis in East London. *Social Science and Medicine*, 30, 131–141.

Hyndman, S. (1998) Making connections between housing and health. In Kearns, R. and Gesler, W. (eds) *Putting Health into Place: Landscape, Identity and Well-Being*. Syracuse, NY, Syracuse University Press. Chapter 10, 191–207.

Iannoccone, P. (2001) Toxicogenomics: 'the call of the wild chip'. *Environmental Health Perspectives*, 109, 1, A8–A11.

IFPRI (1998) *Ghana Seminar on Urban Livelihoods and Food Security*. IFPRI Report 20. Washington, DC, International Food Policy Research Institute.

Imrie, R. (1996) *Disability and the City*. London, Chapman.

Interdepartmental Working Group on Tuberculosis (1996) *The Prevention and Control of Tuberculosis in the United Kingdom. Tuberculosis and Homeless People. A Review and Recommendations for Health Professionals, Local Authorities and Voluntary Support Groups*. Leeds, Department of Health.

Ison, E. (2000) *Resource for Health Impact Assessment*. London, National Health Service National Executive. Available via http://www.doh.gov.uk.

Iversen, L. and Sabroe, S. (1988) Hospital admissions before and after shipyard closure. *British Medical Journal*, 299, 1073–1076.

Jackson, P. (1989) *Maps of Meaning: an Introduction to Cultural Geography.* London, Unwin Hyman.
Jahoda, M. (1992) *Employment and Unemployment.* Cambridge, Cambridge University Press.
James, G. (1991) Blood pressure response to the daily stressors of urban environments: methodology, basic concepts and significance. *Yearbook of Physical Anthropology*, 34, 189–210.
Jarman, B. (1981) *A Survey of Primary Care in London.* RCGP Occasional Paper 16. London, Royal College of General Practitioners.
Jarman, B. and Hirsch, S. (1992) Statistical models to predict district psychiatric morbidity. In Thornicroft, G., Brewin, C. and Wing, J. (eds) *Measuring Mental Health Needs.* London, Gaskell (Royal College of Psychiatrists). Chapter 4, 62–80.
Jarvis, H., Pratt, A. and Cheng-Chong Wu, P. (2001) *The Secret Life of Cities: the Social Reproduction of Everyday Life.* Harlow, Pearson.
Jasmer, R., de Leon, A., Hopewell, P., Alarcon, R., Moss, A., Paz, E., Schecter, G. and Small, P. (1997) Tuberculosis in Mexican born persons in San Francisco: reactivation, acquired infection and transmission. *International Journal of Tuberculosis and Lung Disease*, 1, 6, 536–541.
Jedrychowski, W. (1995) Review of recent studies from central and eastern Europe associating respiratory health effects with high levels of exposure to traditional air pollutants. *Environmental Health Perspectives*, 103, 2, 15–21.
Jeffords, J. and Daschle, T. (2001) Political issues in the genome era. *Science*, 291, 1249–1251.
Jenkins, C. and Campbell, J. (1996) Catchment areas in general practice and their relation to size and quality of practice and deprivation: a descriptive study in one London borough. *British Medical Journal*, 313, 1189–1192.
Jenkins, J. and Scanlan, S. (2001) Food security in less developed countries, 1970–1990. *American Sociological Review*, 66, 5, 718–744.
Jenkins, L., Tarnolpolsky, A. and Hand, D. (1981) Psychiatric admissions and aircraft noise from London Airport: four-year, three hospitals study. *Psychological Medicine*, 11, 765–782.
Jerrett, M., Eyles, J., Cole, D. and Reader, S. (1997) Environmental equity in Canada: an empirical investigation into the income distribution of pollution in Ontario. *Environment and Planning A*, 29, 10, 1777–1800.
Jerrett, M., Burnett, R., Kanaroglou, P., Eyles, J., Finkelstein, N., Giovis, C. and Brook, J. (2001) A GIS: environmental justice analysis of particulate air pollution in Hamilton, Canada. *Environment and Planning A*, 955–973.
Joelson, L. and Wahlquist, L. (1987) The psychological meaning of job insecurity and job loss: results of a longitudinal study. *Social Science and Medicine*, 25, 2, 179–182.
Joffe, M. and Mindell, J. (2002) A framework for the evidence base to support health impact assessment. *Journal of Epidemiology and Community Health*, 56, 2, 132–138.
Johnson, S., Ramsay, R., Thornicroft, G., Brooks, L., Lelliott, P., Peck, E., Smith, H., Chisholm, D., Audini, B., Knapp, M. and Goldberg, D. (eds) (1998a) *London's Mental Health.* London, King's Fund.
Johnson, S., Ramsay, R. and Thornicroft, G. (1998b) Londoners' mental health needs: the sociodemographic context. In Johnson, S., Ramsay, R., Thornicroft, G., Brooks, L., Lelliott, P., Peck, E., Smith, H., Chisholm, D., Audini, B., Knapp, M. and Goldberg, D. (eds) *London's Mental Health.* London, King's Fund. Chapter 3, 15–32.
Joint Health Surveys Unit (1997) *Health Survey for England 1995. Vol. II: Survey Methodology and Documentation.* London, Stationery Office.
Joint Tuberculosis Committee of the British Thoracic Society (1994) Control and prevention of tuberculosis in the United Kingdom. *Thorax*, 49, 1193–1200.

Jones, A., Bentham, G., Harrison, B., Badminton, R. and Wareham, N. (1998) Accessibility and heath service utilization for asthma in Norfolk, England. *Journal of Public Health Medicine*, 20, 3, 312–317.

Jones, A., Bentham, G. and Horwell, C. (1999) Health service accessibility and deaths from asthma. *International Journal of Epidemiology*, 28, 1, 101–105.

Jones, I.R. and Curtis, S. (1997) Health. In Pacione, M. (ed.) *Britain's Cities: Geographies of Division in Urban Britain*. London, Routledge. 218–243.

Jones, J. (1996) Community-based mental health care in Italy: are there lessons for Britain? *Health and Place*, 2, 2, 125–128.

Jones, J. (2000) Mental health care reforms in Britain and Italy since 1950: a cross-national comparative study. *Health and Place*, 6, 3, 171–188.

Jones, K. and Moon, G. (1987) *Health, Disease and Society: an Introduction to Medical Geography*. London, Routledge and Kegan Paul.

Jones, K. and Moon, G. (1993) Medical geography: taking space seriously. *Progress in Human Geography*, 17, 4, 515–524.

Joseph, A. and Phillips, D. (1984) *Accessibility and Utilization: Geographical Perspectives on Health Care Delivery*. London, Harper and Row.

Joshi, H., Wiggins, R., Bartley, M., Mitchell, R., Gleave, S. and Lynch, K. (2000) Putting health inequalities on the map: does where you live matter and why? In Graham, H. (ed.) *Understanding Health Inequalities*. Milton Keynes, Open University Press. 143–155.

Joubert, K. (1991) Size at birth and socio-demographic factors in gypsies in Hungary. *Journal of Biological Science*, 23, 1, 39–47.

Judge, K. (1995) Income distribution and life expectancy: a critical appraisal. *British Medical Journal*, 311, 1282–1287.

Judge, K., Mulligan, J. and Benzeval, M. (1998) Income inequality and population health. *Social Science and Medicine*, 46, 4–5, 567–579.

Judge, K., Mulligan, J. and Benzeval, M. (1999) Reply to Richard Wilkinson (vol. 47, p. 983, 1998). *Social Science and Medicine*, 48, 2, 281.

Kalipeni, E. and Oppong, J. (1998) The refugee crisis in Africa and implications for health and disease: a political ecology approach. *Social Science and Medicine*, 46, 12, 1637–1653.

Karasek, R. and Theorell, T. (1990) *Healthy Work: Stress, Productivity, and the Reconstruction of Working Life*. New York, Basic.

Karlsen, S. and Nazroo, J. (2000) Identity and structure: rethinking ethnic inequalities and health. In Graham, H. (ed.) *Understanding Health Inequalities*. Buckingham, Open University Press. Chapter 3, 38–57.

Kawachi, I., Kennedy, B., Lochner, K. and Prothero-Smith, D. (1997) Social capital, income inequality and mortality. *American Journal of Public Health*, 87, 1491–1498.

Kawachi, I., Kennedy, B. and Glass, R. (1999) Social capital and self-rated health: a contextual analysis. *American Journal of Public Health*, 89, 1187–1193.

Kazmi, J. and Pandit, K. (2001) Disease and dislocation: the impact of refugee movements on the geography of malaria in NWFP, Pakistan. *Social Science and Medicine*, 52, 7, 1043–1055.

Kearns, R. (1991) The place of health in the health of the place: the case of Hokianga special medical area. *Social Science and Medicine*, 33, 4, 519–530.

Kearns, R. (1993) Place and health: toward a reformed medical geography. *Professional Geographer*, 45, 139–147.

Kearns, R. (1994) Putting health and healthcare into place: an invitation accepted and declined. *Professional Geographer*, 46, 111–115.

Kearns, R. (1998) 'Going it alone': place, identity and community resistance to health reforms in Hokianga, New Zealand. In Kearns, R. and Gesler, W. (eds) *Putting Health into Place: Landscape, Identity and Well-Being*. Syracuse, NY, Syracuse University Press. Chapter 12, 226–247.

Kearns, R. and Barnett, R. (1997) Consumerist ideology and the symbolic landscapes of private medicine. *Health and Place*, 3, 171–180.

Kearns, R. and Barnett, R. (1999) To boldly go? Place metaphor and the marketing of Auckland's Starship Hospital. *Environment and Planning D: Society and Space*, 17, 201–226.

Kearns, R. and Barnett, R. (2000) 'Happy meals' in the *Starship Enterprise*: interpreting a moral geography of health care consumption. *Health and Place*, 6, 2, 81–93.

Kearns, R. and Gesler, W. (eds) (1998) *Putting Health into Place: Landscape, Identity and Well-Being*. Syracuse, NY, Syracuse University Press.

Kearns, R. and Joseph, A. (1993) Space in its place: developing the link in medical geography. *Social Science and Medicine*, 37, 6, 711–717.

Kearns, R. and Joseph, A. (2000) Contracting opportunities: interpreting post-asylum geographies of mental health care in Auckland, New Zealand. *Health and Place*, 6, 3, 159–170.

Kearns, R., Smith, C. and Abbott, M. (1991) Another day in paradise: life on the margins in urban New Zealand. *Social Science and Medicine*, 33, 4, 369–379.

Keil, R. (1998) *Los Angeles: Globalization, Urbanization and Social Struggles*. Chichester, Wiley.

Kelleher, D. and Islam, S. (1996) 'How should I live?' Bangladeshi people and non-insulin-dependent diabetes. In Hillier, S. and Kelleher, D. (eds) *Researching Cultural Differences in Health*. London, Routledge. Chapter 10, 220–237.

Kelly, S., Charlton, J. and Jenkins, R. (1995) Suicide deaths in England and Wales, 1982–1992: the contribution of occupation and geography. *Population Trends*, 80, 16–25.

Kelner, M. and Wellman B. (1997) Health care and consumer choice: medical and alternative therapies. *Social Science and Medicine*, 45, 203–212.

Kerr, A. and Cunningham-Burley, S. (2000) On ambivalence and risk: reflexive modernity and the new human genetics. *Sociology*, 34, 2, 283–304.

Kickbush, I. (1993) Health promotion and disease prevention: the implications for health promotion. In Normand, C. and Vaughan, P. (eds) *Europe without Frontiers: the Implications for Health*. Chichester, Wiley. 47–63.

Kisely, S. (1998) More alike than different: comparing the mental health needs of London and other inner city areas. *Journal of Public Health Medicine*, 20, 3, 318–324.

Knapp, K. and Hardwick, K. (2000) The availability and distribution of dentists in rural zip codes and primary health care professional shortage areas (PC–HPSA) zip codes: comparison with primary care providers. *Journal of Public Health Dentistry*, 60, 43–48.

Knowles, C. (2000) Burger King, Dunkin Donut and community mental health care. *Health and Place*, 6, 3, 213–224.

Knox, P. (1978) The intra-urban ecology of primary medical care: patterns of accessibility and their policy implications. *Environment and Planning A*, 10, 415–435.

Koupilova, I., Epstein, H., Holcik, J., Hajioff, S. and McKee, M. (2001) Health needs of the Roma population in the Czech and Slovak republics. *Social Science and Medicine*, 53, 9, 1191–1204.

Kryter, K. (1990) Aircraft noise and social factors in psychiatric hospital admission rates: a re-examination of some data. *Psychological Medicine*, 20, 395–411.

Krzyzanowski, M. (1997) Methods for assessing the extent of exposure and effects of air pollution. *Occupational and Environmental Medicine*, 54, 3, 145–151.

Kuhn, R. and Culhane, D. (1998) Applying cluster analysis to test a typology of homelessness by pattern of shelter utilization: results from analysis of administrative data. *American Journal of Community Psychology*, 26, 2, 207–232.

Kunst, A., Groenhof, F. and Mackenbach, J. (1998) Mortality by occupational class among men 30–64 years in 11 European countries. *Social Science and Medicine*, 46, 11, 1459–1476.

Kunst, A., Groenhof, F., Anderson, O., Borgan, J., Costa, G., Desplanques, G., Filakti, H., Giraldes, M., Faggiano, F., Harding, S., Junker, C., Martikainen, P., Minder, C., Nolan, B., Pagnanelli, F., Regidor, E., Vagero, D., Valkonen, T. and Mackenbach, J. (1999) Occupational class and ischemic heart disease mortality in the United States and 11 European countries. *American Journal of Public Health*, 89, 1, 47–53.

Lacoste, O. (1997) La santé dans le Nord-Pas-de-Calais. *Actualité et dossier en santé publique*, 19, X–XI.

LaMay, C. (1997) Telemedicine and competitive change in health care. *Spine*, 22, 1, 88–97.

Lamont, A., Ukoumunne, O., Tyrer, P., Thornicroft, G. and Slaughter, J. (2000) The geographical mobility of severely mentally ill residents in London. *Social Psychiatry and Psychiatric Epidemiology*, 35, 4, 164–169.

Lang, T. (1999) Diet, health and globalization: five key questions. *Proceedings of the Nutrition Society*, 58, 2, 335–343.

Langford, I. and Bentham, G. (1996) Regional variations in mortality rates in England and Wales: an analysis using multi-level modelling. *Social Science and Medicine*, 42, 6, 897–908.

Lauzardo, M. and Ashkin, D. (2000) Phthisiology at the dawn of the new century: a review of tuberculosis and the prospects for its elimination. *Chest*, 117, 5, 1455–1473.

Lawless, P. (1989) *Britain's Inner Cities*. London, Chapman.

Laws, G. and Dear, M. (1988) Coping in the community: a review of factors influencing the lives of deinstitutionalized ex-psychiatric patients. In Smith, C. and Giggs, J. (eds) *Location and Stigma: Contemporary Perspectives on Mental Health and Mental Health Care*. Boston, MA, Unwin Hyman. Chapter 5, 83–102.

Lawson, K. (2000) Health and illness experiences of African-Caribbean women and men: a study in East London. Unpublished PhD thesis, Queen Mary, University of London.

Learmonth, A. (1988) *Disease Ecology: an Introduction*. London, Blackwell.

Leather, S. and Dowler, E. (1997) Intake of micronutrients in Britain's poorest fifth has declined. *British Medical Journal*, 314, 1412–1413.

Lee, A., Isaac, M. and Janaca, A. (2002) Post-traumatic stress disorder and terrorism. *Current Opinion in Psychiatry*, 15, 6, 633–637.

Lee, K., Buse, K. and Fustukian, S. (eds) (2002) *Health Policy in a Globalising World*. Cambridge, Cambridge University Press.

Lefebvre, H. (1991) *The Production of Space* (1974, trans.). Oxford, Blackwell.

Lehti, A. and Mattson, B. (2001) Health, attitude to care and pattern of attendance among gypsy women: a general practice perspective. *Family Practice*, 18, 4, 445–448.

Levy, C. (2002a) Ingredients of a failing system: a lack of state money, a group without a voice. *New York Financial Times*, 28 April.

Levy, C. (2002b) For mentally ill, death and misery. *New York Financial Times*, 28 April.

Lochner, K., Pamuk, E., Makuc, D., Kennedy, B. and Kawachi, I. (2001) State level income inequality and individual mortality risk: a prospective multi-level study. *American Journal of Public Health*, 91, 3, 385–391.

Löfstedt, R. and Frewer, L. (1998) Introduction. In Löfstedt, R. and Frewer, L. (eds) *Risk and Modern Society*. London, Earthscan. Chapter 1, 3–30.

Loppert, S., Staricoff, R. and Scott, J. (2001) Evidence-based art? *Journal of the Royal Society of Medicine*, 94, 10, 551–552.

Lovell, N. and Celler, B. (1999) Information technology in primary health care. *International Journal of Medical Informatics*, 55, 1, 9–22.

Lowe, J. (1996) Gravity model applications in health planning: analysis of an urban hospital market. *Journal of Regional Science*, 36, 3, 437–461.

Lowry, S. (1991) *Housing and Health*. London, British Medical Journal.

Löytenen, M. (2000) Telemedicine and the geography of health. Paper presented to the 9th International Symposium in Medical Geography, 3–7 July, Montreal.

Löytenen, M. and Arbona, S. (1996) Forecasting the AIDS epidemic in Puerto Rico. *Social Science and Medicine*, 42, 7, 997–1010.

Lucas-Gabrielli, V., Tonnellier, F. and Vigneron, E. (1998) *Une Typologie des paysages socio-sanitaires en France*. Biblio. no. 1220. Paris, Centre de Recherche d'Étude et de Documentation en Économie de la Santé.

Luff, D. and Thomas, K. (2000) 'Getting somewhere', feeling cared for: perspectives on complementary therapies in the NHS. *Complementary Therapies in Medicine*, 8, 4, 253–259.

Lygoshina, T. (1998) Russia. In Davies, P. (ed.) *Clinical Tuberculosis*, 2nd edn. London, Chapman and Hall.

Macintyre, S. (1999) Geographical inequalities in mortality, morbidity and health related behaviour in England. In Gordon, D., Shaw, M., Dorling, D. and Davey-Smith, G. (1999) *Inequalities in Health: the Evidence Presented to the Independent Inquiry into Inequalities in Health, Chaired by Sir Donald Acheson*. Bristol, Policy. 148–154.

Macintyre, S., Maciver, S. and Soomans, A. (1993) Area, class and health: should we be focusing on places or people? *Journal of Social Policy*, 22, 2, 213–234.

Macintyre, S., Hunt, K. and Sweeting, H. (1996) Gender differences in health: are things really as they seem? *Social Science and Medicine*, 42, 617–624.

Macintyre, S., Ellaway, A., Der, G., Ford, G. and Hunt, K. (1998) Do housing tenure and car access predict health because they are simply markers of income or self-esteem? A Scottish study. *Journal of Epidemiology and Community Health*, 52, 10, 657–664.

Macintyre, S., Ford, G. and Hunt, K. (1999) Do women 'over-report' morbidity? Men's and women's responses to structured prompting on a standard question on long standing illness. *Social Science and Medicine*, 48, 1, 89–98.

Maclaran, A. (1981) Area-based positive discrimination and the distribution of well-being. *Transactions of the Institute of British Geographers*, 6, 53–67.

Malmberg, A., Hawton, K. and Simkin, S. (1997) A study of suicide in farmers in England and Wales. *Journal of Psychosomatic Research*, 43, 1, 107–111.

Mangtani, P., Jolly, D., Watson, J. and Rodrigues, L. (1995) Socioeconomic deprivation and notification rate for tuberculosis in London 1982–91. *British Medical Journal*, 310, 963–966.

Manguin, S. and Boussinesq, M. (1999) Remote sensing in public health: application to malaria and other diseases. *Médecine et Maladies Infectieuses*, 29, 5, 318–324.

Marmot, M. and Wilkinson, R. (1999) *The Social Determinants of Health*. Oxford, Oxford University Press.

Marsh, A. (1998) The creation of a global telemedical information society. *International Journal of Medical Informatics*, 49, 2, 173–193.

Martikainen, P. (1990) Unemployment and mortality among Finnish men 1981–1985. *British Medical Journal*, 301, 407–411.

Martin, P. and Lefebvre, M. (1995) Malaria and climate: sensitivity of malaria potential transmission to climate. *Ambio*, 24, 4, 200–207.

Massey, D. (1991) A global sense of place. *Marxism Today*, June, 24–29.

Massey, D. and Jess, P. (eds) (1995) *A Place in the World? Places, Cultures and Globalization.* Oxford, Oxford University Press.
Mattiasson, I., Lindgarde, F., Nilsson, J. and Theorell, T. (1990) The threat of unemployment and cardiovascular risk factors: longitudinal study of quality of sleep and serum cholesterol concentrations in men threatened with redundancy. *British Medical Journal*, 301, 4.
Maturana, H. and Varela, F. (1980) *Autopoiesis and Cognition.* Dordrecht, Reidel.
Maxwell, L. and Evans, G. (2000) The effects of noise on pre-school children's pre-reading skills. *Journal of Environmental Psychology*, 20, 1, 91–97.
May, J. (1958) *The Ecology of Human Disease.* New York, MD.
May, J. (1996) A little taste of something more exotic: the imaginative geographies of everyday life. *Geography*, 81, 1, 57–64.
Mayer, J. (2000) Place, telemedicine and the doctor–patient relationship. Paper presented to the 9th International Symposium in Medical Geography, 3–7 July, Montreal.
Maylath, E., Seidel, J., Werner, B. and Schattmann, P. (1999) Geographical analysis of the risk of psychiatric hospitalization in Hamburg 1988–1994. *European Psychiatry*, 14, 8, 414–425.
Maylath, E., Seidel, J. and Schlattmann, P. (2000) Inequity in the hospital care of patients with alcoholism and medication addiction. *European Addiction Research*, 6, 2, 79–83.
McConnell, L., Koenig, B., Greely, H., Rafine, T. et al. (1998) Genetic testing and Alzheimer disease: has the time come? *Nature Medicine*, 4, 7, 757–759.
McCredie, M., Williams, S. and Coates, M. (1999) Cancer mortality in East and Southeast Asian migrants to New South Wales, Australia, 1975–1995. *British Journal of Cancer*, 79, 7–8, 1277–1282.
McDowell, L. (1983) City and home: urban housing and the sexual division of space. In Evans, M. and Ungerson, C. (eds) (1983) *Sexual Divisions: Patterns and Processes.* London, Tavistock.
McDowell, L. (1993) Space, place and gender relations. Part 1: Feminist empiricism and the geography of social relations. *Progress in Human Geography*, 17, 2, 157–179.
McGuffin, P., Riley, B. and Plomin, R. (2001) Toward behavioural genomics. *Science*, 291, 1232–1233.
McIntyre, L. and Pettigrew, M. (1999) *Methods of Health Impact Assessment: a Literature Review.* Occasional Paper no. 2. Glasgow, MRC Social and Public Health Sciences Unit.
McKeown, T. (1979) *The Role of Medicine: Dream, Mirage or Nemesis.* New York, Tavistock.
McKee, M. and Lang, T. (1997) Food for independence: to distance the sponsors of commercial interests from the defenders of consumer health. *British Medical Journal*, 314, 459–460.
McMichael, A. and Beaglehole, R. (2000) The changing global context of public health. *Lancet*, 356, 495–499.
Mead, L. (1997) *From Welfare to Work: Lessons from America.* London, Institute for Economic Affairs.
Meade, M. and Earickson, R. (2000) *Medical Geography*, 2nd edn. New York, Guilford.
Mechanic, D. (1978) *Medical Sociology*, 2nd edn. New York, Free.
Mechanic, D. (1995) Challenges in the provision of mental health services: some cautionary lessons from US experience. *Journal of Public Health Medicine*, 17, 2, 132–139.
Mechanic, D. (1997) Integrating mental health services through reimbursement reform and managed mental health care. *Journal of Health Services Research and Policy*, 2, 2, 86–93.
Midhet, F., Becker, S. and Berendes, H. (1998) Contextual determinants of maternal mortality in rural Pakistan. *Social Science and Medicine*, 46, 12, 1587–1598.
Miles, A. (1991) *Women, Health and Medicine.* Milton Keynes: Open University Press.
Miller, E. (2001) Telemedicine and doctor–patient communication: an analytical survey of the literature. *Journal of Telemedicine and Telecare*, 7, 1, 1–17.

Milligan, C. (1996) Service dependent ghetto formation: a transferable concept? *Health and Place*, 2, 4, 199–211.
Milligan, C. (2000a) 'Bearing the burden': towards a restructured geography of caring. *Area*, 32, 1, 49–58.
Milligan, C. (2000b) 'Breaking out of the asylum'. Developments in the geography of mental ill-health: the influence of the informal sector. *Health and Place*, 6, 3, 189–200.
Milligan, C. (2001) *Geographies of Care: Space, Place and the Voluntary Sector.* Aldershot, Ashgate.
Mitchell, D. (1995) *City of Bits: Space, Place and the Infobahn.* Cambridge, MA, MIT Press.
Mitchell, J. (1999) The uneven diffusion of telemedicine services in Australia. *Journal of Telemedicine and Telecare*, 5, 1, 45–47.
Mitchell, J., Robinson, P., McEvoy, M. and Gates, J. (2001) Telemedicine for the delivery of professional development of health, education and welfare professionals in two remote mining towns, *Journal of Telemedicine and Telecare*, 7, 3, 174–180.
Mitchell, R., Gleave, S., Bartley, M., Wiggins, D. and Joshi, H. (2000) Do area and attitude influence health? A multilevel approach to health inequalities. *Health and Place*, 6, 2, 67–69.
Mohan, J. (1995) *A National Health Service? The Restructuring of Health Care in Britain since 1979.* Basingstoke, Macmillan.
Mohan, J. (1998a) Explaining geographies of health care: a critique. *Health and Place*, 4, 2, 113–124.
Mohan, J. (1998b) Uneven development, territorial politics and the British health care reforms. *Political Studies*, 46, 2, 309–327.
Mohan, J. (2002) *Planning, Markets and Hospitals.* London, Routledge.
Moon, G. (1990) Conceptions of space and community in British health policy. *Social Science and Medicine*, 30, 1, 165–171.
Moon, G. (2000) Risk and protection: the discourse of confinement in contemporary mental health policy. *Health and Place*, 6, 3, 239–250.
Moon, G. and Curtis, S. (1998) Health and health policy in Europe. In Unwin, T. (ed.) *A European Geography.* London, Longman. Chapter 18, 291–295.
Moon, G. and North, N. (2000) *Policy and Practice: General Medical Practice in the UK.* London, Macmillan.
Moore, M., McCray, E. and Onorator, I. (1999) Cross-matching TB and AIDS registries: TB patients with HIV co-infection, United States, 1993–4. *Public Health Reports*, 114, 3, 269–277.
Moradi, T., Delfino, R., Bergstrom, S., Yu, E., Adami, H. and Yuen, J. (1998) Cancer risk among Scandinavian immigrants in the US and Scandinavian residents compared with US whites 1973–89. *European Journal of Cancer Prevention*, 7, 117–125.
Morse, D. (1994) Multidrug resistance: the New York experience. In Porter, J. and McAdam, K. (eds) *Tuberculosis: Back to the Future.* Chichester, Wiley. Chapter 11, 225–230.
Moser, K., Goldblatt, P., Fox, A. and Jones, D. (1987) Unemployment and mortality: comparison of the 1971 and 1981 Longitudinal Study census sample. *British Medical Journal*, 294, 87–90.
Moss, A., Hahn, J., Tulsky, J., Daley, C., Small, P. and Hopewell, P. (2000) Tuberculosis in the homeless: a prospective study. *American Journal of Respiratory and Critical Care Medicine*, 162, 2, 460–464.
Moss, P. (1999) Autobiographical notes on chronic illness. In Butler, R. and Parr, H. (eds) *Mind and Body Spaces: Geographies Illness, Impairment and Disability.* London, Routledge. Chapter 8, 155–166.

Mossman, D. (1997) Deinstitutionalization, homelessness, and the myth of psychiatric abandonment: a structural anthropology perspective. *Social Science and Medicine*, 44, 1, 71–83.

Mowl, G., Pain, R. and Talbot, C. (2000) The ageing body and the homespace. *Area*, 32, 2, 189–197.

Murcott, A. (2000) Understanding life-style and food use: contributions from the social sciences. *British Medical Bulletin*, 56, 1, 212–232.

Murray, C. (1990) *The Emerging British Underclass*. London, Institute of Economic Affairs.

Mustard, C., Derksen, S., Berthelot, J.-M. and Wolfson, M. (1999) Assessing ecologic proxies for household income and neighbourhood level income measures in the study of population health status. *Health and Place*, 5, 2, 157–171.

Naish, J., Curtis, S., Gilham, V., Gregory, I., McClaren, D. and Ball, C. (2000) *A case study in the application of geographical information systems to inform primary care provision for elderly people*. Paper presented to the 9th International Symposium in Medical Geography, 3–7 July, Montreal.

National Audit Office (2001) *Inappropriate Adjustments to NHS Waiting Lists*. Report by Comptroller and Auditor General, HC 452, session 2001–2, 19 December. London, Stationery Office.

Nazroo, J. (1997) *Ethnicity and Mental Health*. London, Policy Studies Institute.

Nazroo, J. (1998) Genetic, cultural or socio-economic vulnerability? Explaining ethnic inequalities in health. In Bartley, M., Blane, D. and Davey-Smith, G. (eds) *The Sociology of Health Inequalities*. Oxford, Blackwell. Chapter 7, 151–170.

Needleman, C. (1997) Applied epidemiology and environmental health: emerging controversies. *American Journal of Infection Control*, 25, 3, 262–274.

Neelman, J. and Wessley, S. (1999) Ethnic minority suicide: a small area geographical study in South London. *Psychological Medicine*, 29, 429–436.

Nemet, G. and Bailey, A. (2000) Distance and health care utilization among the rural elderly. *Social Science and Medicine*, 50, 1197–1208.

Nesbitt, T., Ellis, J. and Kuenneth, C. (1999) A proposed model for telemedicine to supplement the physician workforce in the USA. *Journal of Telemedicine and Telecare*, 5, 2, s20–s26.

New York City Turning Point Initiative (1998–2000) *Community health profiles: Queens, Bronx, Manhattan, States Island, Brooklyn*. New York, New York City Department of Health.

NHSE (1998) *Tuberculosis Control in London: the Need for Change. A Report for the Thames Regional Directors of Public Health*. London, National Health Service Executive, Thames Region.

NHSE London Region (2000) *Improving TB Control in London*. London, National Health Service Executive, London Region.

Office of National Statistics (2000) *Census News*, 44, December, 24–25.

Ojanlatva, A., Vandenbussche, C., Heldt, H., Horte, A., Haggblom, T., Kero, J., Kahkonen, J., Mottonen, M., Saraste, A. and Turunen, T. (1997) The use of problem-based learning in dealing with cultural minority groups. *Patient Education and Counseling*, 31, 2, 171–176.

Oliver, M. (1990) *The Politics of Disablement*. Basingstoke, Macmillan.

Omran, A. (1971) The epidemiological transition: a theory of the epidemiology of population change. *Milbank Quarterly*, 64, 355–391.

Osterman, P. (1991) Welfare participation in a full employment economy: the impact of neighbourhood. *Social Problems*, 38, 4, 475–489.

Overall, A. and Nichols, R. (2001) A method for distinguishing consanguinity and population substructure using multilocus genetic data. *Molecular Biology and Evolution*, 18, 11, 2048–2056.

Paabo, S. (2001) The human genome and our view of ourselves. *Science*, 291, 1219–1220.

Padgett, D., Struening, E., Andrews, H. and Pittman, J. (1995) Predictors of emergency room use by homeless adults in New York City: the influence of predisposing, enabling and need factors. *Social Science and Medicine*, 41, 4, 547–556.

PAHO (1999) Integrated Management of Childhood Illness (IMCI): 'Healthy Children Goal 2002'. *Epidemiological Bulletin*, 20, 4. PanAmerican Health Organization. http://www.paho.org.

Painter, C. with Clarence, E. (2001) UK local action zones and changing urban governance. *Urban Studies*, 38, 8, 1215–1232.

Palka, E. (1999) Accessible wilderness as a therapeutic landscape: experiencing the nature of Denali National Park, Alaska. In Williams, A. (ed.) *Therapeutic Landscapes: the Dynamic between Place and Wellness*. New York, University Press of America. Chapter 2, 29–52.

Palsson, G. and Rabinow, P. (2001) The Icelandic genome debate. *Trends in Biotechnology*, 19, 5, 166–171.

Panterbrick, C., Todd, A. and Baker, R. (1996) Growth status of homeless Nepali boys: do they differ from rural and urban controls? *Social Science and Medicine*, 43, 4, 441–445.

Parr, H. (1997) Mental health, public space and the city: questions of individual and collective access. *Environment and Planning D: Society and Space*, 15, 435–454.

Parr, H. (1999) Mental health and the therapeutic geographies of the city. In Williams, A. (ed.) *Therapeutic Landscapes: the Dynamic between Place and Wellness*. New York, University Press of America. 123–152.

Parr, H. (2000) Interpreting the 'hidden social geographies' of mental health: ethnographies of inclusion and exclusion in semi-institutional places. *Health and Place*, 6, 3, 225–238.

Parr, H. and Philo, C. (1996) *'A Forbidding Fortress of Locks, Bars and Padded Cells': the Locational History of Mental Health Care in Nottingham*. Historical Geography Research Series no. 32. Historical Geography Research Group.

Pawluch, D., Cain, R. and Gillett, J. (2000) Lay constructions of HIV and complementary therapy use. *Social Science and Medicine*, 51, 2, 251–264.

Pawson, R. and Tilley, N. (1997) *Realistic Evaluation*. London, Sage.

Payne, F. and Jessop, L. (2001) NHS Direct: review of activity data for the first year of operation at one site. *Journal of Public Health Medicine*, 23, 2, 155–158.

Pekkanen, J. and Pearce, N. (2001) Environmental epidemiology: challenges and opportunities. *Environmental Health Perspectives*, 109, 1, 1–5.

Pellegrino, E. (2001) The commodification of medical and health care: the moral consequences of a paradigm shift from a professional to a market ethic. *Journal of Medicine and Philosophy*, 24, 243–266.

Penchansky, R. and Thomas, J. (1981) The concept of access: definition and relationship to consumer satisfaction. *Medical Care*, XIX, 2, 127–140.

Perlin, S., Wong, D. and Sexton, K. (2001) Residential proximity to industrial sources of air pollution: interrelationships among race, poverty, and age. *Journal of Air Waste Management Association*, 51, 4, 406–421.

Pescosolido, B., Gardner, C. and Lubell, K. (1998) How people get into mental health services: stories of choice, coercion and 'muddling through'. *Social Science and Medicine*, 46, 2, 275–286.

Phillips, D. (1979) Spatial variations in attendance at general practitioner services. *Social Science and Medicine*, 13D, 169–181.

Phillips, D. (ed.) (1990) *Health and Health Care in the Third World*. London, Longman.

Phillips, D. and Verhasselt, Y. (eds) (1994) *Health and Development*. London: Routledge.

Philo, C. (1989) Enough to drive one mad: the organization of space in 19th century lunatic asylums. In Wolch, J. and Dear, M. (eds) *The Power of Geography*. Boston, MA, Unwin Hyman. Chapter 12, 258–290.

Philo, C. (1997) Across the water: reviewing geographical studies of asylums and other mental health facilities. *Health and Place*, 3, 73–89.

Philo, C. (2000) *The Birth of the Clinic*: an unknown work of medical geography. *Area*, 32, 1, 11–19.

Picheral, H. (1989) Géographie de la transition epidémiologique. *Annales de Géographie*, 546, 129–151.

Picheral, H. (1998) *Dictionnaire Raisonné de Géographie de la Santé*. Montpellier, Université Montpellier III.

Picheral, H. (2001) La résurgence du lieu: l'émergence de la géographie de la santé. *Technologie et Santé*, 45, 5–17.

Picot, J. (1998) Telemedicine and telehealth in Canada: forty years of change in the use of information and communications technologies in a publicly administered health care system. *Telemedicine Journal*, 4, 3, 199–205.

Pilgrim, D. and Rogers, A. (1999) *A Sociology of Mental Health and Illness*, 2nd edn. Buckingham, Open University Press.

Pinch, S. (1985) *Cities and Services: the Geography of Collective Consumption*. London, Routledge.

Pinch, S. (1997) *Worlds of Welfare: Understanding the Changing Geographies of Social Welfare Provision*. London, Routledge.

Pinfold, V. (2000) 'Building up safe havens . . . all around the world': users' experience of living in the community with mental health problems. *Health and Place*, 6, 3, 201–212.

Poland, G., Coburn, D., Robertson, A. and Eakin, J. (and Critical Social Science Group) (1998) Wealth, equity and health care: a critique of a 'population health' perspective on the determinants of health. *Social Science and Medicine*, 46, 7, 785–798.

Polaschek, N. (1998) Cultural safety: a new concept in nursing people of different ethnicities. *Journal of Advanced Nursing*, 27, 3, 452–457.

Pollard, T. (1997) Urbanism and psychosocial stress. In Schell, L. and Ulizaszek, S. (eds) *Urbanism, Health and Human Biology in Industrialised Countries*. Cambridge, Cambridge University Press. Chapter 12, 231–249.

Popay, J., Bartley, M. and Owen, C. (1993) Gender inequalities in health: social position, affective disorders and minor physical morbidity. *Social Science and Medicine*, 36, 1, 21–32.

Popay, J., Williams, G., Thomas, C. and Gatrell, A. (1998) Theorising inequalities in health: the place of lay knowledge. In Bartley, M., Blane, D. and Davey-Smith, G. (eds) *The Sociology of Health Inequalities*. Oxford, Blackwell. 59–84.

Popkin, B. (2000) Urbanization and the nutrition transition: achieving urban food and nutrition security in the developing world. In *IFPRI 2020 Focus 3*, Brief 7 of 10, Washington, DC, International Food Policy Research Institute.

Porter, J. and Grange, J. (eds) (1999) *Tuberculosis: an Interdisciplinary Perspective*. London, Imperial College Press.

Porter, J. and McAdam, K. (eds) (1994) *Tuberculosis: Back to the Future*. Chichester, Wiley.

Powell, M. (1990) Need and provision in the National Health Service: an inverse care law? *Policy and Politics*, 18, 1, 31–37.

Powell, M. and Moon, G. (2001) Health Action Zones: the 'third way' of a new area-based policy? *Health and Social Care in the Community*, 9, 1, 43–50.

Powell, M., Boyne, G. and Ashworth, R. (2001) Towards a geography of people poverty and place poverty. *Policy and Politics*, 29, 3, 243–258.

Pratt, J., Gordon, P. and Plamping, D. (1999) *Working Whole Systems: Putting Theory into Practice in Organisations*. London, King's Fund.

Pred, A. (1981) Social reproduction and the time-geography of everyday life. *Geografisker Annaler*, 63, 3, 5–22.

Pred, A. (1996) Intefusions: consumption, identity and the practices and power relations of everyday life. *Environment and Planning A*, 28, 11–24.

Preddie, S. and Awai-Boyce, K. (1999) Mental wellbeing in a multi-ethnic society. In Bhui, K. and Olajide, D. (eds) *Mental Health Service Provision for a Multi-Cultural Society*. London, Saunders.

Prior, L. (1993) *The Social Organization of Mental Illness*. London, Sage.

Puentes-Markides, C. (1992) Women and access to health care. *Social Science and Medicine*, 35, 4, 613–617.

Purvis, A. (2001) Is it a hotel? Is it a trendy bar? No, it's a hospital. *The Guardian*, 5 July.

Putnam, R. (2000) *Bowling Alone: the Collapse and Revival of American Community*. New York, Simon and Schuster.

Pyle, G. (1969) The diffusion of cholera in the United States in the nineteenth century. *Geographical Analysis*, 1, 59–75.

Pyle, G. (1979) *Applied Medical Geography*. Washington, DC, Holt, Rinehart and Winston.

Pyle, G. (1986) *The Diffusion of Influenza*. Towota, NJ, Rowman and Littlefield.

Radicova, I. (2001) Poverty of Romanies in connection with the labour force market in the Slovak republic. *Sociologica*, 33, 5, 439–456.

Raleigh, V. and Kiri, V. (1997) Life expectancy in England: variations and trends by gender, health authority, and level of deprivation. *Journal of Epidemiology and Community Health*, 51, 6, 649–658.

Ranade, W. (1998) *Markets and Health Care: a Comparative Analysis*. Harlow, Longman.

Ratner, P.A. (1983) The incidence of wife abuse and mental health status in Edmonton, Alberta. *Canadian Journal of Public Health*, 84, 246–249.

Ritsatakis, A. (2000) Experience in setting targets for health in Europe. *European Journal of Public Health*, 10, 4 supplement, s7–s10.

Rogers, A. and Pilgrim, D. (1996) *Mental Health Policy in Britain: a Critical Introduction*. London, Macmillan.

Rogers, A., Huxley, P., Thomas, R., Robson, B., Evans, S., Stordy, J. and Gately, C. (2001) Evaluating the impact of a locality based social policy intervention on mental health: conceptual and methodological issues. *International Journal of Social Psychiatry*, 47, 4, 41–55.

Rogerson, P. and Han, D. (2002) The effects of migration on the detection of geographic differences in disease risk. *Social Science and Medicine*, 55, 10, 1817–1828.

Rose, G. (1992) *The Strategy of Preventive Medicine*. Oxford, Oxford University Press.

Rosenheck, R., Morrissey, J., Lam, J., Callaway, M., Stolar, M., Johnsen, M., Randolph, F., Blasinsky, M. and Goldman, H. (2001) Service delivery and community: social capital, service systems integration, and outcomes among homeless persons with severe mental illness. *Health Services Research*, 36, 4, 691–710.

Rosenstock, I. (1966) Why people use health services. *Milbank Quarterly*, 44, 94–127.

Ross, C. (2000) Walking, exercising and smoking: does neighbourhood matter? *Social Science and Medicine*, 265–274.

Ross, N., Wolfson, M., Dunn, J., Berthelot, J., Kaplan, G. and Lynch, J. (2000) Relation between income inequality and mortality in Canada and the United States: cross sectional assessment using census data and vital statistics. *British Medical Journal*, 320, 898–902.

Rowland, M., Abdur Rab, M., Freeman, T., Durrani, N. and Rehman, N. (2002) Afghan refugees and the temporal and spatial distribution of malaria in Pakistan. *Social Science and Medicine*, 55, 11, 2061–2072.

Rudat, K. (1994) *Black and Minority Ethnic Groups in England.* London, Health Education Authority.

Ryan, R. and Deci, E. (2001) On happiness and human potentials. *Annual Review of Pscyhology,* 52, 141–166.

Sack, R. (1986) *Human Territoriality, its Theory and History.* Cambridge, Cambridge University Press.

Saks, M. (1992) *Alternative Medicine in Britain.* Oxford, Clarendon.

Sanche Hodge, F. and Fredericks, L. (1999) American Indian and Alaska Native populations in the United States: an overview. In Huff, R. and Kline, M. (eds) *Promoting Health in Multicultural Populations: a Handbook for Practitioners.* London, Sage. Chapter 14, 269–290.

Sassen, S. (1991) *The Global City: New York, London, Tokyo.* Princeton, NJ, Princeton University Press.

Satterthwaite, D. (ed.) (1999a) *The Earthscan Reader in Sustainable Cities.* London, Earthscan.

Satterthwaite, D. (1999b) Sustainable cities or cities that contribute to sustainable development? In Satterthwaite, D. (ed.) *The Earthscan Reader in Sustainable Cities.* London, Earthscan. Chapter 5, 80–106.

Sauderson, T. and Langford, I. (1996) A study of the geographical distribution of suicide rates in England and Wales 1989–1992 using empirical Bayes estimates. *Social Science and Medicine,* 43, 4, 489–502.

Scambler, G. (1997) *Sociology as Applied to Medicine,* 4th edn. London, Saunders.

Schaerstrom, A. (1996) *Pathogenic Paths? A Time Geography Approach in Medical Geography.* Lund, Lund University Press.

Schecter, G. (1997) Supervised therapy in San Francisco. *Clinics in Chest Medicine,* 18, 1, 165.

Schneider, D. and Freeman, N. (2000) *Children's Environmental Health: Reducing Risk in a Dangerous World.* Washington, DC, American Public Health Association.

Schneider, Y., Tikhomirova, E., Shilnikova, I. and Rychkov, Y. (1995) The genetic polymorphism and gene geography of the indigenous populations of the Ural region. 1: The genetic pattern of Ural populations. *Genetika,* 31, 4, 560–572.

Schofield, R. and Reher, D. (1991) The decline of mortality in Europe. In Schofield, R., Reher, D. and Bideau, A. (eds) *The Decline of Mortality in Europe.* Oxford, Clarendon. Chapter 1, 1–17.

Schwartz, S. (1994) The fallacy of the ecological fallacy: the potential misuse of a concept and the consequences. *American Journal of Public Health,* 84, 5, 819–824.

Scott-Samuel, A., Birley, M. and Ardern, K. (1998) *The Merseyside Guidelines for Health Impact Assessment.* Liverpool, Merseyside Health Impact Assessment Steering Group, Liverpool Public Health Observatory.

Scull, A. (1979) *Museums of Madness: the Social Organization of Insanity in Nineteenth Century England.* London, Allen Lane.

Selman, P. (1996) *Local Sustainability: Managing and Planning Ecologically Sound Places.* London, Chapman.

Senior, M. (1991) Deprivation payments to GPs: not what the doctor ordered. *Environment and Planning C,* 9, 79–94.

Shannon, G. (2000) Telemedicine: does distance matter? Paper presented to the 9th International Symposium in Medical Geography, 3–7 July, Montreal.

Shannon, G. and Dever, G. (1974) *Health Care Delivery: Spatial Perspectives.* New York, McGraw Hill.

Sharma, V. (1992) *Complementary Medicine Today.* London, Routledge.

Shaw, C. (2002) Deco style on the Downs: Meadowfield Adult Acute Unit. *HD (Journal for Healthcare Design and Development)*, January, 18–24.

Shaw, M., Dorling, D. and Brimblecombe, N. (1998) Changing the map: health in Britain 1951–91. In Bartley, M., Blane, D. and Davey-Smith, G. (eds) *The Sociology of Health Inequalities.* Oxford, Blackwell. Chapter 6, 135–150.

Shaw, M., Dorling, D., Gordon, D. and Davey-Smith, G. (1999) *The Widening Gap: Health Inequalities and Policy in Britain.* Bristol, Policy.

Shaw, M., Dorling, D. and Mitchell, R. (2002) *Health, Place and Society.* Harlow, Pearson.

Shouls, S., Congdon, P. and Curtis, S. (1996a) Modelling inequality in reported long term illness in the UK: combining individual and area characteristics. *Journal of Epidemiology and Community Health*, 50, 3, 366–376.

Shouls, S., Congdon, P. and Curtis, S. (1996b) Geographic variation in illness and mortality: the development of a relevant area typology for SAR districts. *Health and Place*, 2, 3, 139–156.

Shrader-Frechette, K. (1990) Scientific method, anti-foundationalism and public decision making. *Health, Safety and Environment*, 1, 23–41.

Siahpush, M. and Singh, G.K. (1999) Social integration and mortality in Australia. *Australian and New Zealand Journal of Public Health*, 23, 571–577.

Sibley, D. (1995) *Geographies of Exclusion.* London, Routledge.

Siegel, C., Laska, E., Haugland, G., O'Neill, D., Cohen, N. and Lesser, M. (2000) The construction of community indexes of mental health and social and mental well-being and their application to New York City. *Evaluation Programme Planning*, 23, 3, 315–327.

Siegrist, J. (1996) Adverse health effects of high effort/low reward conditions. *Journal of Occupational Health Psychology*, 1, 27–41.

Sieswerda, L., Soskolne, C., Newman, S., Schopflocher, D. and Smoyer, K. (2001) Toward measuring the impact of ecological disintegrity on human health. *Epidemiology*, 12, 1, 28–32.

Sinnerbrink, I., Silove, D., Manicavasagar, V., Steel, Z. and Field, A. (1996) Asylum seekers: general health status and problems with access to health care. *Medical Journal of Australia*, 165, 634–637.

Sirgy, M. (1998) Materialism and quality of life. *Social Indicators Research*, 43, 3, 227–260.

Sloggett, A. and Joshi, H. (1998) Deprivation indicators as predictors of life events 1981–1992 based on the UK ONS longitudinal study. *Journal of Epidemiology and Community Health*, 52, 4, 228–233.

Slovik, P. (1987) Perception of risk. *Science*, 236, 280–285.

Smallman-Raynor, M. and Cliff, A. (2001) Epidemiological spaces: the use of multidimensional scaling to identify cholera diffusion processes in the wake of the Philippines insurrection, 1899–1902. *Transactions of the Institute of British Geographers*, NS, 26, 3, 288–306.

Smallman-Raynor, M. and Phillips, D. (1999) Late stages of epidemiological transition: health status in the developed world. *Health and Place*, 5, 3, 209–222.

Smallman-Raynor, M., Cliff, A. and Haggett, P. (1991) Civil war and the spread of AIDS in Central Africa. *Epidemiology and Infection*, 107, 69.

Smallman-Raynor, M., Cliff, A. and Haggett, P. (1992) *London International Atlas of AIDS.* Oxford, Blackwell.

Smith, C. (1993) (Over)eating success: the health consequences of the restoration of capitalism in rural China. *Social Science and Medicine*, 37, 6, 761–770.

Smith, C. and Hanham, R. (1981) Proximity and the formation of public attitudes towards mental illness. *Environment and Planning A*, 13, 147–165.

Smith, D. (1977) *Human Geography: a Welfare Approach.* London, Arnold.

Smith, D. (1994) *Geography and Social Justice.* Oxford, Blackwell.

Smith, P., Sheldon, T. and Martin, S. (1996) An index of need for psychiatric services based on in-patient utilisation. *British Journal of Psychiatry*, 169, 3, 308–316.

Smith, R.G. (2000) Windows in time: the spatial experience of chronic illness and disability in men from childhood to adulthood. Unpublished PhD thesis. London, Queen Mary, University of London.

Smith, S., Easterlow, D., Munro, M. and Turner, K. (2003) Housing as health capital: how health trajectories and housing paths are linked. *Journal of Social Issues*, 59, 3, 501–525.

Smith, T. (1997) Racist encounters: Romani 'gypsy' women and mainstream health ethnic inequalities in health. *European Journal of Women's Studies*, 4, 2, 183.

Snider, D. (1994) Tuberculosis: the world situation. History of the disease and efforts to combat it. In Porter, J. and McAdam, K. (eds) *Tuberculosis: Back to the Future*. Chichester, Wiley. Chapter 1, 13–32.

Snow, R., Gouws, E., Omumbo, J., Rapuoda, B., Craig, M., Tanser, F., Le Sueur, D. and Ouma, J. (1998) Models to predict the intensity of *Plasmodium falciparum* transmission: applications to the burden of disease in Kenya. *Transactions of the Royal Society of Tropical Medicine and Hygiene*, 92, 6, 601–606.

Snyder, D., Paz, E., Mohle-Boetani, J., Fallstad, R., Black, R. and Chin, D. (1999) Tuberculosis prevention in methadone maintenance clinics: effectiveness and cost-effectiveness. *American Journal of Respiratory and Critical Care Medicine*, 160, 1, 178–185.

Sooman, A., Macintyre, S. and Anderson, A. (1993) Scotland's health: a more difficult challenge for some. The price and availability of healthy foods in socially contrasting localities in the West of Scotland. *Health Bulletin*, 51, 276–284.

Soskolne, C. and Light, A. (1996) Towards ethics guidelines for environmental epidemiologists. *Science of the Total Environment*, 184, 1/2, 137–147.

Squires, J. (1994) Private lives, secluded places: privacy as political possibility. *Environment and Planning D: Society and Space*, 12, 387–401.

Stafford, M., Bartley, M., Mitchell, R. and Marmot, M. (2001) Characteristics of individuals and characteristics of areas: investigating their influence on health in the Whitehall II Study. *Health and Place*, 7, 2, 117–130.

Stainton-Rogers, W. (1991) *Explaining Health and Illness: an Exploration of Diversity*. Hemel Hempstead, Wheatsheaf.

Stansfeld, S. (1999) Social support and social cohesion. In Marmot, M. and Wilkinson, R. (1999) *The Social Determinants of Health*. Oxford, Oxford University Press. 155–178.

Stansfeld, S., Fuhrer, R., Shipley, M. and Marmot, M. (1999) Work characteristics predict psychiatric disorder: prospective results from the Whitehall II Study. *Occupational and Environmental Medicine*, 56, 5, 302–307.

Stansfeld, S., Haines, M. and Brown, B. (2000) Noise and health in the urban environment. *Reviews on Environmental Health*, 15, 1–2, 43–82.

Stansfeld, S., Haines, M., Curtis, S., Brentnall, S. and Brown, B. (2001) *Rapid Review on Noise and Health for London: a Review to Support the Development of the Mayor of London's Ambient Noise Strategy*. London, Queen Mary, University of London.

Stein, H. (1993) Can supra-national public health strategies improve the health status of national populations? *European Journal of Public Health*, 3, 3–7.

Stellman, S. and Wang, Q. (1994) Cancer mortality in Chinese immigrants to New York City: comparison with Chinese in Tianjin and with United-States-born whites. *Cancer*, 73, 4, 1270–1275.

Stephens, C., Leine, S., Leonardi, G., Chasco, M. and Shaw, R. (2000) Health, sustainability and equity. *Global Change and Human Health*, 1, 1, 44–58.

Sticher, S. and Parpart, J. (eds) (1990) *Women, Employment and the Family in International Division of Labour.* London, Macmillan.

Stoker, G. and Mossberger, K. (1994) Urban regime theory in comparative context. *Environment and Planning C: Government and Policy,* 12, 195–212.

Stone, C. (1993) Urban regimes and the capacity to govern: a political economy approach. *Journal of Urban Affairs*, 15, 1, 1–28.

Subcommittee of the Joint Tuberculosis Committee of the British Thoracic Society (1992) Guidelines on the management of tuberculosis and HIV infection in the United Kingdom. *British Medical Journal*, 304, 1231–1233.

Susser, M. (1996) Choosing a future for epidemiology. II: From black box to Chinese boxes and eco-epidemiology. *American Journal of Public Health*, 86, 674–677.

Sutherland, A. (1992) Gypsies and health-care. *Western Journal of Medicine*, 157, 3, 276–280.

Swanson, B. (1999) Information technology and under-served communities. *Journal of Telemedicine and Telecare*, 5, 2, s3–s10.

Sze, S. (1988) WHO: from small beginnings. *World Health Forum*, 9, 1, 29–34.

Takahashi, L. (1997) The socio-spatial stigmatization of homeless and HIV/AIDS: toward an explanation of the NIMBY syndrome. *Social Science and Medicine*, 45, 6, 903–914.

Takahashi, L. and Gaber, S. (1998) Controversial facility siting in the urban environment: resident and planner perceptions in the United States. *Environment and Behaviour,* 30, 2, 184–215.

Takahashi, L. and Wolch, J. (1994) Differences in health and welfare between homeless and homed welfare applicants in Los Angeles County. *Social Science and Medicine*, 38, 10, 1401–1413.

Tarnopolsky, A., Watkins, G. and Hand, D. (1980) Aircraft noise and mental health. I: Prevalence of individual symptoms. *Psychological Medicine*, 10, 683–698.

Taylor, D. (1998) The natural life of policy indices: geographical problem areas in the US and UK. *Social Science and Medicine*, 47, 6, 713–725.

Thoits, P. (1991) On merging identity theory and stress research. *Social Psychology Quarterly,* 54, 101–102.

Thompson, M., Connor, S., Milligan, P. and Flasse, S. (1996) The ecology of malaria – as seen from Earth-observation satellites. *Annals of Tropical Medicine and Parasitology,* 90, 3, 243–264.

Thouez, J.-P. (1987) *Organization spatiale des systèmes de soins.* Montreal, Presses de l'Université de Montréal.

Thrift, N. (1982) On the determination of social action in space and time. *Environment and Planning D: Society and Space*, 1, 23–57.

Thrift, N. (1992) *Spatial Formations.* London, Sage.

Thrift, N. and Amin, A. (eds) (1994) *Globalization, Institutions and Regional Development in Europe.* Oxford, Oxford University Press.

Tiebout, C. (1956) A pure theory of local expenditure. *Journal of Political Economy,* 64, 416–424.

Tiefenbacher, J. and Hagelman, R. (1999) Environmental equity in urban Texas: race, income, and patterns of acute and chronic toxic air releases in metropolitan counties. *Urban Geography,* 20, 6, 516–533.

Tomlinson, D. (1996) The American, Flemish and British cases of asylum in the community. In Tomlinson, D. and Carrier, J. (eds) *Asylum in the Community.* London, Routledge. Chapter 8, 169–185.

Tomlinson, D., Carrier, J. and Oerton, J. (1996) The refuge function of psychiatric hospitals. In Tomlinson, D. and Carrier, J. (eds) *Asylum in the Community.* London, Routledge.

Townsend, P., Davidson, N. and Whitehead, M. (1988) *Inequalities in Health: the Black Report and the Health Divide*. London, Penguin.

Tsugane, S., Desouza, J., Costa, M., Mirra, A., Gotlieb, S., Laurenti, R. and Wantanabe, S. (1990) Cancer incidence rates among Japanese immigrants in the city of Sao-Paulo, Brazil, 1969–78. *Cancer Causes and Control*, 1, 2, 189–193.

Tually, P., Walker, J. and Cowell, S. (2001) The effect of nuclear medicine telediagnosis on diagnostic pathways and management in rural and remote regions of Western Australia. *Journal of Telemedicine and Telecare*, 7, 2, 50–53.

Tudor-Hart, J. (1971) The inverse care law. *Lancet*, 27 February, 4, 7, 412.

Tulsky, J., White, M., Dawson, C., Hoynes, T., Goldenson, J. and Schecter, G. (1998) Screening for tuberculosis in jail and clinic follow-up after release. *American Journal of Public Health*, 88, 2, 223–226.

Tulsky, J., White, M., Young, J., Meakin, R. and Moss, A. (1999) Street talk: knowledge and attitudes about tuberculosis and tuberculosis control among homeless adults. *International Journal of Tuberculosis and Lung Disease*, 3, 6, 528–533.

Turning Point (2000) *Working Together to Create Healthy Communities. Pilot Bulletin, Issue 1.* New York, New York City Department of Health.

United Nations (2001) *The State of the World's Cities: Report*. New York, UN. http://www.un.org.

United Nations Conference on Environment and Development (1992) *Agenda 21: Action Plan for the Next Century*. Rio de Janeiro, UNCED.

US Department of Health and Human Services (1999) *Mental Health: a Report of the Surgeon General*. Rockville, MD, US Department of Health and Human Services, Substance Abuse and Mental Health Services Administration, Center for Mental Health Services, National Institutes of Health, National Institute of Mental Health.

US Department of Health and Human Services (2001) *Healthy People in Healthy Communities: a Community Planning Guide Using Healthy People 2010*. Pittsburg, US Government Printing Office.

US Environmental Protection Agency (2003) *Toxics Release Inventory (TRI) Program*. http://www.epa.gov/tri.

Vaguet, A. (2000) Editorial. *Espace, Populations Sociétés*, 2, 155–158.

Vahtera, T., Kivimaki, M., Pentti, J. and Theorell, T. (2000) Effect of change in the psychosocial work environment on sickness absence: a seven year follow up of initially healthy employees. *Journal of Epidemiology and Community Health*, 54, 7, 484–493.

Valentine, G. (1999) What it means to be a man: the body, masculinities, disability. In Butler, R. and Parr, H. (eds) *Mind and Body Spaces: Geographies Illness, Impairment and Disability*. London, Routledge. Chapter 9, 167–180.

Van Blerkom, L.M. (1995) Clown doctors: shaman healers of Western medicine. *Medical Anthropology Quarterly*, 9, 4, 462–475.

Van Cleemput, P. and Parry, G. (2001) Health status of gypsy travellers. *Journal of Public Health Medicine*, 23, 2, 129–134.

Van der Gaag, J. and Barham, T. (1998) Health and health expenditures in adjusting and non-adjusting countries. *Social Science and Medicine*, 46, 8, 995–1009.

Veeder, N. (1975) Health services utilization: models for human services planning. *American Institute of Planners Journal*, March, 101–109.

Veenstra, G. (1999) Social capital, SES and health: an individual-level analysis. *Social Science and Medicine*, 50, 619–629.

Verheij, R. (1996) Explaining urban–rural variations in health: a review of interactions between individual and environment. *Social Science and Medicine*, 42, 6, 923–935.

Verwers, A. (1992) Towards a new EC health policy? In *Uniting Health in Europe, Proceedings of Founding Meeting*. Amsterdam, European Public Health Association, 11–14.

Victor, C. (1997) The health of homeless people: a review. *European Journal of Public Health* 7, 398–404.

Vigneron, E. (1997) Santé, société inégalités géographiques en France. *Actualité et dossier en santé publique*, 19, XII–XVI.

Wakefield, S., Elliott, S., Cole, D. and Eyles, J. (2001) Environmental risk and (re)action: air quality, health and civic involvement in an urban industrial neighbourhood. *Health and Place*, 7, 163–177.

Wallace, M. (2001) A new approach to neighbourhood renewal in England. *Urban Studies*, 38, 1, 12, 2163–2166.

Wallace, R. (1990) Urban desertification, public health and public order: 'planned shrinkage', violent death, substance abuse and AIDS in the Bronx. *Social Science and Medicine*, 31, 801–813.

Wallace, R. and Wallace, D. (1991) AIDS deaths in the Bronx 1983–1988: spatiotemporal analysis from a sociogeographic perspective. *Environment and Planning A*, 23, 1701–1726.

Wallace, R. and Wallace, D. (1993) The coming crisis of public health in the suburbs. *Milbank Quarterly*, 71, 4, 543–574.

Wallace, R. and Wallace, D. (1997) Socio-economic determinants of health: community marginalisation and the diffusion of disease and disorder in the United States. *British Medical Journal*, 314, 1341–1345.

Walmsley, D. and Lewis, G. (1984) *Human Geography: Behavioural Approaches*. London, Longman.

Warnes, A. (1999) UK and Western European late-age mortality: trends in cause-specific death rates, 1960–1990. *Health and Place*, 5, 1, 111–118.

Warr, P. (1987) *Unemployment and Mental Health*. Oxford, Clarendon.

Waters, M. (1995) *Globalization*. London, Routledge.

Watkins, G., Tarnopolsky, A. and Jenkins, L. (1981) Aircraft noise and mental health. II: Use of medicines and health care services. *Psychological Medicine*, 11, 155–168.

Watt, I. and Sheldon, T. (1993) Rurality and resource allocation in the UK. *Health Policy*, 26, 19–27.

Watt, I., Franks, A. and Sheldon, T. (1993) Rural health and healthcare. *British Medical Journal*, 306, 358–359.

Weber, M. (1968) *Economy and Society*. London, Bedminster.

West, P. (1991) Rethinking the health selection explanation for health inequalities. *Social Science and Medicine*, 32, 4, 373–384.

Westin, S. (1990) The structure of factory closure: individual response to job loss and unemployment in a 10 year controlled follow up study. *Social Science and Medicine*, 31, 1301–1311.

Whitehead, M. (1995) Tackling inequalities: a review of policy initiatives. In Benzeval, M., Judge, K. and Whitehead, M. (1995) *Tackling Health Inequalities: an Agenda for Action*. London, King's Fund. 22–52.

Whiteis, D. (1998) Third world medicine in first world cities: capital accumulation, uneven development and public health. *Social Science and Medicine*, 47, 795–808.

Whitten, P. and Cook, D. (1999) School-based telemedicine: using technology to bring health care to inner-city children. *Journal of Telemedicine and Telecare* 5, 1, 23–25.

WHO (1946) *Preamble to the Constitution of the World Health Organization as adopted by the International Health Conference, New York, 19–22 June, 1946*. Signed on 22 July 1946 by the representatives of 61 states (Official Records of the World Health Organization, no. 2, 100) and entered into force on 7 April 1948.

WHO (1992) *The International Statistical Classification of Diseases and Related Health Problems, Tenth Revision.* Geneva, World Health Organization.

WHO (1998a) *Malaria.* Factsheet no. 94. Geneva, World Health Organization.

WHO (1998b) *Roll Back Malaria.* Factsheet no. 203. Geneva, World Health Organization.

WHO (1998c) *Health 21. The Health For All Policy for the WHO European Region: 21 Targets for the 21st Century.* Copenhagen, World Health Organization, Regional Office for Europe.

WHO (1998d) *Health For All: Origins and Mandate.* Geneva, World Health Organization. http://www.who.int/archives/who50/en/health4all.html.

WHO (1999a) *Healthy Cities.* Geneva, World Health Organization. http://www.who.int/hpr/archive/cities/index.html.

WHO (1999b) *Healthy Cities: What Is a Healthy City?* Geneva, World Health Organization. http://www.who.int/hpr/archive/cities/what.html.

WHO (1999c) *Healthy Cities: Settings Approach in Public Health.* Geneva, World Health Organization. http://www.who.int/hpr/archive/cities/approach/settings.html.

WHO (2000a) *World Health Report, 2000.* Geneva, World Health Organization.

WHO (2000b) *Tuberculosis.* Factsheet no. 104. Geneva, World Health Organization.

WHO (2001a) *The World Health Report. Mental Health: New Understanding, New Hope.* Geneva, World Health Organization.

WHO (2001b) *Global Tuberculosis Control: WHO Report 2001.* WHO/CDS/TB/2001.287. Geneva, World Health Organization.

WHO (2002) *Scaling Up the Response to Infectious Diseases.* Geneva, World Health Organization.

WHO European Centre for Health Policy (1999) *Health Impact Assessment: Main Concepts and Suggested Approach. Gothenburg Consensus Paper.* Brussels, World Health Organization, Regional Office for Europe. 1–10.

Wiles, J. and Rosenberg, M. (2001) 'Gentle caring experience': seeking alternative health care in Canada. *Health and Place,* 7, 3, 209–224.

Wilkinson, R. (1996) *Unhealthy Societies.* London, Routledge.

Wilkinson, R., Kawachi, I. and Kennedy, B. (1998) Mortality the social environment, crime and violence. In Bartley, M., Blane, D. and Davey-Smith (eds) *The Sociology of Health Inequalities.* Oxford, Blackwell. Chapter 1, 19–38.

Williams, A. (1998) The European Union: cumulative and uneven integration. In Unwin, T. (ed.) *A European Geography.* London, Longman. Chapter 9, 129–147.

Williams, A. (ed.) (1999a) *Therapeutic Landscapes: the Dynamic between Place and Wellness.* New York, University Press of America.

Williams, A. (1999b) Introduction. In Williams, A. (ed.) *Therapeutic Landscapes: the Dynamic between Place and Wellness.* New York, University Press of America. Chapter 1, 1–14.

Williams, N. and Mooney, G. (1994) Infant mortality in an age of great cities: London and the English provincial cities compared 1840–1910. *Continuity and Change,* 9, 2, 185–212.

Williams, R. (1999) Environmental injustice in America and its politics of scale. *Political Geography,* 18, 1, 49–73.

Williams, S. (1995) Theorizing class, health and life-styles: can Bourdieu help us? *Sociology of Health and Illness,* 17, 5, 577–604.

Wilson, D. (1998) Evolutionary epidemiology and manic depression. *British Journal of Medical Psychology,* 71, 375–395.

Wilson, K., Chen, R., Taylor, S., McCracken, C. and Copeland, J. (1999) Socio-economic deprivation and the prevalence and prediction in older community residents. *British Journal of Psychiatry*, 174, 149–553.

Wilson, W. (1987) *The Truly Disadvantaged.* Chicago, IL, University of Chicago Press.

Wilton, R. (1998) The constitution of difference: space and psyche in landscapes of exclusion. *Geoforum*, 29, 173–185.

Winch, P. (1998) Social and cultural responses to emerging vectorborne diseases. *Journal of Vector Ecology*, 23, 1, 47–53.

Wismar, M. and Busse, R. (2000) Targets for health in Germany. *European Journal of Public Health*, 10, 4 supplement, s38–s42.

Wolch, D. and Philo, C. (2000) From distributions of deviance to definitions of difference: past and future mental health geographies, *Health and Place*, 6, 3, 137–158.

Wolch, J. and Dear, M. (eds) (1989) *The Power of Geography.* Boston, MA, Unwin Hyman.

Wood, W. (1994) Forced migration: local conflicts and international dilemmas. *Annals of the Association of American Geographers*, 84, 4, 607–634.

Woods, K., Johnson, J., Kutlar, A., Daitch, L. and Stachura, M. (2000) Sickle cell disease telemedicine network for rural outreach. *Journal of Telemedicine and Telecare*, 6, 5, 285–290.

Woodward, D., Drager, N., Beaglehole, R. and Lipson, D. (2001) Globalization and health: a framework for analysis and action. *Bulletin of the World Health Organization*, 79, 9, 875–881.

World Bank (1993) *World Development Report: Investing in Health.* Oxford, Oxford University Press.

World Commission on Environment and Development (1987) *Our Common Future.* Oxford, Oxford University Press.

Yaganehdoost, A., Graviss, E., Ross, M., Adams, G., Ramaswamy, S., Wanger, A., Frothingham, R., Soini, H. and Musser, J. (1999) Complex transmission dynamics of clonally related virulent mycobacterium tuberculosis associated with barhopping by predominantly human immunodeficiency virus-positive gay men. *Journal of Infectious Diseases*, 180, 4, 1245–1251.

Yi–Fu Tuan (1974) *Topophilia.* London, Prentice Hall.

Young, K. (1997) Gender and development. In Visvanathan, N., Duggan, L., Nisonoff, L. and Wiegersma, N. (eds) *The Women, Gender and Development Reader.* London, Zed.

Index

R

S